A CULTURAL HISTORY OF FIREARMS IN THE AGE OF EMPIRE

A Cultural History of Firearms in the Age of Empire

Edited by

KAREN JONES, GIACOMO MACOLA, DAVID WELCH
University of Kent, UK

Routledge
Taylor & Francis Group
LONDON AND NEW YORK

First published in paperback 2024

First published 2013 by Ashgate Publisher

Published 2016 by Routledge
4 Park Square, Milton Park, Abingdon, Oxon OX14 4RN

and by Routledge
605 Third Avenue, New York, NY 10158

Routledge is an imprint of the Taylor & Francis Group, an informa business

Publisher's Note
The publisher has gone to great lengths to ensure the quality of this reprint but points out that some imperfections in the original copies may be apparent.

British Library Cataloguing in Publication Data
A catalogue record for this book is available from the British Library

The Library of Congress has cataloged the printed edition as follows:
A cultural history of firearms in the age of empire / edited by Karen Jones, Giacomo Macola and David Welch.
pages cm
Includes bibliographical references and index.
ISBN 978-1-4094-4752-8 (hardback)
1. Firearms--Social aspects--Great Britain--History. 2. Firearms--Social aspects--United States--History. 3. Social change--History. 4. Great Britain--Colonies--History. 5. United States--Territorial expansion. 6. Imperialism--History. 7. Great Britain--History, Military. 8. United States--History, Military. I. Jones, Karen R., 1972- II. Macola, Giacomo. III. Welch, David, 1950-
U897.G7C85 2013
683.40941'09034--dc23

2013000829

ISBN 13: 978-1-4094-4752-8 (hbk)
ISBN 13: 978-1-03-292192-1 (pbk)
ISBN 13: 978-1-315-56461-6 (ebk)

DOI: 10.4324/9781315564616

Contents

List of Figures, Maps and Tables

Figures

Maps

Tables

Notes on Contributors

Simon Ball is Professor of International History and Politics at the University of Leeds. His most recent book is *The Bitter Sea: The Struggle for Mastery in the Mediterranean* (2010). His chapter in this volume is the fruit of a project charting the impact of small arms on international politics.

Ian F.W. Beckett is Professor of Military History at the University of Kent. A Fellow of the Royal Historical Society and Chairman of the Council of the Army Records Society, he has previously held chairs in the US as well as the UK. His many publications include *The Victorians at War* (2003) and *Wolseley in Ashanti: The Asante War Journal and Correspondence of Major General Sir Garnet Wolseley, 1873–74* (2009).

Timothy Bowman was educated at Queen's University Belfast and the University of Luton. He is currently Senior Lecturer in Modern British Military History at the University of Kent. His publications include *Carson's Army: The Ulster Volunteer Force, 1910–22* (2007) and, with Mark Connelly, *The Edwardian Army: Recruiting, Training and Deploying the British Army, 1902–1914* (2012).

Jason Bruner is a doctoral candidate in the History Department at Princeton Theological Seminary. He has published articles on the relationship between European imperialism and religious change in Africa and Asia in the nineteenth and twentieth centuries. His dissertation is entitled 'The Politics of Public Confession in the East African Revival in Uganda, ca. 1930–1950'.

Matthew Cragoe is Professor of Modern British History at the University of Sussex. He has written extensively on the history of the Victorian countryside and is currently preparing a new monograph on the long-term effects of eighteenth-century parliamentary enclosure on the communities of Midland England.

Jack Hogan is a doctoral candidate in the School of History of the University of Kent. His thesis – a study of slavery and its abolition in Barotseland, Western Zambia – is due to be completed in 2013.

Karen Jones is Senior Lecturer in History at the University of Kent, and specialises in US and environmental history. She has published widely on environmental issues and the American West specifically, and is currently completing a manuscript for the University Press of Colorado on hunting, nature and the nineteenth-century American West.

Spencer Jones teaches at the University of Birmingham and the University of Wolverhampton. His research focuses on the training, tactics and combat experience of the British Army. His latest work on the subject is *From Boer War to World War: Tactical Reform in the British Army 1902–1914* (2012).

Giacomo Macola is Senior Lecturer in African History at the University of Kent. A specialist in central African history, his latest monograph is entitled *Liberal Nationalism in Central Africa: A Biography of Harry Mwaanga Nkumbula* (2010). He is currently writing a social history of the gun in central Africa to the early twentieth century.

Bill Nasson is a Professor of History at the University of Stellenbosch. His main interest lies in the history of war and society in modern South Africa and in the British Empire. A co-editor of the *Cambridge History of South Africa, Volume Two* (2011), his other recent publications include *Springboks on the Somme: South Africa in the Great War, 1914–1918* (2007), *The War for South Africa: The Anglo-Boer War 1899–1902* (2010), and *South Africa at War, 1939–1945* (2012). He has never fired a gun, in anger or otherwise.

Gianluca Pastori is Lecturer in the History of International Relations at the Catholic University of the Sacred Heart, Milan, Italy. He has written extensively on the history of international politics and on the cultural and ideological foundations of imperial systems. He is currently working on a new monograph on north-west India between the late nineteenth and the early twentieth centuries.

Matthew C. Ward is Senior Lecturer in American History at the University of Dundee. He has written extensively on the history of the early American frontier. His latest monograph is *The Battle for Quebec, 1759* (2005).

David Welch is Professor of Modern History and Director of the Centre for the Study of War, Propaganda and Society at the University of Kent. His books include *Germany, Propaganda and Total War, 1914–1918* (2000) and *The Third Reich: Politics and Propaganda* (2002). He is co-author of *Propaganda and Mass Persuasion: A Historical Encylopedia, 1500 to the Present* (2003) and editor (with Jo Fox) of *Justifying War: Propaganda, Politics and the Modern Age* (2012). He is currently writing a history of propaganda in the twentieth and twenty-first centuries.

Kevin Yuill is Senior Lecturer in American History at the University of Sunderland. He researches race and liberalism in the United States in the nineteenth and twentieth centuries. His most recent book is *Richard Nixon and the Rise of Affirmative Action: The Pursuit of Equality in an Era of Limits* (2006). He is currently preparing a monograph on the 1924 Immigration Act.

Introduction: New Perspectives on Firearms in the Age of Empire

Karen Jones, Giacomo Macola and David Welch

This book steals firearms from the clutches of encyclopaedists and technical enthusiasts, for its central concern is the study of the processes through which firearms and societies have shaped one another across time and space. The deterministic perspectives that dominate most studies of the technological dimension of imperialism are here rejected in favour of approaches that foreground the symbolical value of material things. By moving beyond the mere utility value of guns, the present collection shows that, between the eighteenth and twentieth centuries, the age of modern empires, firearms were much more than weapons of human destruction and/or tools of material production. To be sure, such predictable patterns of gun usage still demand our attention, and it would be unwise entirely to sidestep debates about the different technical properties of successive generations of firearms. But – we contend – it is only by exploring the cultural symbolism of this pervasive technological artefact that its centrality to class, gender and ethnic identities in both the metropolis and the colonies can be accurately assessed. Similarly, the imbrications of guns in racial ideologies and practices of imperial rule are poorly served by utilitarian understandings that do not distinguish between what firearms did and the set of values, meanings and skills that they were taken to embody. Finally, there is a need to explore the manner in which firearms have been projected for different propaganda objectives: from the pragmatic consideration of employing firepower in military strategy and training to the more popular romanticism associated with particular weapons and certain types of fighters. The rest of this introduction is intended to substantiate these claims and illustrate the new perspectives on the history of firearms thrown open by the adoption of culturally sensitive research agendas.[1]

[1] This volume developed out of an international conference held at the University of Kent in May 2011. The editors would like to thank all those who participated in the conference. Sadly, it proved impossible to include all the contributions in this volume.

Environment, Class and Gender on the Imperial Frontier

In *The Wilderness Hunter* (1893), Theodore Roosevelt noted how 'there is endless variety of opinion about rifles'.[2] Champion of 'the strenuous life', pursuer of big game across North America and Africa, and a pioneering conservationist, Roosevelt's remark reflected his own abiding interest in firearms as well as the ubiquity of gun talk around the hunter's campfire. In a broader sense, Roosevelt's comment paid heed to the complex stratifications of meaning surrounding guns on the imperial frontier. As the chapters in part one highlight, firearms served as vital tools in wild country for the harvesting of animal capital: for subsistence, sport and soldiering. But guns were also invested with a set of cultural codes that spoke of their important symbolic value. As signifiers of colonial might, masculine prowess, elitist adventuring and ritualised control over space, the firearm conveyed a power far beyond its ballistics range.

Looming large in this section is a preoccupation with 'the frontier', a label that, in an American context, is festooned with allusions of illustrious national expansion. Often read as a synonym for 'the West', the frontier earned renown courtesy of Frederick Jackson Turner's so-entitled thesis, outlining the line of progress (marching inexorably across the continent) at which point 'civilization' and 'savagery' met, and settlers learned stalwart values of democracy and individualism.[3] In the latter years of the twentieth century, revisionist historians lambasted Turner's thesis as outmoded and jingoistic, a methodological '"F" word' that failed to consider history beyond the frame of the Anglo-Saxon white male. Now, with the 'new Western history' securely installed as a dominant paradigm, an opportunity exists to revisit the folklore of the frontier in the light of new approaches.[4] The mythic West of gunfights and glory may be largely fabricated, but valuable insight can be gained from exploring the mechanics of its construction, dissemination, utility and longevity. Indeed, as chapters here point out (Jones and Ward), the history of firearms on the American frontier is a complex and intriguing (not to mention politically charged) one. In the 'first West' of the trans-Appalachia, guns were not the favoured tools of the toiling subsistence farmer, while the cowboy and the sheriff wandering the streets of the trans-Mississippi cattle town in the 1870s were far from trigger-happy. Statistics on gun ownership and crime (see Ward, this volume, as well as Robert Dykstra's masterful survey of Dodge City) point towards a world in which interpersonal violence was sporadic, inglorious and often enacted by fist, broken bottle or

---

2 Theodore Roosevelt, *The Wilderness Hunter* (New York, 1926; 1st edn, 1893), pp. 370–372.

3 Frederick J. Turner, *The Frontier in American History* (New York, 1921).

4 Patricia N. Limerick, *Legacy of Conquest: The Unbroken Past of the American West* (New York, 1987); Kerwin L. Klein, 'Reclaiming the "F" Word, or Being and Becoming Postmodern', *Pacific Historical Review*, 65/2 (1996), pp. 179–215.

knife, rather than long rifle or Colt '45 – a salient rendition of what historian Elliott West has called the 'longer, grimmer, but more interesting story' of the West.[5]

At the same time, these years saw the crafting of a popular culture of gunplay in which pioneer heroes earned legendary status in literature, art and theatre for their manly swagger and marksmanship skills. Much of this allure – what John Cawelti dubs the 'six gun mystique' – heralded from associations with processes of frontiering, westward expansion, codes of masculine storytelling and democratic rights to bear arms.[6] The iconography surrounding Daniel Boone is instructive (Ward, this volume), as are the adventures of the heroic hunter hero archetype through the nineteenth century and to the Pacific (Jones, this volume). William F. Cody – bison killer and Wild West showman – and George Custer – the flamboyant 'boy general' – became folkloric figures whose status emanated from their credentials as hunters armed with the Winchester repeating rifle.

Beyond the confines of the United States, associations between hunting and gun cultures prove equally pertinent. As chapters here suggest, the idea of the game trail as a place of manly adventuring that staved off the debilitating (and feminising) influences of urban industrial life gained purchase on both sides of the Atlantic (Bruner and Cragoe, this volume). In Africa, just as in North America, the colonial 'big game hunter' skilled in woodcraft and yet technologically advanced, primal and yet honourable in nature, always in possession of a worthy firearm, emerged as a redolent masculine role model.[7] On the imperial home front, such concerns manifested themselves in renewed interest in field sports and outdoor pastimes. Whether enacted in the extra-European theatres of the 'Wild West' or the African savannah – both landscapes

---

5 Robert R. Dykstra, *The Cattle Towns (New York, 1968) and, by the same author,* 'Body Counts and Murder Rates: The Contested Statistics of Western Violence', *Reviews in American History*, 31/4 (2003), pp. 554–563. See also Michael Bellesiles, *Arming America: The Origins of a National Gun Culture* (New York, 2000); Richard Slotkin, *Regeneration through Violence: The Mythology of the American Frontier, 1600–1800* (Middletown, 1973), and *The Fatal Environment: The Myth of the Frontier in the Age of Industrialization, 1800–1890* (New York, 1985); and Henry N. Smith, *Virgin Land: The American West as Symbol and Myth* (Cambridge, MA, 1950).

6 John Cawelti, *The Six Gun Mystique* (Bowling Green, 1999; 1st edn, 1971).

7 On hunting in the British Empire, see John M. Mackenzie, 'The Imperial Pioneer and Hunter and the British Masculine Stereotype in late Victorian and Edwardian Times', in J.A. Mangan and J. Wavin (eds), *Manliness and Morality: Middle Class Masculinity in Britain and* America (Manchester, 1987), pp. 176–195, and *The Empire of Nature: Hunting, Conservation and British Imperialism* (Manchester, 1988); Harriet Ritvo, *The Animal Estate: The English and Other Creatures in the Victorian Age* (Cambridge, MA, 1987). Hunting cultures in Canada are explored in Greg Gillespie, *Hunting for Empire: Narratives of Sport in Rupert's Land, 1840–1870* (Vancouver, 2007); Tina Loo, 'Of Moose and Men: Hunting for Masculinities in British Columbia, 1880–1939', *Western Historical Quarterly*, 32 (2002), pp. 296–319; and Karen Wonders, 'Hunting Narratives of the Age of Empire', *Environment and History*, 11/2 (2005), pp. 269–291.

that were configured as 'hunter's paradises' in the Euro-American imagination (Jones and Bruner, this volume) – or in the deer parks of Scotland and shooting ranges of the home counties, the pursuit of animals for sport was irrevocably bound up with the gun.

Focus on the leisure economy of the hunt should not obscure the fact that outdoor sports, albeit closed in the language of play and adventure, were deeply connected with processes of colonial land seizure and controls over both natural resources and access to firearms. The political economy and, indeed, the ecology of the hunt highlight its role as a locus of imperial power and an important agent of environmental transformation. As Jones points out (this volume), in the American context, the military helped out civilian parties, enacted their own cultures of the hunt as 'target practice' and broadcast a linguistic turn in which indigenous people were cast as enemies or 'bucks'. In various climes, the game trail became a richly discursive landscape in which codes of ritual and masculine proving, technological fetishism and storytelling gained report. Often, the hunting vernacular spoke of gunplay, exploration and masculine renewal, and appreciation of wild landscapes and spiritual immersion in nature. Stories were told about famed marksmen and weapons (here explored by Ward, Cragoe and Jones) and their animal quarries, from grizzly bears and the man-eating lions of Tsavo to wily pigeons. As environmental historian Thomas Dunlap has shown, such tales paid heed to the emerging connections between settlers and the landscapes they chose to inhabit.[8] Hunting on the imperial frontier became a critical conduit through which colonial cultures forged a sense of belonging.

Beneath the rhetoric, tales from the game trail carried forth markers of class, gender, and race – situating the gun (often, but not exclusively) as a tool of empire in the hands of the enfranchised. From belonging came ownership. As Bruner points out (this volume), the mechanics of the hunt (often described by the hunting fraternity in quasi-spiritual terms of connection to nature) allowed for common ground to be found in hunting for lion and 'fishing for men'. With indigenous people serving as guides and porters, opportunities for evangelising around the campfire proved commonplace. Invariably the idea of a missionary 'errand into the wilderness' brings to mind the scholarship of Lynn White, Jr and Roderick Nash on taming the wild and the analogous projects of metaphysical and material conversion that conjoined Christianity and industrial capitalism.[9]

[8] Thomas Dunlap, *Nature and the English Diaspora: Environment and History in the United States, Canada, Australia and New Zealand* (Cambridge, 1999). Other pioneering transnational surveys of empire and environments include: William Beinart and Lotte Hughes, *Environment and Empire* (Oxford, 2007); William Beinart and Peter Coates, *Environment and History: The Taming of Nature in the USA and South Africa* (London, 1995); and Libby Robin and Tom Griffiths (eds), *Ecology and Empire: Environmental History of Settler Societies* (Edinburgh, 1998).

[9] Lynn White, Jr, 'The Historical Roots of our Ecological Crisis', *Science*, 155/3767 (1967), pp. 1203–1207; Roderick Nash, *Wilderness and the American Mind* (New Haven, 1967).

Elsewhere, ritual behaviour, competitive firearms cultures and the political economy of the hunt (Jones and Cragoe, this volume) ensured that patina of glamorous adventuring was underscored by elitist prescriptions that validated certain forms of hunting and limited access to resources on the basis of race, class and regionalism. As Cragoe points out, the community of hunting on the imperial home front also witnessed bifurcations based on configurations of hunting hierarchy – reprising long-standing divisions between the gamekeeper and the poacher with new stratifications based on urban-rural and upper-working class dividers. At home and abroad, the gentleman sportsman communicated a masculine imperial identity based on 'rightful' use of animal resources that allowed him dominion over spaces, peoples and animals.

That said, a monolithic reading of the hunt proves as elusive as the trophy animal 'that got away'. For all the evidence of colonial machismo and assumed authority over natural resources, big game hunters ventured a complicated relationship with those they exerted tutelage over. US Cavalry officers pursued the Lakota as 'bucks', but also showed reverence for their equestrian skills, while civilian hunters on the nineteenth-century western frontier assumed indigenous names, attire, and skills in pursuit of sacred game. As Bruner points out (this volume), the missionary project entailed cultural exchange between imperial agent and subaltern as well as wholesale indoctrination.

The relationship between hunters and conservation also complicates the narrative of the imperial firearms frontier. As many big game animals became scarce at the turn of the nineteenth century, the so-called 'penitent butchers' acted to ensure the preservation of endangered species by lobbying for parks and reserves as well as seasons and gun controls. Others rejected the cult of the gun entirely to hunt with the newest modish technological device – the camera. The rise of the hunter naturalist community – as represented by the Boone and Crockett Club (1887), of which Theodore Roosevelt was a founder member, and the Society for the Preservation of the Wild Fauna of the Empire (1905) – suggested a fresh engagement on the game trail that spoke of 'fair play', gun restraint and an abiding interest in seeing an animal alive as well as a trophy head. Alongside displays of civic altruism and nascent environmental thinking, the sporting community used markers of race and class to cement their own authority over valuable wildlife resources. Guns had enabled the taking of animals on a grand scale, were fetishised in symbol and story and, now, with the wholesale transformation of the imperial frontier, comprised the focus of a colonial conversation about environments and 'rightful users'.[10]

---

10 A stalwart defence of the hunter as conservationist is made in John Reiger, *American Sportsmen and the Origins of Conservation* (Corvallis, 2001; 1st edn, 1975). Critical appraisals that point to racial and class-based restrictions on access include: Thomas Dunlap, 'Sport Hunting and Conservation', *Environmental Review*, 12/1 (1988), pp. 51–59; Karl Jacoby, *Crimes against Nature: Squatters, Poachers, Thieves, and the Hidden History of American Conservation* (Berkeley, 2001); and

Edged Weapons and the Politics of Indigenous Honour

A cultural approach to firearms must be sensitive not only to innovation, but also to continuity. In one way or another, all the chapters included in part two deal with the enduring relevance of edged weapons in the nineteenth and twentieth centuries. In an era commonly portrayed as being characterised by the military and cultural dominance of firearms (both before and after the so-called 'breechloader revolution'), there remained vast numbers of people on the frontier of imperial expansion in Africa and Asia who kept alive earlier tactical and weapons traditions. Sometimes, of course, this was the simple effect of commercial isolation and general lack of economic opportunities. Just as frequently, however, technological disengagement was the result of deliberate choices informed by local social structures and honour cultures.

The Ngoni of eastern Zambia examined by Macola (this volume) wilfully resisted the adoption of firearms for military purposes, as they regarded the new technology as corrupting and emasculating. Firearms threatened hegemonic notions of military bravery constructed around hand-to-hand combat with spear, shield and knobkerrie. In so doing, they also threatened to foreclose the opportunities for individual advancement inherent in the Ngoni's age-set regimental organisation. Technological disengagement, here, amounted to an act of conservation – one that was no less socio-culturally determined than strategies of technological appropriation. Because of KwaZulu-Natal's longer exposure to the new technology from the early nineteenth century, the Zulu's relationship with firearms was more nuanced (Hogan, this volume). Yet, although the Zulu developed greater skills in the tactical deployment of the new weapons than they have commonly been credited with, firearms remained culturally subordinate to the assegai, the central symbol of Zulu masculinity, honour and 'Zulu ideas of the Zulu way of war'.

By foregrounding the importance – and the persuasiveness – of indigenous discourses about weapons, the perspective adopted by Macola and Hogan complicates purely instrumentalist readings of the imperial construction of 'martial races', the subject of a thriving recent literature. The point is borne out by a brief discussion of Timothy Parsons' well-regarded social history of the King's African Rifles (KAR), a colonial regiment in British Kenya, Uganda and

Louis S. Warren, *The Hunter's Game: Poachers and Conservationists in Twentieth Century America* (New Haven, 1997). For Africa and the British Empire, see David Anderson and Richard H. Grove (eds), *Conservation in Africa: People, Policies and Practices* (Cambridge, 1987); Jane Carruthers, 'Tracking in Game Trails: Looking Afresh at the Politics of Environmental History in South Africa', *Environmental History*, 11/4 (2006), pp. 804–829; Martin V. Melosi, 'Equity, Eco-Racism and Environmental History', *Environmental History Review*, 19/3 (1995), pp. 1–16; and R.P. Neumann, 'Dukes, Earls, and Ersatz Edens: Aristocratic Nature Preservationists in Colonial Africa', *Environment and Planning D: Society and Space*, 14/1 (1996), pp. 79–98.

Malawi (Nyasaland).[11] As seen by Parsons, recruitment for the KAR in colonial Kenya (to which the bulk of the book is devoted) consisted of a process in which British authorities automatically categorised as 'martial' every ethnic group that proved willing to respond to calls for enlistment on account of its disadvantaged economic circumstances. In Parsons' understanding, then, 'African societies were most "martial" when taxation and land shortages forced them to seek paid employment, and educational limitations' reduced their options in the labour market.[12] Parsons does not seek to explore the relationships between KAR recruits, their pre-colonial weapons of choice and European appraisals of both. Yet such connections are likely to have been significant, given that the Kamba, the people who eventually emerged as Kenya's prime 'soldiers of the Queen',[13] had excelled in nineteenth-century hunting and trade largely on the strength of their skills with bows and arrows.[14] Before the Kamba began to enlist en masse from the 1930s, northern pastoralists had contributed a large number of recruits to the KAR's Kenyan battalions. To be sure, the latter's home areas were among the least prosperous of the colony; yet they were also regions with a long tradition of inter-communal cattle raiding carried out without the aid of guns.[15]

To posit a link between the absence or rejection of firearms and the attribution of martial qualities by imperial policy-makers and military leaders is not the same as saying that all future 'martial races' were lightly armed on the eve of, and during, the colonial conquest. Indeed, there are instances in which it was rather the abundance of guns that worked towards consolidating the bellicose image of specific groups. The Yao of southern Malawi are a good case in point.[16] Yet, it is clear – if we are allowed to use as many as two neologisms in the same sentence – that 'gunlessness' was an important, though hitherto largely unrecognised, route to 'martialness'. It is certainly not coincidental that the attempt to justify the reverse suffered at Isandlwana led British imperial authorities studiously

---

11 Timothy H. Parsons, *The African Rank-and-File: Social Implications of Colonial Military Service in the King's African Rifles, 1902–1964* (Portsmouth, NH, 1999).

12 Ibid., p. 9.

13 Timothy H. Parsons, '"Wakamba Warriors Are Soldiers of the Queen": The Evolution of the Kamba as a Martial Race, 1890–1970', *Ethnohistory*, 46/4 (1999), pp. 671–701.

14 Edward I. Steinhart, *Black Poachers, White Hunters: A Social History of Hunting in Colonial Kenya* (Oxford, 2006), chapter 3; Edward A. Alpers, *Ivory and Slaves in East Central Africa: Changing Patterns of International Trade to the Later Nineteenth Century* (London, 1975), p. 12.

15 See, e.g., Kennedy A. Mkutu, *Guns and Governance in the Rift Valley: Pastoralist Conflict and Small Arms* (Oxford, 2008), pp. 17–18.

16 The Yao dominated the Central African Rifles and, from 1902, the Nyasaland contingent of the KAR partly because the defeat of their slave-trading chiefs in the 1890s 'had cost hundreds of ... former musketeers their employment. Having formerly considered themselves as warriors, the ex-musketeers probably disdained the idea of resorting to the cultivation of land, which they regarded as women's work'. Risto Marjomaa, 'The Martial Spirit: Yao Soldiers in British Service in Nyasaland (Malawi), 1895–1939', *Journal of African History*, 44/3 (2003), p. 420.

to deemphasise the extent to which firearms had been incorporated into Zulu tactics (Hogan, this volume). Without these discursive elisions, it is unlikely that the Zulu would have emerged as the 'archetypal warriors of the British imperial imagination' and the very epitome of 'ultra-masculinity' (Nasson, this volume). The 'grammar of savagery' that edged weapons came to symbolise was thus profoundly ambivalent, as even the image of the 'treacherous' Afghan on India's North-West Frontier was predicated on a degree of grudging admiration or, at least, an 'inextricable mixture of villainy and valour' (Pastori, this volume).

A more general point stems from the above considerations. A focus on edged weapons and the honour cultures revolving around them help demonstrate that imperial taxonomies – including (but not confined to) the martial race ideology – were never the sole product of European agency and interests. The imperial frontier, as Macola (this volume) points out, was not 'a *tabula rasa* awaiting colonial inscription', and the potential for historical invention – as shown, for instance, by a whole body of works dealing with the creation and/or ossification of ethnic and gender identities in colonial Africa[17] – is never unlimited. However well taken, a focus on imperial fantasies and stereotyping (Pastori, this volume) must not lead us to lose sight of the fact that the history and social organisations of some communities had been more deeply shaped by the experience of warfare and militarism than those of others. The Ngoni of Zambia and Malawi came to be viewed as a 'martial race' between the late nineteenth and the early twentieth centuries, partly because, in their recent past, they had conquered large swathes of Central African territory, incorporating their previous inhabitants into a social system deliberately organised for raiding and warfare. The Sikhs – as even Heather Streets is forced to admit offhandedly – ended up being construed as India's most manly and reliable soldiers not only because they had sided with the British during the Rebellion of 1857, but also because, beginning in about 1700, their religion had been transformed into a 'martial creed' – one that led them to think 'about military service in terms of individual and collective honour' and one that had been able to energise a 'confederacy ... with a large and disciplined army of more than 75,000 men' by the early nineteenth century.[18] The aforementioned Kamba's reputation as 'effective soldiers and police' in colonial Kenya had, no doubt, much to do with British 'prejudices and attitudes ... in looking for members of "warlike tribes." However, we should not ignore the fact that this kind of outdoor life, involving travel, discipline, and a sense of adventure, was very much consonant with Kamba cultural expectations of what

17 Thomas Spear, 'Neo-traditionalism and the Limits of Invention in British Colonial Africa', *Journal of African History*, 44/1 (2003), pp. 3–27.

18 Heather Streets, *Martial Races: The Military, Race and Masculinity in British Imperial Culture, 1857–1914* (Manchester, 2004), pp. 63–64, 212.

was a suitable vocation for young men' socialised into the cult of hunting.[19] Even in circumstances where imperial myth-making was given a free rein, the resulting narratives were never one-way constructs; they were ones to which colonial, or proto-colonial, subjects made a decisive contribution with a view to negotiating for themselves the best possible terms of incorporation into the emerging new structures of domination. Colonial societies have been shown to have been able to project specific images of themselves for the consumption of the new masters.[20] The European fascination for edged weapons in the era of firearms can be safely assumed to have provided abundant grist for the mill of subaltern self-serving stereotyping.

Gun Laws, Race and Citizenship

Martial race ideology continued to influence twentieth-century developments (and its legacy still resonates powerfully in contemporary popular culture). In mature colonial contexts, however, the relationship between guns and race was recast. Internationally, beginning with the Brussels Conference of July 1890, the early years of the twentieth century witnessed a series of diplomatic and military initiatives intended to bring the thriving global trade in small arms under control. Internally, the disarmament of colonial subjects came frequently to be seen as a precondition for asserting the authority of the state and as the symbol of that curtailment of autochthonous citizenship rights on which the edifice of imperial domination was predicated. Even though – as shown by Ball (this volume) with reference to the Arabian frontier before World War I – the resilience of both local and foreign gun smugglers threatened constantly to sabotage colonial efforts at gun control,[21] it would be a mistake to underestimate the social and ideological significance of these same efforts, which, especially in settler-dominated regimes, took on the function of delimiting the boundaries of race.[22] Segregationist South Africa is a good case in point. There, not only were black Africans deprived of the right to own firearms, but even the arming of

---

19 Steinhart, *Black Poachers, White Hunters*, pp. 56–57.

20 See, e.g., Gwyn Prins, *The Hidden Hippopotamus: Reappraisal in African History. The Early Colonial Experience in Western Zambia* (Cambridge, 1980), and Patrick Harries, *Butterflies and Barbarians: Swiss Missionaries and Systems of Knowledge in South-East Africa* (Oxford, 2007).

21 The long history of gun trafficking along the border between present-day Angola, Zambia and Congo is examined by both Giacomo Macola, 'Reassessing the Significance of Firearms in Central Africa: The Case of North-Western Zambia to the 1920s', *Journal of African History*, 51/3 (2010), pp. 301–321, and Jean-Luc Vellut, 'Garenganze/Katanga – Bié – Benguela and Beyond: The Cycle of Rubber and Slaves at the Turn of the 20th Century', *Portuguese Studies Review*, 19/1–2 (2011), pp. 133–152.

22 William K. Storey, *Guns, Race and Power in Colonial South Africa* (Cambridge, 2008).

African policemen was viewed with profound anxiety. Such concerns, however, clashed with the state's dependence on 'black intermediaries to provide the control and specialized knowledge of Africans that a vast settler project entailed'. The end-product of this tension – as Keith Shear has noted – was a series of measures that 'emasculated black policemen institutionally and socially'.[23] Not the least significant features of such precautionary regime were the more and more stringent curbs placed on black policemen's access to firearms.[24]

Similar contradictions underpinned the workings of the South African Army (Union Defence Forces), from which, as Nasson (this volume) shows, black Africans were excluded in peace time. When world-war mobilisation made the enlistment of Africans necessary, a gun colour-bar was implemented that resulted in black Africans being wholly confined to unarmed labour units. Moreover, during World War II, race-based regulations were further tightened so as to include even such Coloured Cape Corps as had mustered combat battalions and served in East Africa and Palestine during World War I. The surreal image of black South African non-combatant soldiers pointlessly guarding airfields and prisoner-of-war camps armed solely with their knobkerries and assegais might be taken as symbolising the endpoint of the trajectory of edged weapons in the era of formal empires. From markers of agency and of the integrity of indigenous military systems on the eve of the colonial conquest, the stabbing spear – the prime cultural signifier of manliness in Zulu and Zulu-inspired societies – had become a symbol of degradation and rights deprivation in an oppressive and racially exclusive political order. The emasculating implications of the gun colour-bar – which South African authorities sought to justify in neo-traditional terms by appealing to increasingly obsolete arguments about culturally appropriate tactics and weapons – were not lost on Mshiyeni ka Dinzulu, the Acting Zulu Paramount, who demanded the formation of a 'Zulu Military Regiment trained in the manipulation of Big Guns', or on the Communist Party of South Africa, which launched an unsuccessful 'Give Him a Gun, NOW' campaign (Nasson, this volume).

These debates about firearms and citizenship – and the reactions they elicited from racially oppressed groups – also found a distorted echo in the United States between the nineteenth and the twentieth centuries (Yuill, this volume). In colonial America, race-specific legislation had made it possible sporadically to curtail black gun ownership. Such legislation offered a template for the southern states' infamous 'Black Codes' in the immediate aftermath of the Civil War. The passing of the Fourteenth Amendment to the US Constitution, however,

---

23 Keith Shear, '"Taken as Boys": The Politics of Black Police Employment and Experience in Early Twentieth-Century South Africa', in L. Lindsay and S. Miescher (eds), *Men and Masculinities in Modern Africa* (Portsmouth, NH, 2003), pp. 109–110.

24 Ibid., p. 118.

circumscribed the white supremacists' room for legal manoeuvring and forced them to change tactics. Late nineteenth-century gun control laws in the southern states were now couched in an ostensibly non-racial idiom; yet enforcement was far from being colour-blind, as the same laws were applied selectively with a view to disarming African-Americans, as part of a concerted effort to curtail the latter's newly obtained civil rights and liberties. In this context, the fight for the right to bear firearms became a synecdoche for the broader fight for black equality, and most African-American civil rights and community leaders resisted attempts at legal or extra-legal disarmament. However unpalatable this may sound to present-day liberal sensitivities, the fact remains that, for oppressed African-Americans in the southern states, living under the constant threat of white violence, disarmament meant helplessness and emasculation. The parallel with South Africa is again appropriate, for a similar link between access to firearms and the promise of freedom and emancipation also came to prominence during that country's liberation struggle, as powerfully attested, for instance, by the semi-mythical status attributed to Nelson Mandela's hidden Makarov pistol.[25]

Firearms in Popular and Military Cultures

The final section of the volume deals with firearms within military and paramilitary cultures and the manner in which guns were celebrated and utilised for propaganda purposes. Ian Beckett traces the history of a specific rifle, the Martini-Henry, exploring, *inter alia*, the process through which popular culture turned a sometimes problematic weapon into an icon. The Martini-Henry first entered service in 1871; it was a breech-loading single-shot lever-actuated rifle adopted by the British Army and used throughout the British Empire for 30 years. During this period, the British were involved in a number of colonial wars, notably the Anglo-Zulu War of 1879. Martini-Henry rifles were used at Rorke's Drift, where 139 soldiers successfully defended themselves against thousands of Zulu fighters. Considered at the time to be the state of the art in weaponry, the Martini-Henry was referred to by Rudyard Kipling in his poems and in his short story 'The Man Who Would be King' (1888). The defence of Rorke's Drift was subsequently immortalised in the British film *Zulu* (1964), starring Stanley Baker and Michael Caine, while the Martini-Henry also featured in Barrie Hughes's 1978 novel *The Martini-Henry*. In order to uncover the myths associated with the weapon, Beckett traces its roots and antecedents, beginning

[25] Garth Benneyworth, 'Armed and Trained: Nelson Mandela's 1962 Military Mission as Commander in Chief of Umkhonto we Sizwe and Provenance for his Buried Makarov Pistol', *South African Historical Journal*, 63/1 (2011), pp. 78–101.

with the crushing defeat suffered by the British at Isandlwana. The defeat – which took place at the same time as the Rorke's Drift encounter and which resulted in the greatest single day's loss of British troops between 1815 and 1914 – was put down, at the time, to the failure to open the Boxer-Henry ammunition boxes. Once again this myth has been perpetuated in British popular culture by a number of films (*The Man Who Would be King* [1975], *Zulu Dawn* [1979] and *The Four Feathers* [2002]). Beckett ably demonstrates that the present-day status of the Martini-Henry (not least among specialists and collectors) is almost entirely due to the cinema and to Michael Caine's order to 'Fire' in *Zulu*!

In his chapter on the Second Anglo-Boer War of 1899–1902, Spencer Jones points out that during the Victorian period the British Army fought over 230 wars. Most of these were 'small' wars against poorly equipped colonial foes. The exceptions to the rule were the Afrikaners of South Africa, who were well-equipped with modern rifles and with a military culture that emphasised firepower. The experiences of the British Army in conflict with the Boers confirmed the reputation of the Afrikaners as formidable opponents. What particularly impressed the British was their enemies' individual marksmanship.

The British Army had already suffered a humiliating defeat in the First Anglo-Boer War (1880–1881), and observers had pointed to a stark contrast between skilful Boer shooting and cumbersome British musketry. According to Jones, however, it was the Second Anglo-Boer War – a conflict that dwarfed earlier engagements in terms of scale, duration, cost and intensity – that left a lasting impression upon the British Army. Firepower proved to be an important battlefield factor, and the popular press were quick to attribute success to natural Boer skills. Jones re-examines British impressions of Boer marksmanship during this major conflict by focusing on three key elements that contributed to Boer firing accuracy, namely terrain, culture and equipment, demonstrating how they combined to produce unusually effective rifle fire. In the aftermath of the war, British admiration of Boer musketry would, according to Jones, play an important role in the British Army's musketry reforms in the years leading up to World War I and partly explains the favourable rifle skills of the British Expeditionary Force in the opening battles of 1914.

While the British Army were reflecting on military strategy and undergoing new forms of rifle training in the light of its experiences in the Boer wars, a different type of gun culture was emerging in Ireland. Between 1910 and 1921, a number of paramilitary forces (both Unionist and Nationalist) were established. Although such organisations drew on a long tradition of physical violence, in the initial period of their formation, both the Ulster Volunteer Force and the Irish Volunteers were largely unarmed. According to Bowman (this volume), despite increasing demands for arms from elements of their respective rank and file membership, it was not really until 1914 that an identifiable 'gun culture' emerged. Indeed, Bowman shows how the Ulster Volunteer Force (still wedded to parliamentary opposition) was actually attempting to withhold arms from its militant rank and file. The Irish Volunteers and the Irish Republican Army also comprised large numbers of inactive members. It was the threat of conscription early in 1918 that acted as a catalyst, leading to a sizeable expansion in membership.

Of course, it should also be recognised that a vibrant gun culture existed in Edwardian Britain. Firearm ownership, as Bowman points out, was increasing, allied to concepts of individual liberty (even though crime rates were falling). Before 1920 there were, in fact, few legal restraints on carrying firearms in Britain. When the government forbade the importation of arms and ammunition into Ireland in December 1913, it met with little success. Not surprisingly, gun-running became a major feature of the ensuing period in Ireland, with both Nationalists and Republicans claiming propaganda coups. Bowman argues that it played a significant part in the Easter Rising in 1916 – all of which has served to embroider a deep sense of romanticism that has been associated with gun culture and the smuggling of firearms. And yet, although there was a strong paramilitary and also civilian gun culture in Ireland in these years, not all paramilitaries enthusiastically endorsed the gun. Bowman's chapter thus helps correct the romanticised impression of the armed 'freedom fighter' that has entered into the Irish popular psyche. Indeed, in a final ironical twist, Bowman points out that the worst armed paramilitary force of this period proved to be the most militant.

* * *

Firearms have been studied by historians mainly as means of human destruction and material production. Yet, throughout their history, they have been invested with a whole array of additional social meanings. This collection places these latter at the centre of the analysis and extends the study of guns beyond the confines of military history and the examination of their impact on specific events and outcomes. By bringing cultural perspectives to bear on the subject, the contributors explore the densely interwoven relationships between firearms and broad processes of social change. They, moreover, do so within a global frame of reference. By bringing together historians of different periods and regions, *A Cultural History of Firearms in the Age of Empire* overcomes traditional compartmentalisations of historical knowledge and encourages the drawing of novel and illuminating comparisons across time and space.

PART I

Adopting Guns: Environment, Class and Gender on the Imperial Frontier

Chapter 1

Guns, Violence and Identity on the Trans-Appalachian American Frontier

Matthew C. Ward

In 1860, residents of Frankfort, the state capital of Kentucky, sought to honour the memory of the most famous pioneer of their state, Daniel Boone. Boone had died in 1820 in Missouri, but 25 years later his body was reinterred in Frankfort. Now, another 15 years later, the town's residents sought to mark his grave with a fitting memorial. The choice of memorial was significant – a 15-foot stone column with relief panels on each of its sides portraying scenes from Boone's life. One panel depicted his wife, Rebecca, in a typical feminine domestic role milking a cow. The other three reliefs showed Boone in various masculine activities: in one Boone the hunter sits with a slaughtered deer at his feet, a rifle at his side; in another Boone stands offering advice to a fellow settler, a rifle in his hand; in the final relief he battles desperately with a Native American warrior, parrying his opponent's blows with a rifle. That in every relief, and in almost every portrait ever painted, Boone is depicted bearing a Kentucky rifle says much about the association between the rifle and the image of Kentucky settlers in nineteenth-century America.[1]

By the early nineteenth century the rifle had become symbolic of the 'First West', the region across the Appalachian Mountains settled during and in the immediate wake of the Revolutionary War. Hunters, armed with rifles, were viewed as the archetypal settlers of the region; militiamen sniping from the cover of the woods, armed with rifles, were viewed as the region's principal heroes of the Revolutionary War. However, it is surprising that a weapon based upon the German Jaeger, developed by gunsmiths in Lancaster County, Pennsylvania, and known to residents of early America as the Pennsylvania rifle, should become known by the early nineteenth century as the Kentucky rifle and have become intimately associated with the development of the Trans-Appalachian frontier.[2]

---

1 Meredith Brown, *Frontiersman: Daniel Boone and the Making of America* (Baton Rouge, 2008), pp. 269–271; John M. Faragher, *Daniel Boone: The Life and Legend of an American Pioneer* (New York, 1993), pp. 356–360.

2 M.L. Brown, *Firearms in Colonial America: The Impact on History and Technology, 1492–1792* (Washington DC, 1980), pp. 28–30, 261–269; James B. Whisker, *Arms Makers of Colonial America* (Selinsgrove, 1992).

Figure 1.1 The Life of Daniel Boone from his memorial at Frankfort Cemetery, Kentucky
Courtesy of Matthew C. Ward.

In few places has the gun become so central to the imagery and identity of a region. Yet frontiersmen were not always so closely associated with their weapons and in many ways the association of the frontiersman and the rifle is highly ironic, for guns were more important to the Native American foes of early western settlers than they were to the settlers themselves. The arrival of firearms completely transformed the nature of Native American warfare. Before the arrival of European guns, Native Americans had fought large open battles sometimes even with the opposing armies lined up in ranks. The heavy causalities inflicted by firearms meant that this battle tactic was abandoned and replaced with what would become known as the 'skulking way of war'. Not only did the arrival of firearms transform Native American warfare, but it also had a more profound effect on Native American society by transforming the nature and role of hunting. Pre-contact native cultures had relied on a mix of agriculture, hunting and gathering. By the eighteenth century hunting had become essential to the Native American economy, and the fur trade now supplied many necessities of life that had previously been produced in Indian villages. Gender roles were transformed, as women's work, particularly agricultural labour, became less important. Competition between Native American peoples for skins, and between Native Americans and Europeans for land, dramatically increased conflict on the frontier and meant that guns were also essential for protection while hunting and travelling. Consequently, by the middle of the eighteenth century, Native Americans were almost certainly the most heavily armed people on earth.[3]

By contrast guns seem to have had a much slower impact on the life of Euro-Americans. Frontiersmen were not always skilled woodsmen. Before the outbreak of the Seven Years War in 1754, residents of the backcountry, particularly in the middle colonies and in Virginia, had little reason to use or possess firearms. Some early settlers were pacifist Quakers or Mennonites for whom the use of firearms was anathema; for others who did not have religious scruples against gun ownership, a gun was an expensive item to be purchased only if there was a clear need for its use for hunting or defence. In the middle of the eighteenth century a Pennsylvania rifle cost around four pounds sterling at a time when the average wages for an agricultural labourer were around two shillings and six pence per day.[4] It therefore represented over one month's income for a

---

3 Kathryn E. Holland Braund, *Deerskins and Duffels: The Creek Indian Trade with Anglo-America, 1685–1815* (Lincoln, 1993), pp. 121–163; Colin Calloway, *New Worlds for All: Indians, Europeans and the Remaking of Early America* (Baltimore, 1997), pp. 92–93, 191; Armstrong Starkey, *European and Native American Warfare, 1675–1815* (London, 1998), pp. 20–24; Daniel K. Richter, *Facing East from Indian Country: A Native History of Early America* (Cambridge MA, 2001), pp. 174–175.

4 Walter S. Dunn, *The New Imperial Economy: The British Army and the American Frontier, 1764–1768* (Westport, 2001), p. 98; Donald R. Adams, Jr, 'Prices and Wages in Maryland, 1750–1850', *Journal of Economic History*, 46/3 (1986), p. 632.

typical settler. In the colonies of New Jersey, Pennsylvania and Virginia in the mid-eighteenth century there seemed little need to possess a gun for defence. There had been no major Indian conflict on the frontiers of these colonies since the 1670s. A few settlers did hunt to supplement their meagre incomes and ventured into the woods. While English laws had forbidden hunting, in the North American colonies, hunting laws were steadily liberalised, although there were still some restrictions placed on what game could be hunted and in which season. A Virginia law of 1734, for instance, restricted the hunting of deer to the autumn, although there were few restrictions placed on those who needed to hunt 'for the necessary subsistence of himself or family'.[5]

However, most of those who participated in the fur trade did not themselves head out west to hunt, but instead exchanged agricultural surpluses, particularly alcohol, directly with Indians who passed by their door. These small-scale trading activities were not encouraged, particularly when they involved, as they often did, the sale of alcohol, and the colonial authorities made repeated attempts to regulate liquor traders. Despite such attempts, the small-scale trade remained widespread and involved hundreds of families. However, while these families may have been involved in the fur trade, there was little need for them to own a gun.[6]

Settlers who sought principally to provide meat for their family may also have found more effective means than hunting with a gun. Before the 1770s, most settlers would have only had access to a musket. Muskets might have been of some use in hunting larger game, such as deer or bison, but for the smaller game that most settlers might have hunted in the colonial backcountry muskets were of limited use. They were heavy, weighting over 15 pounds, had a very short range, and the noise and smoke would scare away any game for miles. Except perhaps for the few settlers who lived on the very furthest western fringes of settlement, where larger game was more plentiful, if a backcountry settler wished to feed his family, it was far more efficient to trap small animals and birds or to fish, rather than to attempt to use an expensive gun in search of small game.[7]

---

5 William W. Hening (ed.), *The Statutes at Large: Being a Collection of all the Laws of Virginia, From the First Session of the Legislature in the Year 1619* (13 vols, Richmond Samuel Pleasants, 1809), vol. 5, p. 61; P.B. Munsche, *Gentlemen and Poachers: The English Game Laws, 1671–1831* (Cambridge, 1981).

6 Proceedings of Criminal Court, Box 6, 1750–1759, Cumberland County Historical Society, Carlisle; Samuel Hazard (ed.), *Minutes of the Provincial Council of Pennsylvania: from the Organization to the Termination of the Proprietary Government* (16 vols, Harrisburg, 1838–1853), vol. 5, p. 628, 749, vol 6, p. 149; Stephen H. Cutcliffe, 'Indians, Furs and Empires: The Changing Policies of New York and Pennsylvania, 1674–1768', unpublished PhD thesis, Lehigh University, 1976, p. 221.

7 Whisker, *Arms Makers of Colonial America*, pp. 11–22; Brown, *Firearms in Colonial America*, pp. 91–95.

During the Seven Years War complaints about settlers' incompetence in the use of firearms were commonplace and were made by both British and American officers. For instance, William Parsons, a commander of the Pennsylvania provincial forces, reported that his men had great difficulty using their guns and were 'generally as much afraid to fire them, as they would be to meet an Indian'.[8] George Washington complained repeatedly about the Virginia militia's lack of martial skills and in particular the ways in which instead of ranging quietly through the woods in search of the enemy, they would dash 'hooping' and 'hallooing', scaring off game and warning the enemy of their presence.[9] Such comments can be found throughout the journals and letters of both British and American officers who served on the frontier during the Seven Years War. Much of the reason for this lack of familiarity with firearms seems to have been that many settlers did not own guns. Swiss-born British officer Colonel Henry Bouquet reported in 1758 that, as the Pennsylvania troops mustered, only 'half ... have their own Arms, the rest Walks with Sticks'.[10]

The Seven Years War, however, marked a turning point in the relationship between settlers and guns. Native American raiding parties had wrought havoc on the colonial frontier during the course of the war. Over 2,000 settlers were killed, even more captured, and settlers abandoned an area of around 30,000 square miles along the frontier from New Jersey to North Carolina. In 1763 war erupted again on the frontier in Pontiac's War. Tales of Indian attacks and fear of Indian cruelty permeated backcountry settlements and became part of backcountry folklore. Whereas there had been no conflict on the frontiers for 80 years, the Seven Years War marked the first stage of continual conflict that some historians have termed 'The Sixty Years' War'. For the first time there were compelling reasons for backcountry settlers to possess a gun for self-defence.[11]

The Seven Years War and Pontiac's War also dramatically increased the number of settlers who hunted in the west and literally opened up the west to settlers. Thousands of backcountry settlers had served in the provincial forces or had worked as teamsters or road-builders for the army and had seen the west first-hand. Daniel Boone, for instance, served as a teamster on Braddock's expedition. Many were impressed by the fertility of the soil and the abundance of land in the west, and after the war sought to return to the west to hunt. The

---

8 William Parsons to Gov. Morris, 14 July 1756, Northampton County Records: Miscellaneous Papers 1:209, Historical Society of Pennsylvania, Philadelphia.

9 George Washington to Gov. Dinwiddie, 9 November 1756, in W.W. Abbot, D. Twohig and P.D. Chase (eds), *The Papers of George Washington* (10 vols, Charlottesville, 1983), vol. 4, p. 1.

10 Col. Bouquet to Sir John St Clair, 3 June 1758, in L. Waddell et al. (eds), *The Papers of Henry Bouquet* (6 vols, Harrisburg, 1951–1994), vol. 2, p. 23.

11 David C. Skaggs, 'The Sixty Years' War for the Great Lakes, 1754–1814: An Overview', in D.C. Skaggs and L.L. Nelson (eds), *The Sixty Years' War for the Great Lakes, 1754–1814* (East Lansing, 2001), pp. 1–20.

British Army facilitated the growth of hunting by constructing a network of roads and stations to move supplies and troops into the region. The new British posts in the west, such as Fort Pitt and Fort Vincennes, all paid hunters for any meat that they could supply, and therefore hunters could profit not only from the sale of their skins, but also from the sale of meat.[12]

By the 1760s, some 'long-hunters', such as Daniel Boone, were even venturing across the Appalachians on hunts lasting many months. James Dysart, for example, had emigrated from Ireland to Philadelphia in 1762. Over the next several years, Dysart gradually made his way to the western settlements of Pennsylvania and began to learn to hunt and to use the rifle. Towards the end of 1769 he enrolled in a party of around 40 men bound for Kentucky to hunt for up to a year. For these few men, hunting was both a viable commercial activity and, as Steven Aron has demonstrated, an opportunity to escape from farm work and emasculating dependence upon their fathers. The number of men who began to participate in such hunting expeditions was substantial. Angus McDonald wrote from Fort Burd, on Pennsylvania's western frontier, 'Here Comes Such Crowds of Hunters out of the Inhabitence as fills those woods at which the Indians seems very much disturbed and say the white people Kills all there Deer.'[13] Obviously for hunters travelling across the Appalachians in search of game, there was a clear necessity to possess and be skilled in the use of firearms.

The American Revolution in the west further intensified the need for ordinary settlers to own guns for protection from attack. As pioneers crossed the Appalachian Mountains and founded new settlements in Kentucky, they faced almost constant attack from Indians, earning Kentucky the moniker of the 'dark and bloody ground'. Attacks on frontier settlements could be sudden and unexpected, and there was every reason for settlers to keep their firearms close at hand. Joseph Smothers reported how a party of Indians had managed to surprise one frontier settlement and to get undetected into a cabin through the chimney, scalping one of the women inside and taking the other two captive before anyone even knew of the attack. Settlers in Kentucky had very pressing

---

[12] Faragher, *Daniel Boone*: pp. 69–70; Matthew C. Ward, *Breaking the Backcountry: The Seven Years' War in Virginia and Pennsylvania, 1754–1765* (Pittsburgh, 2003), pp. 71–72; Gen. Forbes to Gen. Amherst, 26 January 1759 National Archives of the UK, Kew, London, WO34/44; Col. Bouquet to Gen. Stanwix, 26 April 1760, in Waddell et al., *Papers of Henry Bouquet*, vol. 4, pp. 541–543.

[13] Angus McDonald to Col. Bouquet, 25 October 1761, in Waddell et al., *Papers of Henry Bouquet*, vol. 5, p. 840; Richard Peters to Conrad Weiser, 21 February 1760, Conrad Weiser Papers, 2:169, Historical Society of Pennsylvania; John B. Dysart to Lyman Copeland Draper, 27 March 1849, Boone Papers, 5C61 Draper Manuscript Collection, Wisconsin Historical Society, Madison; Stephen Aron, *How the West Was Lost: The Transformation of Kentucky from Daniel Boone to Henry Clay* (Baltimore, 1996), p. 26.

reasons to own guns, and particularly rifles, for defence, and had every reason to be skilled in their use.[14]

By the 1790s firearms had become vitally important to the lives of frontier and backcountry settlers in the Trans-Appalachian West. However, it is difficult, if not impossible, to gauge the exact extent of backcountry settlers' experiences and familiarity with the use of guns, and the extent of gun ownership in early America remains a highly contentious issue. In *Arming America*, Michael Bellesiles has argued that 'only a small percentage of the American population' owned guns before the middle of the eighteenth century.[15] Bellesiles's work, particularly his analysis of probate records, has been subject to intense scrutiny and largely discredited, but even if more than 'a small percentage' of early Americans owned guns, gun ownership was certainly far from universal. Estate inventories from Virginia in the 1720s reveal that just over half of the estates contained firearms. By 1774 this percentage had risen to almost two-thirds. This was somewhat higher than in Massachusetts, where fewer than half the inventories on the eve of the Revolution contained firearms, and substantially higher than in England, where fewer than one in seven inventories in London contained firearms.[16]

A preliminary analysis of estate inventories in Mason and Muhlenberg Counties, Kentucky, at the beginning of the nineteenth century suggests a similar pattern to Virginia, with around two-thirds of estate inventories containing firearms. However, many of these weapons were described as old or broken. Firearms do seem to have been much more common in North America than they were in Great Britain. However, in Kentucky and Virginia, they were possessed by far fewer households than were horses. In Christian County in 1805, for instance, 11 out of 12 households possessed a horse.[17]

Although guns may not have been as ubiquitous as horses in early Kentucky, for ordinary settlers they became more symbolic of western life than horses because of the connotations of the two and in particular their association with

---

14 Interview with Joseph Smothers, Kentucky Papers, 12CC96-97, Draper Manuscript Collection; Brown, *Frontiersman*, pp. 91–103.

15 Michael Bellesiles, *Arming America: The Origins of a National Gun Culture* (New York, 2000), p. 149. For a broad discussion of Bellesiles's findings, see 'Forum: Historians and Guns', *William and Mary Quarterly*, 59/1 (2002), pp. 203–268.

16 Of 298 Virginia estate inventories from the 1720s, 51 per cent contained firearms. By 1774 the percentage had risen to 60 per cent of a sample of 141 Virginia inventories. In Massachusetts 45.6 per cent of a sample of 298 estate inventories in 1774 contained firearms, whereas in England only 13.4 per cent of a sample of 172 London inventories and 10.4 per cent of a sample of 260 Worcestershire inventories contained firearms. Carole Shammas, *Wealth, Household Expenditure, and Consumer Goods in Preindustrial England and America, 1550–1800* [computer file]. Colchester, Essex: UK Data Archive [distributor], May 1993. SN: 2994, http://dx.doi.org/10.5255/UKDA-SN-2994-1.

17 Mason County Will Books, Muhlenberg County Will Books, Kentucky State Archives; Christian County Tax List, 1805, Kentucky State Archives.

the construction of masculinity. Emigration across the Appalachian Mountains challenged the masculinity of western men in several ways. Most basically, many emigrant men found that they were unable to provide for and protect their families. Women and children were exposed to the constant fear, if not reality, of Indian attacks. Sarah Graham, for instance, remembered how as a child in Kentucky her family had been forced to take shelter in frontier forts and she had witnessed first-hand Indians scalping the dead. Frontier stations and forts could not always provide protection and on occasion they fell to Indians and news of the fall of a frontier fort spread terror through the western settlements. The end of the Revolutionary War did not end the attacks on Kentucky, which in many ways intensified after the massacre of the Delaware community at Gnadenhütten in 1782. Twice after 1788 settlers in Lexington, the largest and oldest town in Kentucky and a major settlement, were forced to take shelter in their forts and in 1794 a raiding party even attacked the state capital of Frankfort, stealing horses from the middle of town and appearing at the State House.[18]

Living with such tensions and fear could have long-term implications for settlers. Sarah Graham reported that she knew 14 settlers who committed suicide. Others spent years trying to reconstruct their families. James McIlvaine reported that Joe Young had 'hunted for (3 yrs.) [for] his w[ife] before he found her. Found her at last on the Ohio River. The Indians had traded her to the French. [He] Always cried freely, talking about it.'[19] Providing safety and protection for one's family was the most basic definition of masculinity in the late eighteenth century, yet many men who emigrated to Kentucky with their families had failed to do that.

Those settlers who arrived later in the Trans-Appalachian West, while they may not have had to face up to Indian attacks, still faced disappointment and failure that challenged their identity as providers for their families. Settlers were lured to Kentucky by the belief that land was all but free for the taking. Early settlers spoke of 'tomahawk rights' in the woods, a right to land to each settler who could clear a plot and establish his family there. But such rights, while widely recognised by the settlers themselves, had no firm legal basis. Many settlers who arrived in Kentucky, expecting to be able to claim lands simply by

---

[18] Interview with Sarah Graham, interview with Daniel Bryan, Kentucky Papers 12CC46 22C27-28, Draper Manuscript Collection; John May to Samuel Beall, April 15, 1780, Beall-Booth Family Papers, A/B365, Filson Historical Society, Louisville; Fredrika J. Teute, 'Land, Liberty, and Labor in the Post-Revolutionary Era: Kentucky as the Promised Land', unpublished PhD thesis, Johns Hopkins University, 1988, pp. 205–206.

[19] Interview with Sarah Graham, interview with James McIlvaine, Kentucky Papers, 12CC47 12CC58, Draper Manuscript Collection

settling upon them, soon found themselves sorely disappointed and unable to acquire land.[20]

Even those who could afford the costs of buying land and registering their claims soon found that they faced innumerable challenges. The system of recording land claims in Kentucky using the ancient 'metes and bounds' tradition, where surveyors recorded the direction and distance of the plot by relation to geographic features such as streams, rocks and trees, proved completely inadequate in a territory where settlers arrived so quickly and there were so many competing claims. In 1797 the state's Surveyor General reported to the legislature that 'there has been Lands granted to sundries in the State above 24 Millions of Acres, [but] that all the Counties contain only 12,476,116 Acres, so that some persons will fall short'.[21] Twice as many acres had already been granted and surveyed than there were total acres in the state.

This opened up a new route of speculation. Land-speculators found that they if they could identify tracts of land where there were several competing claims, they could buy up some of the claims and take any remaining claimants to court. John Preston's agent, John Smith, specifically sought out tracts of land where the settlers' claims were in doubt, to buy up disputed claims and promising Preston that 'those that will not become tenants I shall eject'.[22] The ability of the wealthy elite to use the courts to monopolise land in early Kentucky meant that many settlers found their dreams of economic security in the west quickly dashed, yet the lure and reputation of Kentucky was so strong that settlers kept on travelling to the state. Moses Austin, travelling through Kentucky in 1796, was appalled at the poverty that he saw. He was horrified to see women and children

> with out Shoe or Stocking, and barely as maney rags as covers their Nakedness, with out money or provisions ... Ask these Pilgrims what they expect ... the Answer is Land. have you any. No, but I expect I can git it. have you anything to pay for land, No ... can any thing be more Absurd than the Conduct of man[?] here is hundreds Travelling hundreds of Miles, they Know not for what Nor Whither, except its to Kentucky ... when arriv.d at this Heaven in Idea what do they find? a goodly land I will allow[,] but to them [a] forbiden Land. exhausted and worn down with distress and disappointment they are at last Oblig'd to become hewers of wood and Drawers of water.[23]

---

20 Joseph Doddridge, *Notes on the Settlement and Indian Wars of the Western Parts of Virginia and Pennsylvania, from 1763 to 1783* (Pittsburgh, 1912), p. 81.

21 Thomas Perkins Abernethy, *Western Lands and the American Revolution* (New York, 1937), p. 228.

22 John Smith to John Preston, 3 December 1793, Preston Family Papers, Joyes Collection, Box 1, Folder 3, Filson Historical Society.

23 Moses Austin, 'The Journal', *American Historical Review*, 5/3 (1900), pp. 525–526.

By the 1790s, fewer than half of the settlers in Kentucky had acquired land, while the wealthiest elites monopolised thousands of acres. In Harrison County in 1797, 425 out of 813 households or 52.3 per cent of all households owned no land, while half the land in the county was owned by just 22 settlers. For many, failure in the west was viewed as the result of some inadequacy of character. Most viewed those who failed to succeed in the west as somehow morally deficient; failure was the result of their own personal shortcomings. William Priest claimed that the successful planter was 'sober and industrious: but when a man of an opposite description makes such an attempt, he often degenerates into a demisavage'.[24] Timothy Dwight described settlers on the frontier who failed to acquire land as 'too idle; too talkative; too passionate; too prodigal; and too shiftless; to acquire either property or character'.[25] For the majority of male settlers in Kentucky, who owned no land, what had their migration to the west meant and what did this mean about their status as 'men'? They had exposed their families to the privation of an arduous journey and the hardships of the settlement of a new country; they had braved Indian attacks and resolutely clung on. And now they had nothing. In an era when reputation was vital to identity, by failing to provide support and protection for their families, these frontiersmen had failed to meet the most basic definition of a 'man'.[26]

Lacking such claims to masculinity, frontiersmen may have become increasingly conscious of threats to their honour. Robert Shoemaker has noted how in eighteenth-century London new concepts of politeness and gentility challenged traditional patterns of masculinity and encouraged those who did not meet those concepts to 'reassert distinctive masculine traits'. In the Trans-Appalachian West, frontiersmen participated in the creation of a new model of masculinity that Anthony Rotundo has defined as the 'masculine primitive'. In so doing, they rooted their heritage, their claim to land and, most importantly, their claim to masculinity in the mythic frontiersman hunter and Indian fighter.[27]

---

[24] William Priest, *Travels in the United States of America; Commencing in ... 1793 and Ending in 1797, with the Author's Journals of His Two Voyages across the Atlantic* (London, 1802), p. 42; Lee Soltow, 'Kentucky Wealth at the End of the Eighteenth Century,' *Journal of Economic History*, 43/3 (1983), p. 620; Harrison County 1797 Tax List, Kentucky State Archives.

[25] Quoted in John R. Van Atta, '"A Lawless Rabble": Henry Clay and the Cultural Politics of Squatters' Rights, 1832–1841', *Journal of the Early Republic*, 28/3 (2008), p. 349.

[26] Ed Hatton, '"He Murdered Her Because He Loved Her": Passion, Masculinity, and Intimate Homicide in Antebellum America', in C. Daniels and M.V. Kennedy (eds), *Over the Threshold: Intimate Violence in Early America* (New York, 1999), pp. 111–134.

[27] Robert Shoemaker, 'Male Honour and the Decline of Public Violence in Eighteenth-Century London', *Social History*, 26/2 (2001), p. 200. See also Karen Harvey, 'The History of Masculinity, circa 1650–1800', *Journal of British Studies*, 44/2 (2005), pp. 296–311; E. Anthony Rotundo, *American Manhood: Transformations in Masculinity from the Revolution to the Modern Era* (New York, 1993), pp. 227–232.

In eighteenth-century Europe, the presentation of the body and issues of honour and gender were all closely related. In North America, presentation was particularly important as a means of judging the social status of individuals. In Europe, dialect or manners could easily betray an individual's status; in North America, where immigrants from across Europe with different speech patterns and different manners mingled, and where social mobility could transform Benjamin Franklin from a poor printer's apprentice to a member of the high elite, how individuals presented themselves was a primary means of judging social status. Dress and fashion became an important measure of judging status. In such a social milieu the image of the body became extremely important. Elias Fordham, for instance, commented that during his travels in the west he had 'not seen an effeminate, or a feeble man, in mind or body'.[28]

While horses would later become symbolic of the western cowboy, they were not seen as symbolic of early western pioneers largely because they were used by the Kentucky elite to enforce their own construction of gentility. In the early nineteenth century, the Kentucky elite used horse-racing and the breeding of thoroughbred horses as a means of constructing their own image of gentrified masculinity. By closely monitoring the pedigree of their horses and replacing the quarter race run over a quarter of a mile straight track with distance races run over several miles, Kentucky elite horse-owners excluded poorer Kentuckians from participation in the sport and relegated them to spectators. Horses and horse-racing thus symbolised in many ways elite masculinity, and while horses may have been central to the lives of western settlers, they did not become symbolic of their lives.[29]

If horse-racing represented elite gentility, so too did fighting with firearms, and ordinary settlers did not use firearms in combat as a means of displaying their masculinity. Although the image of the late nineteenth-century frontier is dominated by images of guns and gunslingers, guns seem to have played a relatively unimportant role in interpersonal violence in the early nineteenth century. Whereas Kentucky court records are full of accusation of horse theft and disputes over horse-ownership, firearms are surprisingly absent from the court records and, if settlers possessed guns, they do not seem to have used them widely in disputes with one another. For instance, at the end of the eighteenth century, western Kentucky was terrorised by two outlaws known as the Harpe

---

28 Elias Pym Fordham, *Personal Narrative of Travels in Virginia, Maryland, Pennsylvania, Ohio, Indiana, Kentucky; and of a Residence in the Illinois Territory: 1817–1818* (Cleveland, 1906), p. 129; Pieter Spierenburg (ed.), *Men and Violence: Gender, Honour, and Rituals in Modern Europe and America* (Columbus, 1998), p. 4; T.H. Breen, '"Baubles of Britain": The American and Consumer Revolutions of the Eighteenth Century', *Past and Present*, 119 (1988), p. 81.

29 Catriona Paul, '"The Horsemen Had the Start": Horse Ownership and Advantage in Early Kentucky, 1770–1830', unpublishd PhD thesis, University of Dundee, 2012; Rhys Isaac, *The Transformation of Virginia, 1740–1790* (New York, 1988), pp. 98–102.

Brothers, Micajah and Wiley Harpe. For most of 1799 the Harpes conducted a reign of terror across western Kentucky, with often-exaggerated reports of their crimes and murders appearing in the *Kentucky Gazette.* They joined with Indians in attacking settlers travelling down the Ohio River, they pushed people off cliffs, and they mutilated bodies. By the time a posse caught up with the Harpe Brothers in September 1799, newspaper reports attributed as many as 40 murders to them. Although nineteenth-century images of the Harpe Brothers invariably show them as armed to the teeth with several guns, the Harpes do not seem to have used guns as their weapon of choice. Of the almost 40 victims that were killed by the Harpes, only three were definitely shot. Indeed, the Harpes' 'signature' method of murdering was cutting up bodies with knives and then filling the bodies with stones and dumping them in water. It was this brutality, the cutting up of bodies, filling them with stones and dumping them in a river, that was the hallmark of the Harpe Brothers, not the use of guns.[30]

The Harpe Brothers do not seem to have been alone in resorting to other means of violence before guns. In two western Kentucky counties, Christian and Muhlenberg, between 1800 and 1810, the local county courts investigated four murders. Two of the cases involved knives. Of the other two, one, the murder of Tobias Penrod in Muhlenberg County, was clearly premeditated, as Penrod was shot in the head by a rifle while ploughing his field by his brother Peter. The only other murder to involve a gun was in Christian County when Michael Dillingham murdered John Fisher. The two had been drinking and playing cards and soon began to quarrel. Dillingham threatened to 'whip' Fisher, who responded that 'he would fight Dillingham any way with gun or pistol'. Fisher then picked up his gun and paced around the room before Dillingham grabbed his knife and rushed at him, fatally stabbing him. It was the knife, not the gun, that was the fatal weapon.[31]

Guns were more likely to result in deaths than knives or other weapons, so the proportion of murders involving guns could be expected to be higher than the proportion of mere assaults. Indeed, the involvement of guns in assaults is even more limited. In many instances courts did not bother to record the weapons used in an assault and this makes any meaningful analysis of the frequency with which guns were used in assaults almost impossible. However, Frederick County in Virginia has relatively detailed records for such cases, although from a slightly earlier period. Between 1745 and 1755, of 46 court cases in which weapons were specifically mentioned in the court record, guns were mentioned only in 5 cases, or just over 10 per cent. By contrast knives were mentioned in

---

30 Otto A. Rothert, *The Outlaws of Cave-in-Rock: Historical Accounts of the Famous Highwaymen and River Pirates* (Cleveland, 1924), pp. 89–90; *Kentucky Gazette*, 15 August 1799.

31 Commonwealth of Kentucky vs Penrod, Muhlenberg County Case Files, Box 4. Kentucky State Archives; Christian County Circuit Court Order Book, 2:152, Kentucky State Archives.

25 cases, or 54 per cent, and whips were mentioned in four cases, or just under 9 per cent – mainly reflecting cases brought by indentured servants. Records for the Trans-Appalachian West are scarcer, but suggest a similar pattern. For the 19 cases of assault in which a weapon was specifically named in Washington County, Ohio, and Muhlenberg and Christian Counties, Kentucky, in the first decade of the nineteenth century, only one case mentioned a gun, while seven mentioned knives, six fists or feet, and one assault with a chair.[32]

Indeed, the type of fighting with which the early west was most identified by contemporaries was not 'gun-slinging', but 'rough and tumble fighting' or gouging, where settlers attempted to gouge out their opponent's eyes or remove certain body parts. Irish traveller Fortescue Cuming maintained that Kentuckians

> fight for the most trifling provocations, or even sometimes without any, but merely to try each others prowess, which they are fond of vaunting of. Their hands, teeth, knees, head and feet are their weapons, not only boxing with their fists ... but also tearing, kicking, scratching, biting, gouging each others eyes out by a dexterous use of a thumb and finger, and doing their utmost to kill each other.[33]

Another traveller Isaac Weld commented

> Whenever these people come to blows, they fight just like wild beasts, biting, kicking, and endeavouring to tear each other's eyes out with their nails. It is by no means uncommon to meet with those who have lost an eye in a combat, and there are men who pride themselves upon the dexterity with which they can scoop one out. This is called *gouging* ... If ever there is a battle, in which neither of those engages loses an eye, their faces are however generally cut in a shocking manner with the thumb nails, in the many attempts which are made at gouging.[34]

If Kentuckians used a weapon in such fights it was rarely a gun, but a dirk, or short knife. Traveller Elias Fordham commented that 'there are a number of dissipated and desperate characters, from all parts of the world, assembled in these Western

---

32 Frederick County Order Books, Volumes 1–6, Frederick County Ended Causes, 1744–1755, Library of Virginia, Richmond; Muhlenberg County Criminal Case Files, Muhlenberg Circuit Court Order Books, 1799–1817, Christian County Circuit Court Order Books, 1803–1809, Kentucky State Archives.

33 Fortescue Cuming, 'Cuming's Sketches of a Tour to the Western Country, 1807–1809', in R.G. Thwaites (ed.), *Early Western Travels* (4 vols, Cleveland), vol. 4, p. 137; Elliott J. Gorn, '"Gouge and Bite, Pull Hair and Scratch": The Social Significance of Fighting in the Southern Backcountry', *American Historical Review*, 90/1 (1985), pp. 18–43.

34 Isaac Weld, *Travels through the States of North America, and the Provinces of Upper and Lower Canada, during the years 1795, 1796 and 1797* (London, 1799), p. 192.

States; and these, of course, are overbearing and insolent. It is nearly impossible for a man to be so circumspect, as to avoid giving offence to these irritable spirits; who, in fact, do not always wait for provocation to be insolent. The Kentuckians on these occasions use their dirks.' When the General Assembly of Kentucky revised the state's laws relating to violence in 1797, no mention was made of guns, but the state did feel it necessary to include a specific clause for those who 'shall unlawfully cut out or disable the tongue, put out an eye, slit the nose, ear or lip, or cut off or disable any limb or member, with intention in so doing to maim or disfigure such person, or shall voluntarily, maliciously, and of purpose, pull or put out an eye while fighting'.[35]

The use of knives and dirks may simply reflect the fact that before the advent of mass-produced pistols, guns and rifles could not be carried as concealed weapons. However, there may also have been other factors influencing the choice of fighting style. Peter Spierenberg has suggested that, during the course of the seventeenth and eighteenth centuries, knife-fighting in Amsterdam became increasingly ritualised and was conducted principally with the aim of disfiguring the opponent by which 'the participants built up a self-image of a tough, non-effeminate man'.[36] Similarly, fighting in the early Trans-Appalachian West seems to have been peculiarly fashioned to mutilate and disfigure, and nothing could be better designed to disfigure than gouging.

If guns were not central to interpersonal violence, they still did play a central role in the fashioning of a self-image for Kentucky settlers and in the construction of a new image of masculinity. Some of the roots of this new construction of masculinity lay in a strange place: John Filson's *Discovery, Settlement and Present State of Kentucky*, first published in 1784. Filson specifically aimed to use this work to encourage settlers to move to the west. To achieve this, Filson had to explain away the dangers of Indian warfare about which so much had been reported. He could not deny that frontier settlers lived in danger, but he had to make this danger less of a barrier, perhaps even attractive. As a means of doing this, Filson chose to append a 'biography' of Daniel Boone, purportedly written by Boone himself, but actually crafted by Filson. Boone was portrayed by Filson as an adventurer who did not follow the etiquette of eastern society but lived in a simpler and less refined way, in tune with nature. He stressed that Boone lacked any formal education but relied on his own quick wits and particularly his skill with a rifle. Filson also helped to craft the image of Boone as an Indian fighter. This was an image with which Boone was very uncomfortable, himself admitting

---

35 Fordham, *Personal Narrative of Travels*, pp. 148–149; William Littell, *The Statute Law of Kentucky* (2 vols, Frankfort, 1809), vol. 2, p. 13.

36 Pieter Spierenburg, 'The Taming of the Noble Ruffian: Male Violence and Dueling in Early Modern and Modern Germany', in Spierenburg, *Men and Violence*, p. 117.

to having certainly killed only one Indian during his life and continuing to show affection for the Indian family that had adopted him during his captivity.[37]

Filson's image of Boone became an overnight success, and the work was widely republished across both North America and Europe, making Boone an international sensation. By the early nineteenth century, the image of the rifle-carrying frontiersmen, encapsulated by Boone, had become symbolic of the Trans-Appalachian West, not only in North America but also in Europe. When James Fenimore Cooper sought a hero to embody the frontiersman for his Leatherstocking Tales, he modelled his character Nathaniel Bumppo quite explicitly on Daniel Boone. Bumppo's claim to fame lay in his marksmanship, which earned him the epithet Deerslayer, and his proudest possession was his trusty rifle. Cooper's Bumppo drew strength and prestige from his lack of formal education and his simplicity. He was innocent and naïve, acquiring his reputation solely from his wilderness experience. Towards the end of his life, like the real-life Boone, pushed from his home by the hordes of incoming settlers and prosecuted in the courts, he fled across the Mississippi to Missouri. The image of the long-hunter, exemplified by Boone and Bumppo, provided a perfect model for many settlers who had crossed the Appalachians and failed to acquire land and wealth.[38]

The image of the long-hunter had particular appeal because it explicitly rejected elite conceptions of masculinity. But it could not have appealed without the special place that the long-hunter had played in early Kentucky's history. For the first decade of Kentucky's history it had been hunters and backwoodsmen who had defended the settlements. The first settlers who crossed the Appalachian Mountains in the early 1770s were mainly long-hunters in search of deer and bison skins. In 1774 they were followed by the first permanent settlers, but the outbreak of the Revolutionary War in the following year saw Indian raiding parties threaten the very existence of the settlements and made the activities of hunters essential for the survival of Kentucky. Unable to plant or harvest crops without fear of Indian attacks, those settlers who remained in Kentucky came to rely on meat acquired by hunters as the primary component of their diet. Benjamin Allen recalled that due to the coarse texture of buffalo meat it was 'a good deal like corn bread' and was used as a substitute on many occasions. The Kentucky settlements needed a consistent supply of meat to feed the settlers and this meant that the skilled hunters were much in demand and could quickly develop a respectable reputation. Without the food and protection offered by

---

37 Interview with Nathan and Olive Boone, in Neal O. Hammon (ed.), *My Father, Daniel Boone: The Draper Interviews with Nathan Boone* (Lexington, 1999), pp. 29–32; John Filson, *The Discovery, Settlement, and Present State of Kentucky* (London, 1793); Richard Slotkin, *Regeneration through Violence: The Mythology of the American Frontier, 1600–1860* (Middletown, 1973), pp. 268–348.

38 William P. Kelly, *Plotting America's Past: Fenimore Cooper and The Leatherstocking Tales* (Carbondale, 1983); Slotkin, *Regeneration through Violence*, pp. 484–507.

the hunters, early Kentucky would not have endured, and consequently it is not surprising that their role was celebrated by many Kentuckians.[39]

The overwhelming majority of Kentucky's settlers, however, arrived after the end of the Revolutionary War, when hunting was no longer a commercially viable activity and when communities were no longer dependent upon the supplies gathered by hunters. In 1783 Kentucky's population had numbered a few thousand, scattered principally around the forts and stations in the central Bluegrass region; by the census of 1790, it had risen to almost 74,000. The expansion of western settlement meant that bison had disappeared and deer had become too scarce for commercially profitable hunting. By 1790 commercial hunters had moved across the Mississippi in search of richer hunting grounds. But these later settlers still sought to emulate their forebears, and skills with a gun could still compensate for lack of material possessions and became central in demonstrating how, in a world where so many men failed to acquire property and provide for their families, they could still remain men. Kentucky men of varying social status paid increasing attention to hunting as a leisure activity. For instance, wealthy, aspiring Kentucky lawyer and future governor Robert Breckinridge McAfee spent much of his leisure time as a young man hunting squirrels. Boys were encouraged to develop their hunting skills at an early age. Charles Drake remembered how as a boy 'father purchased me a little old shot gun, and I circum-perambulated the little field with the eye of a hunter and the self-importance of a sentinel on the ramparts of a fortress'.[40] Keeping squirrels and crows off the corn-fields of a Kentucky farm was sufficient to provide an adolescent boy with a sense of manhood and shooting the fast-moving vermin was a good way to demonstrate skill. Learning to fire a gun and hunt was often viewed as more important than receiving a formal education. Joseph Doddridge noted that western settlers placed a 'higher value on physical than on mental endowments, and on skill in hunting and bravery in war than on any polite accomplishments'.[41]

While the image of the hunter played a central role in shaping the perception of guns and tied guns to images of masculinity and independence, the role of the militia also informed these images, linking guns to concepts of citizenship and independence. During the Revolutionary War, isolated across the Appalachian Mountains and with little support from the regular Continental Army, settlers were forced to rely almost exclusively on the militia for defence. Militia officers such as Daniel Boone and Simon Kenton became central figures of the heroic

39 Interview with John Hedge, interview with Benjamin Allen, Kentucky Papers 11CC19-23, 11CC69, Draper Manuscript Collection.

40 Daniel Drake and Emmet F. Horine, *Pioneer Life in Kentucky 1785–1800* (New York, 1948), p. 51; Robert Breckinridge McAfee Journal, Filson Library; Craig T. Friend, *Kentucke's Frontiers* (Bloomington, 2010), pp. 130, 68.

41 Doddridge, *Notes on the Settlement of Western Virginia*, p. 121.

age of Kentucky's defence. For eight years the militia was the principal bulwark protecting the homes and families of Kentucky settlers. Even after the end of the Revolutionary War, the Kentucky militia remained an important institution serving in campaigns against the Ohio Indians and in the War of 1812.[42]

The image of the western rifle-carrying militiaman became central to the western perception of the frontier wars. While not all frontier militiamen were equipped with rifles, many were. Possession of rifles allowed the western militia to serve as sharp-shooters, sniping at opposing troops from cover, particularly targeting enemy officers and disrupting advancing lines. Western militiamen themselves made significant changes to how the rifle was used in combat, adopting methods used by western hunters. In particular, western militiamen placed a greased patch over the bullet, making reloading much faster by enabling the bullet to fit snugly into the barrel without having to drive it down with a mallet, while the grease trapped the expanding gasses from the gunpowder more efficiently, giving the bullet greater velocity and providing both greater accuracy and inflicting greater damage. Western sharp-shooters fought not only in the Ohio Valley but also in campaigns in the east, perhaps most famously at the Battle of King's Mountain in 1780, gaining a national reputation. Indeed, such a reputation did these militiamen gain that British chronicler of the war Charles Stedman wrote of the 'wild and fierce inhabitants of Kentucky' who opposed the British troops in the South.[43]

The performance of the western militia did not, however, always impress officers, and militia units were notoriously fickle in battle. In 1778 Virginia General Andrew Lewis concluded that the western militia were 'lost to all sence of [duty] & self preservation'. The limitations of the militia seemed most apparent when it was engaged in the offensive operations north of the Ohio River. Militiamen were reluctant to be away from their homes and crops for prolonged periods and, on occasion, militia units deserted en masse. During General Arthur St Clair's expedition against the Ohio Indians in 1791, the desertion of an entire company on the eve of battle compelled St Clair to detach 300 regulars in pursuit, and then, when battle commenced, the militia 'fled through the main army without firing a gun', causing panic and mayhem and contributing to the defeat and destruction of his entire army. However, in many

---

[42] Malcolm J. Rohrbough, *Trans-Appalachian Frontier: People, Societies, and Institutions, 1775–1850* (Bloomington, 2008; 1st edn, 1978), pp. 31–36.

[43] Charles Stedman, *The History of the Origin, Progress and Termination of the American War* (2 vols, London, 1794) vol. 1, p. 221; Matthew C. Ward, 'The American Militias: "The Garnish of the Table"', in R. Chickering and S. Förster (eds), *War in an Age of Revolution, 1775–1815* (Cambridge, 2010), pp. 159–176; Neal Hannon and Richard Taylor, *Virginia's Western War 1775–1786* (Mechanicsburg, 2002), p. 55; Robert Middlekauff, *The Glorious Cause: The American Revolution, 1763–1789* (Oxford, 1982), p. 461.

ways, it was that very indiscipline and independence that appealed to western settlers; militiamen did not always follow the orders of their superiors.[44]

The defeat of the Kentucky militia at the Battle of Blue Licks in northern Kentucky in August 1782 symbolised both the weaknesses and the appeal of the western militia. Although experienced leaders such as Daniel Boone had urged caution, accusations of cowardice, exemplified by Hugh McGary's taunt to his fellow officers 'they that ain't cowards follow me', coerced the militiamen into rushing headlong into disastrous ambush.[45] The defeat at Blue Licks symbolised the bravery and individualism of the militia and, despite its reputation for indiscipline, the militia became a heroic icon to many western settlers. The independence and informal nature of the militia appealed to Kentuckians seeking to define a new form of identity and masculinity. Militiamen were encouraged to fight not by fear of discipline, but for their regard and respect for their comrades and commanders. Indeed, George Washington counselled his western commanders that their militiamen were 'not to be governed by military laws, but must be held by the ties of confidence and affection to their leader'.[46]

Pride in the militia and the belief in its accomplishments continued to grow throughout the late eighteenth and early nineteenth centuries. On the eve of the War of 1812, William Henry Harrison bemoaned the belief throughout the west 'that the untutored rifleman is the most formidable of all warriors ... [and] undisciplined militia armed with rifles, are superior to regular troops'.[47] Taking such pride in the militia, western settlers liked to display publicly their connections with it. Travellers to Kentucky commented on the ubiquity of the public use of military titles and dress. Perhaps most infamously, Fanny Trollope commented that 'the eternal recurrence of their militia titles is particularly ludicrous, met with, as they are, among tavern-keepers, market-gardeners, &c.'. One settler Trollope encountered was referred to by all as the 'General'. When someone questioned what role he had played in the militia, they were informed that he was 'not in the army ... but he was surveyor-general of the district'.[48]

44 Gen. Andrew Lewis to Col. William Preston, 8 June 1778, in L. Phelps Kellogg (ed.), *Frontier Advance on the Upper Ohio, 1778 –1779* (Madison, 1916), p. 79; Report of a Special Committee of the House of Representatives on the Failure of the Expedition against the Indians, 27 March 1792, in W.H. Smith (ed.), *The St. Clair Papers: The Life and Public Services of Arthur St. Clair ... with his Correspondence and other Papers* (2 vols, Cincinnati, 1882), vol. 2, p. 295.

45 Interview with Jacob Stevens, Draper Manuscript Collection 12CC134; Darren Reid, 'Walking the Line of Fire: Violence, Society and the War for the Kentucky and Trans-Appalachian Frontier, 1774–1795', unpublished PhD thesis, University of Dundee, 2011, pp. 161–166.

46 George Washington to Col. Daniel Brodhead, 29 December 1780, in L. Phelps Kellogg (ed.), *Frontier Retreat on the Upper Ohio, 1779–1781* (Madison, 1917), p. 312.

47 William Henry Harrison to Charles Scott, 17 April 1810, in L. Esarey (ed.), *Messages and Letters of William Henry Harrison* (2 vols, New York, 1975), vol. 1, p. 412.

48 Frances Milton Trollope and Pamela Neville-Sington, *Domestic Manners of the Americans* (London, 1997), pp. 137–138.

The militia not only played a significant social role in Kentucky but also played an important political role. Militia companies became important sources of political influence in the early west and engendered a link between militia membership and the concept of membership of the body politic. In both Kentucky and western Pennsylvania the militia became a means through which ordinary settlers could lobby for reform. In Kentucky, militia companies served as a focus for radical canvassing during the fierce debates about the nature of the state's Constitution on the eve of statehood in 1792. In particular, the Bourbon County militia emerged as a proponent of radical ideas attacking the 'golden key which can open those doors of access' and which provided 'the admission of immoral men into places of power and trust'.[49] Across the west, in Pennsylvania, Virginia and Kentucky, militia companies formed the focus for opposition to the Whiskey Excise in 1793 and 1794. While property requirements deprived poorer men of a vote in state elections, particularly in Kentucky, the militia was a way in which those deprived of a vote could make their views heard. Silenced by the powerful elite, protests by militiamen could give voice to the poor and landless in society, and militiamen were not reluctant to impose their views upon members of the elite. On occasion this could lead to violent clashes. Elias Fordham reported that in a tavern that he visited 'some of the young men armed themselves with Dirks ... to resist the intrusion of the Militia, as the vulgar are contemptuously called. Unluckily one of our party was electioneering, and treated some hunters in the bar room with rum.'[50] That Fordham should conflate the presence of the militia with the hunters speaks to the extent to which the two concepts had merged.

By the 1790s in the Trans-Appalachian West a link had been forged between militia membership and popular participation in politics. The link between the militia, guns, political activity and most importantly citizenship was made most tangible through the passage of the Second Amendment to the United States Constitution. In stating that 'A well regulated militia being necessary to the security of a free state, the right of the people to keep and bear arms shall not be infringed', the Second Amendment codified the relationship between the militia, citizenship and arms. In the West the link was very apparent. In Kentucky slaves were prohibited from carrying all weapons. Kentucky law stated, 'no slave shall keep any arms whatever ... Arms in possession of a slave contrary to this prohibition shall be forfeited to him who will seize them.' At the same time, militiamen were required by law to own arms and to muster regularly and train with those arms. The requirement for adult white men to be armed, while

---

49 *Kentucky Gazette*, 15 October 1791; Harry S. Laver, 'Rethinking the Social Role of the Militia: Community-Building in Antebellum Kentucky', *Journal of Southern History*, 68/4 (2002), pp. 777–816.

50 Fordham, *Personal Narrative of Travels*, p. 219. Scrap Book 1793–1815, pp. 60–62, James Veech Manuscripts, Historical Society of Western Pennsylvania, Pittsburgh.

simultaneously depriving those not deemed to be citizens of arms, reinforced the connection between citizenship and the ownership of arms. The possession of arms thus represented both membership of the militia and the body politic.[51]

* * *

While guns may not have been intimately associated with the American Appalachian frontier in the middle of the eighteenth century, by the end of the century, a clear link had been forged in the west between guns and both citizenship and masculinity. In part, this lay in the image of the Kentucky rifle as a tool of the long-hunter. To many Trans-Appalachian settlers, success in hunting quickly came to demonstrate 'manliness' and success even if a man had few material possessions. Hunting implied independence from the restrictions of authority, both state and patriarchal. Its celebration rejected much of the definition of status based on the acquisition of land and property espoused by the early Kentucky elite. Because this image was crafted as a reaction to elite conceptions of masculinity, horses did not become a central component of the image of the masculine settler. Similarly, the elite practice of duelling, which was still central to the construction of elite masculinity in the early Trans-Appalachian West, meant that guns did not become a major feature of interpersonal violence in the early west, but knives, dirks and fists remained symbolic of western masculinity. The ownership of guns and firearms demonstrated that an individual was a member of the body politic and implied images of the masculine hunter and backwoodsman. For poorer settlers, who failed to get land in Kentucky and who struggled to provide for and protect their families, the image of the western hunter and of the heroic militiaman provided sources of identity and masculinity alternative to the elite construction of the polite gentleman, and central to the image of both the hunter and the militiaman was the Kentucky rifle.[52]

---

51 Littell, *Statute Law of Kentucky*, vol. 1, p. 243. Nathan R. Kozuskanich, 'Pennsylvania, the Militia, and the Second Amendment', *Pennsylvania Magazine of History and Biography*, 133/3 (2009), pp. 119–147; Robert H. Churchill, 'Gun Regulation, the Police Power, and the Right to Keep Arms in Early America: The Legal Context of the Second Amendment', *Law and History Review*, 25/1 (2007), pp. 139–175; Saul Cornell, 'Early American Gun Regulation and the Second Amendment: A Closer Look at the Evidence', *Law and History Review*, 25/1 (2007), pp. 197–204; Robert E. Shalhope, 'The Ideological Origins of the Second Amendment', *Journal of American History*, 69 (1982), pp. 599–614; Robert E. Shalhope and Lawrence D. Cress, 'The Second Amendment and the Right to Bear Arms: An Exchange', *Journal of American History*, 71 (1984), pp. 587–593.

52 Friend, *Kentucke's Frontiers*, p. 42; Stephen Aron, '"Rights in the Woods" on the Trans-Appalachian Frontier', in A.R.L. Cayton and F.J. Teute (eds), *Contact Points: American Frontiers from the Mohawk Valley to the Mississippi, 1750–1830* (Chapel Hill, 1998), p. 183; W.J. Rorabaugh, 'The Political Duel in the Early Republic: Burr v. Hamilton', *Journal of the Early Republic,* 15/1 (1995), pp. 1–23; Dick Steward, *Duels and the Roots of Violence in Missouri* (Columbia, 2000).

When Illinois-born Methodist preacher William Milburn attacked the gentlemen he saw around him on the east coast, he decried the 'dapper, diminutive thing, which seems to possess some features of both sexes ... [who] talks magniloquently of first circles and old families ... [and] aspires to become a connoisseur of horse-flesh, an amateur in cigars, brandy-smashes, and gin cocktails'. He contrasted this with another vision of masculinity that 'was all muscle, nerve, [and] backbone': the young man in Kentucky with 'his rifle ready'.[53] Just like the residents of Frankfort in commemorating Boone, when a western Methodist preacher sought to portray an image of masculinity in early America, it was unthinkable without a rifle in hand.

[53] William Henry Milburn, *The Rifle, Axe, and Saddle-bags, and Other Lectures* (New York, 1857), pp. 42–43.

Chapter 2

Guns, Masculinity and Marksmanship: Codes of Killing and Conservation in the Nineteenth-Century American West

Karen Jones

As broadcast in colourful fashion by western movies from *High Noon* to *The Wild Bunch*, the gun looms large in the culture and mythology of the American frontier. In the movie *Winchester '73*, 'the gun that won the West' earned top billing alongside James Stewart, its biography serving as the basis for an expansive cinematic narrative. In 2011, the Colt single-shot pistol – the so-called 'Peacemaker' – was formally recognised by the Arizona legislature as an official 'state weapon'. Machinations of celluloid and modern politics aside, firearms have long been valorised as iconic relics of the nineteenth-century West, a continuation of earlier frontiering motifs but imprinted with a monumental and histrionic geography of continental acquisition, identity politics and a national creation story wrought through Manifest Destiny. From dime novels to dioramas, lithographs to live shows, firearms were celebrated as agents of American might and right, gleaming markers of technological supremacy, individualism and masculinity in the broader project of westward expansionism. According to hunter and 'Wild West' showman, Buffalo Bill Cody, a man expertly placed to comment on the fantasy architecture of the frontier and its most striking props, the gun (alongside the axe, the Bible and the schoolbook) carried the aspirations and the power of the American nation across the plains to the Pacific. Trapper Jim Bridger remained unconvinced by spiritual assistance, yet effusively confident in the almighty providence of firepower as an armament to the frontier project: 'the grace of God won't carry a man through these prairies, it takes powder and ball.' In the American theatre, the gun boasted a decidedly western genealogy.[1]

In frontier gun culture, the 'six-gun mystique', as historian John Cawelti dubs it, the paragons of American masculinity and marksmanship are typically

[1] 'Buffalo Bill's Wild West and Congress of Rough Riders of the World', n.d., p. 22, held at the Autry Museum of the American West, Los Angeles (hereafter cited as Autry); Bridger, quoted in Charles G. *Worman, Gunsmoke and Saddle Leather: Firearms in the Nineteenth-Century American West* (Albuquerque, 2005), p. 59.

represented by a triumvirate of swaggering heroes – the lawman, the outlaw and the cowboy – who collectively roam the West, all guns blazing. In recent years, historians have deconstructed the cult of the gun to discover a far less glamorous (yet, arguably, more interesting) nineteenth-century frontier in which occasions of *High Noon* style inter-personal violence were sporadic, atypical and confined to the early years of notorious hell-raising cattle towns, such as Dodge City and Ellsworth. The 'code of the West' notwithstanding (extrapolated by Richard Maxwell Brown as a set of social conventions that encouraged no duty to retreat and the defence of honour, family and property as part of the frontier condition), guns were more likely deployed to scare off stray animals than face off against black-hat-wearing bad guys. Hunting, of course, weighs into this equation in an intriguing and instrumental way. Although Wyatt Earp may have spent more time putting out chimney fires and rounding up hogs than pistol-whipping bandits in the dusty streets of Tombstone, it remains much harder to conceive of the history of hunting in the West without the imprint of the gun. The Euro-American hunter utilised firearms as essential tools in the collection of animal capital – whether the goal was subsistence, sport, or to satisfy market demands for fur, hide and meat. While the residents of Dodge City were settling into a 'civilised' phase courtesy of civic boosterism, bureaucratic infrastructures and gun ordinances (corralling the town's reputation for gunplay to history and the realms of tourism), the cult of the gun was alive and well on the plains and in the Rockies. As this chapter illustrates, sportsmen (so-called 'sports'), settlers and market hunters alike brandished Winchesters, Springfield and Sharps with abandon, harvesting the animal capital of the faunal frontier on an industrial scale, as well as imprinting their chosen weapons with considerable symbolic purchase.[2]

An important part of the firearms culture of the nineteenth-century West saw the valorisation of the hunter hero – a man (and overwhelmingly they were, although the performing frontier did allow space for the likes of Calamity Jane and Annie Oakley) configured in various iterations from rough and ready trappers, settlers or guides, and sharpshooting bison hunters (of which William F. Cody was the most iconic), to flamboyant well-to-do sporting tourists and soldier-heroes with a love of the chase in the mould of George Armstrong Custer. Obscured by other masculine archetypes in the twentieth century, the hunter assumed a leading role as a signifier of manly authority in

[2] John Cawelti, *The Six Gun Mystique* (Bowling Green, 1999; 1st edn, 1971). For qualifications on the nature of violence and extent of gun usage in the West, see Robert R. Dykstra, *The Cattle Towns* (New York, 1968) and 'Body Counts and Murder Rates: The Contested Statistics of Western Violence', *Reviews in American History*, 31/4 (2003), pp. 554–563; Michael Bellesiles, *Arming America: The Origins of a National Gun Culture* (New York, 2000); Roger D. McGrath, *Gunfighters, Highwaymen, and Vigilantes: Violence on the Frontier* (Berkeley, 1984); Richard M. Brown, *Strain of Violence: Historical Studies of American Violence and Vigilantism* (New York, 1975).

the nineteenth. Hunters demonstrated skills with horse and gun, capacities for wilderness woodcraft, and a manly air that appealed to a populace keen to both make sense of, and celebrate, the mechanics of westward conquest. According to writer Cecil Hartley, here stood the self-sufficient, everyman patriot: 'The early pioneers of the West were all hunters. They acquired in the pursuit of the bear, the panther, and the bison, those habit of courage, coolness, presence of mind, and indifference to danger, which made them such formidable enemies to the Indians, and such efficient defenders of the infant settlements.' By the latter years of the 1800s, a growing fraternity of upper and middle class easterners invested the hunter with additional currency as a vigorous role model for a generation of men apparently suffering under the feminising and degenerative influences of urban industrialism. Leading advocate of the 'strenuous life', Theodore Roosevelt lionised the hunter as a man of hardihood and reliance, and conspired in the creation of a sporting economy of the West through his writings and advocacy for the Boone and Crockett Club (1887), an organ founded to promote 'manly sport with the rifle'. In cultural production – art, literature, and particularly in the numerous biographies and autobiographies depicting action-packed adventures on the western game trail – the hunter stood tall as a leading man. A photograph of Roosevelt in full buckskin attire, grasping his Winchester, steely gaze directed to the camera, exemplified the cultural import of the masculine hunter, a leitmotif that Teddy rode all the way to the White House.[3]

Beneath the folkloric gaze of patriotic pioneering and sporting flourish, the gun was a material agent of change and irrevocably connected to structures of political authority. Evident in the hunting and firearms cultures of the frontier army, the gun represented a tool of empire, albeit a colourful one, which was deployed by the enfranchised to control contested spaces. The right to hunt emerged as a marker through which different groups demonstrated territorial claims, using ritual codes, axioms of class, race and nationhood, and technological advantage to demonstrate economic, political and environmental provenance. As this chapter illustrates, hunting dynamics on the trans-Mississippi frontier saw the creation of a martial ecology, a place for playing war and for its actualisation. In a final twist to the story, with game animals increasingly scarce by the latter 1800s, fresh takes on firearms and western game(s) emerged in the shape of the hunter naturalist (who stressed personal integrity and rightful mastery of the game trail through conservation and restraint with the gun)

---

[3] Cecil Hartley, *Hunting Sports of the West* (Philadelphia, 1865), pp. 8–9; Theodore Roosevelt, *The Wilderness Hunter* (New York, 1926) and *Hunting Trips of a Ranchman* (New York, 1885). On masculinity, see Michael S. Kimmel, *Manhood in America: A Cultural History* (Oxford, 2006); Gail Bederman, *Manliness and Civilization: A Cultural History of Gender and Race in the United States, 1880–1917* (Chicago, 1995); Mark C. Carnes and Clyde Griffen, *Meanings for Manhood: Constructions of Masculinity in Victorian America* (Chicago, 1990); E. Anthony Rotundo, *American Manhood: Transformations on Masculinity from the Revolution to the Modern Era* (New York, 1993).

and the 'camera-hunter' (a heretical figure who rejected the rifle in favour of a photographic approach).

The Technology of the Hunting Trail

Leaving aside thorny debates of technological detachment versus determinism (seen in modern arguments over gun control and the relative culpability of the object over the user), the gun remained an essential part of the armoury of the chase in the nineteenth-century West. The fact that the period witnessed nothing short of a revolution in firearms technology altered the dynamics of the hunt substantially. Compare the pursuit of bison by horse-bound Lakota armed with bows and arrows, with a contingent of sporting types armed with breech-loading rifles and transported by railroad: a vastly different environmental transaction. Developments in firearms technology allowed the hunter to shoot further, faster and with more reliable and powerful weapons – a technological boon that allowed the harvest of animal capital on an unprecedented scale. At the dawn of the nineteenth century, hunters wielded the Kentucky rifle, a smoothbore flintlock that was heavy, liable to jam, slow to restock and difficult to maintain. On his expedition to the Red River in 1806–1807, Zebulon Pike failed to make inroads into nearby bison herds with his small bore rifle. Fifty years on, Captain Randolph Marcy cautioned prospective emigrants that, when greeted with potentially hostile Indians, the gun was useful *only* as a visual prop:

> he should halt, turn around and point his gun at the foremost, which will often have the effect of turning them back, but he should never draw a trigger unless he finds that his life depends upon the shot; for, as soon as his shot is delivered, his sole dependence, unless he have time to reload, must be on the speed of his horse.[4]

The situation had altered radically by the time the Census Bureau declared the frontier closed (and thus the West 'won') in 1890. From the 'Hawken's rifle', a modified Kentucky rifle with a re-bored barrel and percussion ignition that catered specifically to the fur trade, to the invention of the breech-loader rifle (the Sharps, 1848, which fired four times as fast as the muzzle loader), self-contained metallic cartridges (such as those used in the Springfield Joslyn, 1865), and repeating arms (the Henry, Spencer, Winchester and the Colt Revolving Rifle), advances in firearms technology were nothing short of revolutionary. Meanwhile, mass production methods, industrial techniques, developments in print culture and advertising techniques ensured a lively and competitive trade

[4] Randolph Marcy, *The Prairie Traveler: A Handbook for Overland Expeditions* (New York, 1859), p. 188.

between manufacturers, including Remington (1816), Colt (1836), Marlin (1870) and the Winchester Repeating Arms Company (1866). The West emerged as a critical market – both in terms of the demand for guns from settlers, marketeers, and from the community of 'sports' eager to take the latest arms on their hunting adventures in pursuit of grizzly bears, bison and other charismatic animals of the faunal frontier. Where Jacob and Samuel Hawken of St Louis produced 100 rifles annually in the 1820s, Winchester was selling 51,000 by 1885. The Winchester '73 – famed for 'the Strength of its parts, the Simplicity of its construction, the Rapidity of its fire', available in carbine, rifle and musket variants, and at the bargain price of $20 for a basic model – sold 720,000 units to emerge as a favourite weapon on the frontier. According to Montana rancher Granville Stuart (who ordered special editions for himself and friends), '[i]f poor Custer's heroic band had been armed with these rifles they would have covered the earth with dead Indians for 500 yards around'. Residents of a new settlement in Idaho named it 'Winchester' in honour of the famous rifle. By 1912, the company was making $11 million from arms sales. Winchester won the West. And the West made Winchester.[5]

From Tool to Totem: The Firearms Culture of the Hunt

Hunting for the pot, the market, or for recreation, demanded a cache of weapons. A.J. Leach, a settler in Nebraska, recalled that most of his neighbours retained a firearm for the pursuit of small game as well as for self-defence. According to popular sporting journal *Forest and Stream*, the gun represented the 'constant companion' of the western man out for sport and for subsistence hunting. As well as practical tools, guns were also symbolic objects – signifiers of power, belonging and identity – and invested with significant social capital. Granville Stuart fondly remembered learning to shoot with his father using a flintlock and a small bore rifle, while Irish aristocrat Sir George Gore packed 75 rifles and 12 shotguns (and hired a man specifically to maintain them) for his extended safari across the plains in 1854–1857. Firearms were part of the imaginative arsenal of the masculine hero, played a critical role in the storytelling culture of the hunt and, at times, seemed to take on an agency all of their own.[6]

Narratives of the hunt depicted a romanticised vision of the West as a realm where nature was savage and resplendent, its animals feisty and fearsome, and

[5] Marble Arms Company, catalogue no. 18, 1911, Autry; Stuart, quoted in Worman, *Gunsmoke*, p. 387.

[6] A.J. Leach, *Early Day Stories* (Norfolk, NE, 1916), p. 94; 'Dr George Bird Grinnell, Buffalo Hunter', *Rod and Gun News* (c. 1876), clipping 35/198, George Grinnell Papers, HM223, Manuscripts and Archives, Yale University Library; Granville Stuart, *Forty Years on the Frontier* (Cleveland, 1925), vol. 1, p. 33.

the all-conquering hero manly and energetic. Firearms played a vital role in this discursive landscape of contest, threat and masculine triumph. The challenge proved contingent on environmental hazards, the canny tactics of prey animals and the wilderness woodcraft and sharpshooting attributes of the hunter hero. As Rocky Mountain hunter William Pickett exclaimed: 'I had acquired such skill in the use of my rifle and its manipulation, and such confidence in myself, that I did not fear an encounter with any of the wild animals to be met with.' On a practical level, firearms seemed a useful tool travelling in a landscape of potentially dangerous critters (notably the grizzly) as well as American Indians. Montana settler William Allen described his gun, somewhat succinctly, as 'my life preserver'. Firearms conferred a material sense of security and a semblance of psychological comfort courtesy of technological power. Charged by a pugnacious grizzly, Malcolm Mackay found succour in the reliability of his Winchester, as well as his proficiency in using it: 'This time I knew he had to be stopped for good or he would be shaking hands with me and he wasn't a friendly looking animal by a long shot. But I was an old shotgun shooter and the nearer he came the better I liked it, and the safer I felt.' One might be forgiven for thinking that technological advance might *detract* from the repute of the hunter hero, but actually the opposite proved true. Knowing how to use a gun, and, more importantly, how to use it *well*, represented a critical signifier of masculine authority on the frontier. As Theodore Roosevelt motioned: 'It is the man behind the rifle that counts after the weapon has reached a certain stage of perfection.'[7]

Among other things, the stories told by hunters paid heed to the atavism of the game trail experience and the critical purchase of the firearm in this formative transaction with the wild. According to William Allen, the appeal of hunting was in no small part down to his 'innate love' of the gun. In the fatal environment of the American West – a landscape configured as a space of mortal danger and visceral pleasure – firearms became a critical vector through which ideas of heroic contest and personal transformation were communicated. A 'machine in the garden', the gun was apprehended both practically and fantastically. Hunters spent significant time talking about the use and maintenance of their weapons, trading in technical specifications, and engaged in the pursuit of the perfect weapon and the perfect shot. Loquacious comments on the aesthetics of the gun, the 'feel' of the stock and barrel, the aural pleasure of gunfire and the mesmerising qualities of bullet trajectory attested to a technological fetishisation of the hunt. Horace Edwards recalled banter with English sportsmen in 1884 near Elk Creek on the merits of their armoury of Winchesters, Sharps and an English

[7] William D. Pickett, Diary for 1880, p. 13, William Pickett Diary, SC1436, Montana Historical Society, Helena, Montana (hereafter MHS); William Allen, *Adventures with Indians and Game, or Twenty Years in the Rocky Mountains* (Chicago, 1903), p. 108; Malcolm Mackay, *Cow Range and Hunting Trail* (New York, 1925), p. 167; Roosevelt, *Wilderness Hunter*, pp. 370–372.

Express rifle, while guidebooks and manuals such as *Hitting Versus Missing with the Shotgun* (1898) illuminated sporting interest in the mechanics of the gun and its proper use. This 'knowledge trade' in arms fostered a vibrant culture of the gun both on the game trail, and of it. On occasion, the West seemed a landscape of experiment, an animate firing range for the display of a modish and ever more destructive arsenal. Big game hunter Frederick C. Selous modified his trusty Mannlicher elephant gun used in Africa for a trip to the American frontier, while Pickett regarded his hunting sojourn in the fall of 1878 as a prime occasion for the testing of the new Express bullet. Such examples highlighted both the power and the fascination attached to the gun, as well as complicating the narrative of the sportsman naturalist: a lover of industrial technology as well as of wild nature.[8]

Being well-armed and demonstrating skill with the gun marked the true hunter hero apart. Unsurprisingly, the storytelling culture of the West as hunter's paradise (an important locus in which protagonists advertised their masculine prowess) saw firearms elevated as companions, trophies and characters in possession of an agency of their own. A. Wislizenus noted how 'one gets habituated to his rifle as to a trusty travelling companion'. Described as loyal subalterns and feisty agents of environmental transaction, guns were given names and stories to become integral elements in the descriptive turn of the trail. The names given to favourite rifles conveyed their powerful, trustworthy nature, and communicated not only the intimate relationship between hunter and weapon but also schematics of conquest and racial and gender hierarchies. The 'explosive express' of Theodore Roosevelt or George Shields' 'Old pill driver' conveyed a sense of martial authority, while Bill Cody's .50 Springfield, 'Lucretia Borgia', found historical referent for violent tendency. Aside from the politics of nomenclature, the practical and symbolic significance of the gun was demonstrated by its trophy status. Not just devices to secure animal heads and horns, the gun became an object of veneration and conspicuous display in its own right. Firearms featured prominently in big game hunting photography, typically wielded by confident looking hunter heroes who held weapons in strident pose, pointed them at game in a visual re-enactment of the hunt, or arranged trusty rifles to rest on bodies of vanquished animals. Trophy rooms and dens, the principal architectures of hunting display, featured animal shields and gun racks aplenty. William A. Baillie-Grohman spoke of visiting the house

---

8 Leo Marx, *Machine in the Garden: Technology and the Pastoral Ideal in America* (Oxford, 2000; 1st edn, 1964); Allen, *Adventures*, p. 186; Horace Edwards to Henry, 28 December 1875, folder 6, Horace Edwards Correspondence, Western Americana Collection, Beinecke Rare Book and Manuscript Library, Yale University; William D. Pickett, Diary for 1878, pp. 10–13, William Pickett Diary, SC1436, MHS.

of a frontiersman, whose walls were arrayed with firearms, each named and with their own story of western provenance.[9]

According to *The Complete Sportsman* (1893): 'the song of the wild bird is sweet from the thorn, but the gun hath more music than these.' Such comment communicated the gun as a powerful totem of industrial pornography, but it also alluded to a further vocal function: the gun as agent and 'speaker' in its own right. Hunting black-tail deer between Bozeman and Miles City, Allen described the process by which he killed a doe as an act in which 'my Winchester spoke to her, and as the smoke cloud cleared away, I saw her lying on the ground, her neck broken just below the ear'. Allen's notation romanticised the exchange, paid heed to the abject violence invoked by the gun, and issued it a form of agency. The euphemism of the vocal gun added to the 'historical moment' of the kill, related the mechanics of the act in an intimate and gentle way, and granted the gun a form of personhood. Others, too, used the metaphor of the gun to communicate a communion between the hunter hero and vanquished nature – at once a literary device, marker of its implied power, and a signal of the power of language to distance an act of violence from its less savoury connotations.[10]

Beyond the agency of the gun, the use of militarised language also proved a feature of the descriptive landscape of the hunt. Such was most common among sports, but also seen in the testimonies of settler and market hunters. William Allen spoke of the hunting 'battlefield', while Shields gloried in 'the music of our artillery' and 'the smoke of battle'. Montana rancher and buffalo hunter David Hilger recalled 'fierce battles' with 'old Mr Bull', who was not 'going to surrender his authority without a contest'. Of course, the practical attributes of the hunting encounter lent themselves to military comparison – of chase, contest and kill. A substantial number of sports also had a service background, which partly explained their linguistic frames of reference. In addition, the storytelling codes of the hunt favoured a literary formula based around the quest – adventuring, suspense and masculine triumph – which found useful grist in martial connotations. According to Anthony Rotundo, the use of such metaphor served a sociological function, with 'martial ideals and images as a way to focus their vision of a manly life'. Moreover, for many of the aspiring hunter heroes keen to hone their self-created identities as vigorous fighting men, the prospect of war with nature promised an attractive military service, the West as a grand theme park for the exercise of martial ambition and patriotic fire without the problematic associations of history – a 'virgin land', in the parlance of Henry Nash Smith. Unlike the bloodletting of the Civil War, the

9 A. Wislizenus, *A Journey to the Rocky Mountains in the Year 1839* (St Louis, 1912), p. 122; Roosevelt, quoted in 'Charlie Marble and Teddy Pals', 22 November 1937, Montana News Association Inserts, MHS; George O. Shields, *Hunting in the Great West (Rustlings in the Rockies)* (Chicago, 1888), p. 39; William A. Baillie-Grohman, *Camps in the Rockies* (New York, 1910), p. 142.

10 Howland Gasper, *The Complete Sportsman* (New York, 1893), p. 1; Allen, *Adventures*, p. 135.

contest with the assembled fauna of the West was heroic, violent and largely absent of moral quandary: playing war had its appeal. Of course, this distinction between leisure culture and processes of westward conquest was somewhat contrived. In assuming the animal spoils of the frontier as rightfully theirs, Euro-American hunters advanced processes of colonial contest, their guns as smoking instruments of empire wielded against a faunal citizenry of grizzly bears, elk, deer and bison.[11]

The US Army and the West as Martial Ecology

The relationship between firearms, the hunting trail and processes of colonial exchange was ably illuminated by the US Army, whose various encounters with western game animals and those who pursued them saw the construction of the West as a martial ecology, a landscape of the gun in which access to environmental resources depicted a broader contest for the American frontier. Most notably, the 'right' to game became a point (geographic and conceptual) of conflict between military forces and the American Indian.

Under the Fort Laramie Treaty (1868), plains tribes were granted the right to hunt in 'unceded Indian territory', a swathe of land stretching from the Black Hills to the Bighorn Mountains, and encompassing the ancestral bison grounds of various tribes, including the Lakota, Blackfeet, Mandan and Cheyenne. Peace, though, was short-lived, as increasing numbers of Euro-American prospectors and homesteaders encroached on the plains to challenge rightful possession of the land and its resources. The hunt became a locus of contention and marker of western contest. As Bill Cody remarked, the 'desperate redskins ... had made up their mind never to give up that great hunting range'. Access to the game trail meant subsistence, but also freedom of manoeuvre. Elizabeth Custer, for one, assumed that the warriors who were out 'hunting' when her husband toured the tribal camps were actually engaged in something resembling domestic terrorism. Accordingly, the US Army regarded restrictions on guns, game and territory as effective measures to pacify the Great Plains.[12]

As well as policing the plains – preventing access to the hunt, skirmishing with 'hostiles' and mandating confinement to reservations – the army encouraged other Euro-American cults of the gun as a corollary mechanism of empire on the trails. Market hunters armed with high-powered Sharps rifles were given encouragement in the form of official statements of support, and provided with

---

11 Allen, *Adventures*, p. 118; Shields, *Hunting*, pp. 124–125; David Hilger, 'The last of the buffalo', David Hilger papers, SC854: 5, MHS; Rotundo, *American Manhood*, p. 232.

12 William Cody, 'Famous Hunting Parties of the Plains', *The Cosmopolitan*, 17/2 (June 1894), p. 137; Mary Burt (ed.), *The Boy General: Story of the Life of Major General George A. Custer as told by Elizabeth Custer* (New York, 1901), p. 115.

ammunition, logistical aid and security. The dollar and the imperial project aligned. General Sheridan was quoted as saying to hide hunter John Cook: 'let them kill, skin and sell until the buffaloes are exterminated.' Military testimony read the killing of the herds as an act of patriotic honour and the gun as an able architect of progress. Lieutenant General John Schofield, commander of the Department of the Missouri (1869–1870), illustrated the opinion of many when he ventured the following comment: 'With my cavalry and carbined artillery encamped in front, I wanted no other occupation in life than to ward off the savage and kill off his food until there should no longer be an Indian frontier in our beautiful country.'[13]

Assistance, supply and protection were likewise extended to sport hunters. The US Cavalry served, in many cases, as de facto tour guides and hunting outfitters for the legion of American eastern elites and European noblemen who ventured to the plains in pursuit of big game. The Earl of Dunraven showed up at Fort McPherson in 1868, armed with a letter of introduction from General Sheridan and grand plans for bison hunting. The officer at the fort duly assigned horses, men, guns and wagons for the aristocrat's hunting sojourn. The army proved exemplary hosts, and dispatched their duties with efficiency and flair. Scout Bill Cody remembered how visiting sports were issued with guns, and many 'favored the Springfield standard issue'. The military provided further information in the shape of reconnaissance, telling hunting parties where to find bison herds and where to watch out for American Indians. Those who played the hunting game on the western frontier found the army a worthy host. As sportsman George Shields pointed out, a day of fine sport and an evening of fine venison and twilight tales delivered by officers on such subjects as 'frontier life, Indian Warfare, hunting yarns' made the hunting experience truly special. More than that, the tourists gained a further bargain from travelling with the military in allowing their martial inclinations full rein. Shields, for one, gloried in his assumed role as a 'sentry' and the preparations for a 'warpath' made against the buffalo.[14]

The Culture of the Gun in Military Life: Subsistence, Soldiering and Sport

For the frontier army in the West, firearms served a multitude of functions. Aside from the obvious use of guns and ammunition as technologies of attack and defence, the military machine had its sights firmly locked on animal life. Much

13 John Cook, *The Border and the Buffalo* (Topeka, 1907), pp. 663–664; John Schofield, *Forty-Six Years in the Army* (New York, 1897), p. 428.

14 Cody, 'Famous Hunting Parties', pp. 137–139; Shields, *Hunting*, pp. 145–152. See also David D. Smits, 'The Frontier Army and the Destruction of the Buffalo: 1865–1883', *Western Historical Quarterly*, 25/3 (1994), pp. 312–338.

as in the civilian sphere, guns were valuable as tools of subsistence, sport and storytelling. For a frontier army on the march – the Yellowstone campaign of 1876 for instance covered 1,200 miles – or for a regiment holed up in remote and inhospitable country, the function of the gun as a means of securing subsistence was critical. The army purchased meat from market hunters and settlers, and also commissioned special scouts and hunters known for their shooting skills to procure supplies for the troops. As Elizabeth Custer noted: 'The best shots in a company were allowed to leave the column and bring in game for the rest.' As well as fulfilling a basic provision for food, the addition of game meat on the army ration was greeted with pleasure by men who had grown tired of low quality fare and small portions. Major A. F. Mulford of the 7th Cavalry recalled the typical marching menu of 'hard tack, bacon and coffee for breakfast, raw bacon and tack for dinner, fried bacon and hard bread for supper', and added, enthusiastically, 'if our hunters have good luck ... we feast on antelope meat'. The nutritional value of game was also a boon, while the presence of bison, deer and the like was a literal lifeline when supplies fell scarce or domestic animals died off. Without food, the troops were more prone to recalcitrance and insubordination. In the field, as Walter Schuyler noted, the army needed 'to eat to live'.[15]

Beyond subsistence concerns, the officer class embraced opportunities for sport hunting in the western theatre, sometimes for several weeks at a time. General Crook committed an entire month to the chase in 1886, instructing his second in command to 'act on such cases according to your best judgment without sending [telegrams] to me, unless they involve some important question in which you are in doubt about'. The allure of the West as a hunter's paradise occupied the hearts and minds of officers, who wrote fervently and religiously in their journals of hunting adventures, favourite weapons and skirmishes with triumphal encounters – reprising many of the storytelling codes of the civilian canon. Kinzie Bates, writing at Camp Niabara in October 1875, suggested that thoughts of sport dominated the daily roster:

> What shall I write about? I get up in the morning at 7 o'clock, breakfast, perhaps Edmunds takes the mounted party on a scout, dinner at 12 o'clock if he gets back, take supper at 5 o'clock, and spend the rest of the evening until dark on a bar in the river waiting for geese and ducks. We come home and settle down in one of the tents with a good fire and talk, talk, talk.[16]

---

15 Elizabeth Custer, *Following the Guidon* (New York, 1898) p. 34; A.F. Mulford, *Fighting Indians in the 7th United Stated Cavalry* (Corning, 1878), p. 79; Walter Scribner Schuyler to George Schuyler, 1 November 1876, Fort Laramie, Wyoming, Papers of Walter Scribner Schuyler, Mss WS87, Huntington Library, San Marino, California.

16 Quoted in Kevin Adams, *Class and Race in the Frontier Army: Military Life in the West, 1870–1890* (Norman, 2009), p. 85; Kinzie Bates, letter to wife, Camp at Niabara, 8 October 1875, Papers of Kinzie Bates (1863–1929), Mss HM60325-60354, Huntington Library.

On one level, the interest of the officer cadre in hunting sports suggested a distraction from military duties and requisite levels of dedication and battle readiness. Custer on one occasion became so excited by the prospect of running his hounds in pursuit of a bison spotted near the column that he gathered up aides and dashed off in hot pursuit, only to put a bullet through the brain of his horse. Left alone wandering the plains, defenceless, the general expressed sadness at the death of one his favourite steeds, before noting the greater predicament: what if American Indians found him before the guidon did? That said, for all the risks of 'buck fever', the love of the chase actually befitted military priorities in several critical ways. According to Custer, there was 'nothing so nearly resembling a cavalry charge as a buffalo chase'. Chasing the buffalo herd over unfamiliar terrain encouraged skills in reconnoitre, scouting and the mapping of space. The chance of a hunt also kept morale high and encouraged a state of alertness. Most significantly, its gunplay, vigorous exercise and equestrian pursuits provided valuable training for warfare. The General saw bison hunting as ideally placed to 'break the monotony and give horses and men exercise', and many regiments developed a competitive gun culture in relation to the hunt, including rifle competitions and horse races.[17]

Allied to practical uses, the flamboyant culture of martial aristocracy on the game trail demonstrated American authority and power. According to Katherine Fougera, wife of an officer, the hunt gave 'a glamor to army life that nothing ever quite equaled'. The hunting codes of military sports helped to craft a cult of the gun in which animal pursuit and cavalry flourish added to the romance of the military project in the West. Embedded in the illustrious gaming excursions and attendant kill tallies were codes of masculinity, class, race, imperial power and technology. The accoutrements of the chase, the maintenance of hounds, lavish picnics and parties, and glorying in the relationship between the elite hunter, his horse and his gun, added to the sense of pomp and confirmed Euro-American provenance over a landscape and its spoils. Conspicuous consumers of the sporting game on the plains, generals Crook, Custer, Sheridan, and Miles became lionised as hunter heroes, 'knights of the plains' renowned for their gunplay, equestrianism and performative posturing. As Elizabeth Custer fondly recalled, there was nothing quite like the 'jingling spurs, rattling arms, and impatient, stamping horses' and the 'warlike preparations' of the hunt. Beneath the codes of entertainment lay gesture politics of might and right.[18]

---

[17] Custer, *Following*, p. 33; George Armstrong Custer, *My Life on the Plains* (New York, 1874), p. 47.

[18] Katherine Gibson Fougera, *With Custer's Cavalry* (Caxton, 1942), p. 78; Custer, *Following*, pp. 264, 195.

A Frontier Education: Training the Troops for Fighting the 'Bucks'

Among the regular troops, encounters with hunting and firearms proved quite different. While the officer class galloped around playing hunter heroes, the hierarchy of military life ensured no such favours for those in the lower ranks. Elitism and regimental code, combined with security issues, meant the right to hunt remained an officer dictate. Troops earned service in the hunting theatre as subalterns, gun carriers and aides. Permissions were needed to discharge arms or undertake hunting pursuits. As A.F. Mulford pointed out, 'We have strict orders not to fire without orders from the commanding officer. Exercising restraint when greeted with the faunal assemblage of the West nonetheless proved hard.' As Mulford added: 'it is enough to provoke a Deacon to see so much game on every hand, and not be allowed to take a shot at it.' The requirements of discipline, defence and the saving of ammunition were cited as motivation. As James Calhoun, First Lieutenant of the 7th Cavalry and Custer's brother-in-law, elucidated in one of his journal entries, an attack of the 'buck fever' among the men had potentially drastic ramifications:

> Some of the soldiers are very careless in shooting across the column. Antelope were plentiful. Some came within 25 yards of the command, and the soldiers were firing in all directions. The excitements at one time became so great as to cause a stampede with one of the artillery carriages ... this caused a circular to be issued prohibiting shooting at game.[19]

Environmental circumstances dictated opportunities for hunting. As did risk assessments of Indian attack. With hunting rights contingent on obtaining a pass, the requirements of food acquisition or the sporting proclivities of officers had to be weighed up against determinations of threat to the hunting party. Commanders, then, in more hostile locations found opportunities for leaving camp in pursuit of game severely curtailed – a salient illustration of the political economy of hunting on a martial frontier. Keim issued consternation that 'the manly sport, the chase, was almost excluded from pastimes', as 'the intervening country was a barren waste, traversed by roaming bands of savages, closely watching every movement of the invaders of their lands, and ready to pounce upon small parties should they leave camp'. Beyond that, when the 'all clear' for a hunt was given, many among the lower ranks clamoured for service. A reward system typically operated whereby dispensations for good behaviour or a reputation as a 'sharpshooter' governed selection processes. Once out on the trail, the troops seemed to relish the freedom, immersion in nature, and

---

19 Mulford, *Fighting Indians*, p. 80; Lawrence A. Frost (ed.), *With Custer in '74: James Calhoun's Diary (*Provo, 1979), p. 25.

adventuring hi-jinks that appealed for officer sports and civilians alike. The idea of fun and target practice infused commentary, as did the promise of escape from regimental authority. Dispatched to a wood-chopping camp in the woods for three weeks with a team of soldiers, H.H. McConnell recalled: 'during most of which the weather was delightful, was like a "picnic", no military duty to perform, our time at our disposal, after the quota of logs were cut, which was generally completed by noon, and in the afternoon and evening we hunted.'[20]

Sitting alongside the rubric of military hierarchy and the demands of frontier security, however, was the need to train troops in the art of warfare on the plains. Many enlistees were woefully inexperienced in firearms training – a fact made evident when the 'buck fever' proved too great a temptation and the troop line fired on wandering bison without hitting the target (no small feat considering the herd sometimes took a full half hour to pass by). Drill training and weapons practice thus became part of daily routines, as mandated by Army General orders 8, 9 and 10 (1867), which made provisions for exercises and target practice. Here the masculine ideal-type of the frontiersman hunter hero gained purchase as a role model for the lower ranks. A frontier education, and hunting activities in particular, duly emerged as appropriate methods to school the cadre of recruits into 'fighting men'. In *Thirty Years of Army Life on the Border* (1866), Captain Randolph March presented the case for hunting as training regime to channel the masculine vigour of the hunter hero in a martial context: 'I know of no better school of practice for perfecting men in target-firing, and the use of firearms generally, than that in which the frontier hunter receives his education.' The skills of the hunting trail – knowledge of terrain, woodcraft, a healthy constitution and gun skills – ably translated to the needs of the military. As Marcy urged, 'all of which will be found serviceable in border warfare'. Meanwhile, an illustration of the interchangeable categories of 'enemy', as well as an assumption of the American Indian as a fixture of 'savage' nature, was evident in military reports. Scouting journals often described the hunting of indigenous warriors as 'bucks', while in an article for *Galaxy* magazine (Custer wrote avidly for the sporting press, another signal of his stature as a soldier-hunter hero), the general talked of chasing 'prowling' Sioux on the Yellowstone as if they were scarpering game animals, glorying too in the technological promise of the Remington and its 'leaden messages' charged with a 'deadly errand'. In the martial ecology of the West, the gun delivered a message of imperial authority and rightful claims over the landscape, to animals and indigenous peoples alike.[21]

---

20 Quoted in Adams, *Class and Race*, p. 90; H.H. McConnell, *Five Years a Cavalryman, or Sketches of Regular Army Life on the Texas Frontier, 1866–1875* (Norman, 1996), p. 55.

21 Randolph B. Marcy, *Thirty Years of Army Life on the Border* (New York, 1866), pp. 283–285; George A. Custer, 'Battling with the Sioux on the Yellowstone', *Galaxy*, 22 (July 1876), pp. 94–97.

Conclusion: Guns, Conservation and the Hunter Hero Heretic

From the expeditions of Lewis and Clark (1804–1806), explorers who convincingly shot their way across the continent for the purposes of nation-building, natural history and their own sustenance, to the legion of hunter-autobiographers wandering game trails for sport, the nineteenth-century West was presented as a cornucopian landscape, a place where buffalo blackened the prairie in countless numbers and fearsome grizzlies wandered the forests of the Rockies with impunity. By the end of the 1800s, however, that landscape had changed, courtesy of industrial technology, settlement, market forces and prerogatives of national expansion. The gun played an important part in that story. Firearms served as commanding tools for the hunters, both civilian and military, that blazed trails across the American West in the name of sport, subsistence, pecuniary gain and Manifest Destiny. On a material level, firearms exerted an influence, courtesy of their technological prowess, complicity in processes of political and economic power-broking and sheer capacity for environmental transformation. Guns did not do this alone, of course. The hunters of the West brought not only Winchesters and Sharps, but also an armoury of assistive forces, including: animal companions in the shape of dogs and horses; industrial technology, from railroads to chemical preservatives; and an imperial rubric that brought both a bureaucratic infrastructure and a value system asserting supreme rights to the West's natural resources. Guns emerged as intensely symbolic objects that found a way into the storytelling codes of the frontier and gained stature as iconic agents of conquest – a fact best demonstrated by the Winchester '73: 'the gun that won the West'. Standing equally proud behind the sights was the hunter, whose idealised status as a paragon of American masculinity loomed large in campfire yarns, dime novels, visual culture and theatre. From Theodore Roosevelt's tales of sport and the strenuous life to the 'Wild West' show antics of sharpshooter Buffalo Bill Cody, the cult of the gun, manly adventuring and the pursuit of charismatic animals conspired to create a powerful, and popular, mythology of the American West. The hunter hero boasted a western genealogy.

Dominant narratives of the 'winning of the West' spoke of triumph, progress and American Exceptionalism, most evident in Frederick Jackson's Frontier Thesis (1893). Set against this, however, was an anxiety about the foreseen end to hunter's paradise and the promise of game trail gunplay. With the faunal frontier under threat, new iterations of the hunter hero emerged, together with a revised approach to firearms culture that stressed restraint and fair play. Hunter-naturalists (of which Roosevelt was a prime example) lofted a new relationship to the hunt, its animals and their weapons cache – one based around codes of honourable ethics, bag limits and wilderness lore over trigger itch. This cadre of sportsmen emerged as a powerful conservation lobby that sought to preserve wild animals and wild landscapes for future generations (and, some might say, for their own shooting interests). Touting codes of ritual and honour, the sportsmen-conservationists effected a schism in the ranks of the hunting fraternity, divorcing themselves from market, pot and Indian hunters, who, they claimed, represented a shooting culture founded on crass materialism, waste and bloodlust. The 'true' hunter hero gave prey a fighting chance, avoided shooting female or young animals and killed humanely and sparsely. As *The Dead Shot: The Complete Guide for Sportsmen* (1882), written by the mysteriously yet appositely titled 'A. Marksman', motioned, the pot-hunter was not only a bad shot but 'more or less a poacher'. Forged in a new era of scarcity, these emerging conservationist sensibilities spoke of associations between wild nature and national identity, critical vantages on modern industrialism, and the ciphers of race, class and empire in deciding access to natural resources. They also pointed to the multifaceted identity of the hunter as both a consumer and a protector of the natural world, as well as to the complex nature of firearms culture in the American West (and indeed elsewhere).[22]

The prospect also loomed in the late 1800s of a new technological weapon to take to the game trail: the camera. In its practical techniques of stalk and pursuit, as well as a shared lexicon of loading, aiming, shooting and the final 'capture' of subjects, photography and firearms inhabited similar terrain. According to Susan Sontag, the technologies shared a colonial modus: the camera as an artefact of industrial modernity and a 'death weapon' of imperial tutelage. Technological fetishism made for an easy segue between the worlds of the rifle and the camera. Just as hunters got excited about calibres, cartridges and stock lengths, so too did many talk enthusiastically about shutters, mounts and exposure plates. Sometimes the technical connect was starkly obvious – as in the case of novelty products such as the 'gun camera' invented by Massachusetts based photographer and inventor B.W. Kilburn, who used a traditional gun stock as the mount for 4 x 5 compact picture camera (1883) to allow him to take pictures in rugged country and take advantage of the gun sight gaze for processes of photographic capture. For some, the camera and the gun were able companions on the trail. Others, however, ventured a more heretical position, consigning their guns to a decorative position above the mantle and assuming a new identity as 'camera hunters'. Such individuals (who, while relatively vocal, represented a minority among the hunting fraternity) argued that the demise of hunter's paradise in the West mandated a new position and a new weapons culture more in tune with an industrial age of game scarcity. New Englander A.W. Dimock saw his photographic vocation as that of a 'fugitive' and, in common with many of his peers, as one of atonement: 'I like to forget the brutal bags of game I made in the long ago, but the thought of each camera shot brings pleasure. The life history of birds and animals as pictured by the camera contrast curiously with the game

22 'A. Marksman,' *The Dead Shot: The Complete Guide for Sportsmen* (London, 1882), p. 126.

bag product of the fowling piece and the bloody trophies of the rifle.' Moreover, the dangers and difficulties associated with stalking ferocious beasts armed not with a loaded 'life preserver' but a device of image capture alone seemed the ultimate in demonstrations of manly adventuring. As conservationist William Hornaday remarked: 'any duffer with a good check book, a professional guide, and a high-powered repeating rifle can kill big game, but it takes good woodcraft, skill and endurance of a high order ... to secure a really fine photograph.' The Winchester may have 'won the West' in the nineteenth century, but for the new breed of 'camera hunters' it seemed a weapon less suited to the twentieth than the Kodak.[23]

[23] Susan Sontag, *On Photography* (New York, 1977), pp. 13–14; A.W. Dimock, *Wall Street and the Wilds* (New York, 1915), pp. 444–453; William Hornaday, *A Wild Animal Round Up* (New York, 1925), p. 331.

Chapter 3

Fishers of Men and Hunters of Lion: British Missionaries and Big Game Hunting in Colonial Africa

Jason Bruner

'Come after me, and I will make you to become fishers of men.'

– The Gospel of Mark 1:17

'Your adversary the devil, as a roaring lion, walketh about, seeking whom he may devour.'

– 1 Peter 5:8[1]

From the time of the earliest Christian communities, missionaries viewed themselves as fishers – a metaphor referring to the gathering of people into Christian churches through the proclaimed message of the Christian gospel. But as one examines the lives of many British missionaries to colonial Africa, it is clear that they did a lot of hunting as well. And yet missionary hunting has not been analysed with respect to the missionaries' most basic intention: the religious conversion of non-Christians to Christianity. Missionaries often spoke of their hunting exploits in religious terms, either as a form of spiritual retreat with other missionaries, or as a way to demonstrate to African onlookers God's providence in the natural world.

This chapter examines missionary hunting, primarily in East Africa and with reference to evangelical British missionaries. It argues that missionaries found hunting and other British sporting activities to be in accordance with their aim of converting non-Christians to Christianity and employed them towards these ends. In both hunting a lion that Africans believed to be demon-possessed, or in climbing a mountain on which Africans believed ancestral or malevolent spirits dwelled, missionaries were intentionally combining British practices of outdoor leisure and adventure with their religious vocation. This resulted from the missionaries' conviction that Africans had a fundamentally fearful disposition to the natural world. They, on the contrary, wished their converts to understand

[1] Biblical references used throughout the chapter are quoted in the King James (or Authorised) Version because this was the version most commonly used by the British missionaries under discussion.

nature as displaying God's benevolent providence, that is, the working-out of events or processes in the world that are only understood through the eyes of the Christian faith. Hunting, therefore, was one means by which missionaries attempted to cultivate in their converts what they thought to be a proper understanding of the relationship between divine or metaphysical beings and the natural world.[2]

The picture of missionary hunting that emerges from this study is not simply one of British expatriates flaunting an imperial martial masculinity, though that was undoubtedly part of the story. Rather, an exploration of the religious and spiritual dimensions of hunting in the colonial era brings into focus a range of meanings, responses and interactions between Western and African conceptions of the natural world and of its relation to the supernatural. Quite often European missionaries hunted on terms established by Africans, including participating in protective hunting rituals and ceremonies, or even committing physical and metaphysical violence at Africans' behest.

This chapter begins by focusing on missionary views of the natural world and the missionaries' experiences of hunting in the imperial era. Next, it moves to an examination of the ambiguous interactions between missionary and African understandings of hunting and the natural world. It concludes by discussing the relationship between missionary-African hunting and religious conversion in the imperial era. The overall narrative, as a result, is one that attempts to highlight the agency of Africans and African Christians, who, far from being the passive victims of imperial European sport or religion, actively conscripted missionaries into their own agendas and worldviews as they creatively responded to new socio-religious contexts.

The Calling of the Natural World

During the First World War, Algernon Stanley Smith and Leonard Sharp served in Kenya as medical officers with the King's African Rifles. These Cambridge-educated men came from upper-middle-class families. As evangelical Christians, they were imbued with the spirit of the annual Keswick holiness conference in England's Lake District and held rigorously to their belief in the verbal inerrancy of the Christian Bible, among other conservative evangelical doctrines. It might, then, come as no surprise that, while they were stationed in Kenya, these men felt called to missionary careers in Africa.[3] But one would expect this missionary

---

[2] Sujit Sivasundaram, *Nature and the Godly Empire: Science and Evangelical Mission in the Pacific, 1795–1850* (Cambridge, 2005), chapter 4.

[3] J.A. Mangan and Callum McKenzie, '"Duty unto Death" – the Sacrificial Warrior: English Middle Class Masculinity and Militarism in the Age of the New Imperialism', *International Journal of the History of Sport*, 25/9 (2008), pp. 1080–1105.

calling to have come, as it did for so many other evangelicals, at the Keswick conference itself, or perhaps from reading a biblical text such as the 'Great Commission' issued by Jesus Christ to his disciples at the conclusion of the Gospels of Matthew and Mark.[4] Their missionary vocation did come while they were reading. However, it was not the Bible that was before their eyes, but rather the Duke of Mecklenberg's travelogue, *In the Heart of Africa*.[5] It was through this book, which details the duke's adventures through the mountainous wilderness of eastern and central Africa, that these young doctors said God had spoken to them, calling them to initiate an Anglican mission to Rwanda.[6] Interestingly, Mecklenberg's account scarcely mentioned African traditional religions, and it did not contain any semblance of a missionary call. But it did include a great deal of stunning pictures, mostly landscapes of interlacustrine hills – quite similar, in fact, to those found in England's Lake District. It seems that these aesthetic elements connoted overlapping senses of adventure, spiritual rejuvenation and redemptive exertion for the would-be missionaries.

This example indicates an intriguing and under-explored relationship between evangelical missionaries, the natural world and the divine. This dialectic is made all the more interesting in light of Matthew Engelke's 2007 monograph: *A Problem of Presence*. Engelke argues that the simultaneous presence and absence of God is a recurring tension in Christian theology. He defines the problem of presence as follows: 'how a religious subject defines and claims to construct a relationship with the divine through the investment of authority and meaning in certain words, actions and objects.'[7] He further states that what distinguished conservative Protestants from other Christians, especially since the early twentieth century, was their tendency to identify the 'Word of God' as being constituted exclusively by the precise words of the Bible itself.[8] However, the evangelical doctors' divine missionary call to Rwanda came as they looked at pictures of bucolic hills of eastern Africa. They heard God's voice through adventurous accounts of mountain exploration and mysterious peoples virtually unknown to Europeans. Images of the natural world (namely, landscapes), therefore, indicated God's providential direction for them within a divine plan for the spiritual redemption of all peoples. When missionaries believed God to

---

[4] The Gospel of Matthew 28:19–20 reads: 'Go ye therefore, and teach all nations, baptizing them in the name of the Father, and of the Son, and of the Holy Ghost: Teaching them to observe all things whatsoever I have commanded you: and, lo, I am with you always, even unto the end of the world. Amen.'

[5] Adolphus F. of Mecklenberg, *In the Heart of Africa*, trans. G.E. Maberly-Oppler (New York, 1910).

[6] A. Stanley Smith, *Road to Revival: The Story of the Ruanda Mission* (London, 1946).

[7] Matthew Engelke, *A Problem of Presence: Beyond Scripture in an African Church* (Los Angeles, 2007), p. 9.

[8] Ibid., pp. 1–48.

be present in the natural world, that presence was constituted by its redemptive purposes; and when missionaries interacted with the natural world, they often did so within this redemptive framework. While this does not get missionaries around Engelke's recurring tension of the simultaneous presence and absence of God, it does broaden the scope of the investigation to include the evangelical missionaries' extra-biblical experiences of God's imminence.

Imperialism, Athleticism and Missionary Hunting

British evangelical missionary societies saw a demographic shift in their recruits from the early to late nineteenth century, and the two doctors mentioned above are indicative of this shift. In general, by the end of the nineteenth century, a recruit to the evangelical Anglican Church Missionary Society (CMS) would more likely come from an upper-middle class background and have a public school and university education. Perhaps no other missionaries exemplified this shift better than the much-lauded 'Cambridge Seven'.[9] This new breed of missionary underwent a remarkably similar formation as other young men who would venture to the frontier in the high imperial era, and they, in fact, did share many of the same manly sporting ideals.[10] J.A. Mangan, John MacKenzie and John Tosh, among others, have thoroughly described the games ethic that so pervaded English public schools and universities, creating a paradigmatic 'martial male' dedicated to duty, self-sacrifice and fair play. Furthermore, the missionaries' incorporation into colonial educational structures gave them a central role in the diffusion of Western games, athletics and notions of sportsmanship to the non-Western world.[11] A growing body of literature on missionaries of the high imperial era describes them as 'global Tom Browns' – international purveyors of muscular Christianity, a movement that, generally speaking, recognised the confluence of personal moral character, Protestant Christianity and physical health.[12] From this description, it is relatively safe to assume that missionaries

[9] See Jason Bruner, 'The Cambridge Seven, Late Victorian Culture, and the Chinese Frontier', *Social Sciences and Missions* (forthcoming).

[10] John Tosh, *Manliness and Masculinities in Nineteenth- Century Britain: Essays on Gender, Family and Empire* (New York, 2005), chapters 7–8.

[11] See, e.g., 'The Making of Men in Kashmir', *Times*, 11 August 1949; Hamad S. Ndee, 'Western Influences on Sport in Tanzania: British Middle-Class Educationalists, Missionaries and the Diffusion of Adapted Athleticism', *International Journal of the History of Sport*, 27/5 (2010), pp. 905–936; and J.A. Mangan, 'Christ and the Imperial Playing Fields: Thomas Hughes's Ideological Heirs in Empire', *International Journal of the History of Sport*, 23/5 (2006), pp. 777–804.

[12] Boria Majundar, 'Tom Brown Goes Global: The "Brown" Ethic in Colonial and Post-Colonial India', *International Journal of the History of Sport*, 23/5 (2006), pp. 805–820; John M. MacKenzie, 'The Imperial Pioneer and Hunter and the British Masculine Stereotype in Late Victorian and Edwardian Times', in J.A. Mangan and J. Walvin (eds), *Manliness and Morality: Middle-Class Masculinity in Britain*

hunted for many of the same reasons as did other men in the empire. It was manly, it showed pluck and courage, it required discipline and familiarity with the natural world, and it also provided excitement – especially when converts were hard to come by.[13]

Since missionaries often sought out peoples on the peripheries of colonial territories, they frequently placed themselves in good hunting grounds on the frontier of European expansion.[14] Some missionaries consciously utilised their position in the wilderness to attract colonial hunting expeditions, which in turn brought revenue to support the mission's work. Such was the case from Roman Catholic missionaries in the Jos Plateau of Nigeria to evangelical faith missionaries in Mongolia.[15] It was, in fact, the missionary's ambition to be a pioneer, forging through virgin forests and reaching peoples not yet contacted by Westerners, that likewise required many of them to be skilled marksmen (and markswomen) for the sake of food and protection. These values and skills led Robert Baden-Powell to characterise missionaries, along with explorers, as 'peace scouts' in his classic *Scouting for Boys*.[16] Missionaries, however, were poised to have unique interactions with indigenous people, as they went abroad for religious purposes and lived closer to non-Westerners than did many other colonists. They were also people who were often deeply attuned to their own spirituality – often focused on the person of Jesus Christ; and their religious vocation and personal spirituality frequently impacted on their hunting experiences.

African big game hunts led by Europeans were often massive affairs, with African helpers employed as porters, trackers, cooks, cleaners, tanners, among others. Missionaries occasionally took advantage of these circumstances to evangelise their African helpers. For example, in southern Africa, Joseph Dupont,

---

and America, 1800–1940 (Manchester, 1987), pp. 176–196; J.A. Mangan, 'Britain's Chief Spiritual Export: Imperial Sport as Moral Metaphor, Political Symbol and Cultural Bond', in J.A. Mangan (ed.), *The Cultural Bond: Sport, Empire and Society* (London, 1992), pp. 1–10.

13 Brian Stanley, '"Hunting for Souls": The Missionary Pilgrimage of George Sherwood Eddy', in P.N. Holtrop and H. McLeod (eds), *Missions and Missionaries* (Woodbridge: The Boydell Press, 2000).

14 David Livingstone, *Missionary Travels and Researches in South Africa* (London, 1857), pp. 116–117. For a discussion of Livingston and lions, see ibid., pp. 136–142. See also John Barker, 'Where the Missionary Frontier Ran ahead of Empire', in N. Etherington (ed.), *Missions and Empire* (Oxford, 2005), pp. 86–106.

15 Andrew Barnes, 'Catholic Evangelizing in One Colonial Mission: The Institutional Evolution of Jos Prefecture, Nigeria, 1907–1954', *Catholic Historical Review*, 84/2 (1998), pp. 240–262; Erik Sidenvall, *The Making of Manhood among Swedish Missionaries in China and Mongolia, c. 1890–c. 1914* (Leiden, 2009), chapter 4.

16 Robert Baden-Powell, *Scouting for Boys: A Handbook for Instruction in Good Citizenship* (New York, 1994; 1st edn, 1908), 13; J.A. Mangan and Callum McKenzie, 'Imperial Masculinity Institutionalized: The Shikar Club', *International Journal of the History of Sport*, 25/9 (2008), pp. 1218–1242.

of the Roman Catholic White Fathers, would intentionally bring along 200 or more African porters for his hunts, so that at night he could tell them Bible stories around a bonfire. Others, like Stuart Watt, in Kenya and Tanganyika, tried giving Bible readings while his porters marched along. Given the ethnic and linguistic diversity of many of these safaris, one wonders what, exactly, was communicated by these kinds of attempts. And the response to such large-scale attempts was tepid, even by the missionaries' own accounts.[17]

The personal spiritual dimension of hunting is illuminated further when one considers that some missionaries, like Leonard Sharp and Joe Church of the Church Missionary Society's Ruanda Mission, used small hunting excursions as a form of spiritual retreat. After a famine in the region left them spiritually depleted and physically exhausted, they arranged a small hunting safari. Joe Church described the experience as follows:

> As Len and I stalked the buffalo together and faced the danger of their cunning, and talked about the deep things of God in these surroundings of incredible beauty, I think we found depths of oneness in Christ that were very profound. We did not always see eye to eye in the methods of big game hunting or in the problems of revival later on, but these times of hunting and danger bound us together in a very deep way.[18]

While his description of hunting as a crucible for male bonding is hardly unique, what is distinctive here is Church's emphasis upon the spiritual benefits that were uniquely received through hunting. This is not in the sense of communing or spiritually bonding with nature itself, but of the peculiarly Christian fellowship experienced in a particular setting: the stalking of big game. These examples begin to open up new doors of inquiry into how British missionaries mediated their religious identity and vocation with Western hunting practices. They briefly indicate that missionaries did attempt to capitalise on hunts and safaris to proselytise a captive audience. They also show that hunting could have a deep personal and spiritual significance. But how did hunting itself

---

[17] Rachel Watt, *In the Heart of Savagedom* (London, n.d. [1924?]), 39; Bengt Sundkler and Christopher Steed, *A History of the Church in Africa* (Cambridge, 2000), pp. 605–606; Thomas Beidelman, 'Contradictions between the Sacred and the Secular Life: The Church Missionary Society in Ukaguru, Tanzania, East Africa, 1876–1914', *Comparative Studies in Society and History*, 23/1 (1981), pp. 73–95.

[18] Joe Church, *Quest for the Highest: An Autobiographical Account of the East African Revival* (London, 1981), p. 30. The revival mentioned in this quote is known as the East African Revival or Balokole Revival; this emerged in the early 1930s in a network of CMS mission stations between northern Rwanda and Uganda. It eventually spread into all neighboring colonies and mandate territories and endured into the era of early independence.

factor into missionary evangelism, that is, the sharing of the Christian message of salvation through Jesus Christ?

Hunting as Evangelism

Large, formal big game hunts were the exception for missionaries. More common were small-scale affairs, with the missionary and a handful of Africans tracking an animal either for food or protection. And, as these kinds of hunting were far more common, they offer a better means of understanding how missionaries utilised hunting for their goal of the religious conversion of Africans.

Missionaries frequently talked about the Africans' seemingly insatiable appetite for meat and their expectations that the missionary, with his superior hunting technology, should provide them with it. Whether on the painfully long safari from Mombasa to Kampala or around their stations, missionaries seemed to think that their weapons' effectiveness at securing meat endeared them to their African helpers.[19] In truth, it does appear that missionaries with guns (and their many African companions and assistants) were able to curb the devastating effects of famine in some cases. For missionaries, hunting for sustenance was thus an opportunity to demonstrate God's gracious (and efficient) provision of food, particularly in a time of need.[20]

And hunting for food often produced the same thrilling stories that were the stuff of aristocratic sportsmen's epic travelogues. While out on a food-acquiring hunt in the midst of a famine, Stuart Watt was chased by an antelope and dropped his gun. While running, he turned, grabbed the antelope's horn and said a prayer. As his brief prayer was completed, he happened upon his dropped gun, which he somehow managed to pick up. He then turned and fired, instantly killing the antelope. Some years later, Watt is also said to have stared down a lion only three metres away. His wife's account of the incident is enlightening: 'The lion's courage, which knows no dread and heeds no repulse, can only be subdued, under the providence of Almighty God, by exercising the power that rests in the human eye which was bestowed upon man by the Omniscient Creator.'[21] Importantly, dozens of Africans witnessed this encounter with the lion and were reportedly overawed by the power displayed, as they had been earlier with Watt's uncanny ability to level a large beast with a tiny bullet. Rachel Watt's commentary does not sacralise nature, nor is the lion possessed by anything but the nature that God gave it. For Watt, the lesson to be learned from

19 Edward I. Steinhart, *Black Poachers, White Hunters: A Social History of Hunting in Colonial Kenya* (Oxford, 2006), pp. 107–109, 136–137, 211.

20 Watt, *Heart of Savagedom*, pp. 80, 87, 151, 341.

21 Ibid., 367.

the encounter is one about nature. It is only through the proper knowledge of the nature of man and the nature of the lion within a divine providential system that the encounter could be understood properly. But one does wonder what her commentary might have been had her husband not been so fortunate.

Since they had the best guns and lived in close contact to indigenous peoples, missionaries were frequently called upon to hunt and dispose of predatory animals that had killed individuals on mission stations or in safari caravans.[22] If British males perceived big game hunting in general to be the ideal forum (outside of actual warfare) in which a man could prove his mettle in the midst of danger, the tracking of a particular homicidal animal took this danger to an even higher level. This was a duty that missionaries took up with much trepidation, as they frequently hunted at night to catch the beasts just before they took their next victim.[23] It was not uncommon for these night-time campaigns to end in severe injury or death. One missionary, Joe Church, was plagued by a recurring nightmare in which he met a charging leopard he had been called upon to kill. In the nightmare, his worn-out gun misfired, resulting in his being mauled (which, ironically, he was some years later, though he survived the attack). His anxieties were resolved a few weeks later, when, in what he considered an act of divine providence, a new gun was miraculously provided to him free of charge – a 10.75 mm Mauser rifle.[24] And often God's workings were simply serendipitous. When Joe Church and Stanley Smith simultaneously shot at a pouncing lion that had terrorised a village, one bullet went directly between the lion's eyes, with the other hitting its shoulder. In a conflation of providence and Shikar Club ideals, Church reflected: 'Neither of us felt worthy of that incredible shot in the brain when it was charging, so we just praised God and shared the trophy.'[25] God, therefore, arranged for the missionary to continue to carry out his duty of protecting fearful villagers from predatory beasts.

The above examples indicate what missionaries hoped to communicate to African onlookers: that the Christian God manifested himself providentially to provide and protect. For missionaries, provision came in the form of good hunting results in time of want, which showed God's goodness to humanity. Protection was understood within the same framework, for missionaries believed God was somehow present even in particular gunshots, or in the several cases where they themselves were saved from a predator by an African hunting companion. Missionaries then used these stories of providential personal

---

22 Steinhart, *Black Poachers, White Hunters*, pp. 107–109; Church, *Quest*, 39; Ruth B. Fisher, *On the Borders of Pigmy Land* (London, n.d. [1905?]), pp. 56–58.

23 See, e.g., Church, *Quest*, and Allen and Lillian Bilderback, *Our African Journal, 1945–1950* (Puyallup, 1993), pp. 101–104.

24 Church, *Quest*, 56.

25 Ibid, 42, Mangan and McKenzie, 'Imperial Masculinity Institutionalized'.

protection to support their belief that God had sent them to Africa – and preserved them in Africa – to preach a message of salvation.

These examples throw light on how missionaries viewed the relationship of the Christian God to the physical world. But how did this perspective interact with African understandings of the relationship between the physical and the metaphysical? It is here that we begin to understand how hunting was a factor, not only in evangelism, but also in conversion, the turning from one religion or belief system to another.

Hunting for Converts

Missionaries often thought of Africans as fearful people. They were afraid – so common missionary perceptions went – because they were superstitious. And they were superstitious because they did not adequately understand the correct relationship between the natural and the divine.[26] In the missionaries' belief system, gods did not inhabit caves, nor did potentially malevolent spirits or ancestors dwell on particular mountains. One missionary asserted that the reason why African witchcraft had no effect upon Europeans was because it was the fear in the heart of the bewitched African that killed him. The antidote to witchcraft was simply a pithy quote from Jesus: 'Fear not, little flock.'[27] Ludwig Krapf, a pioneer Protestant missionary to East Africa in the 1860s, put it succinctly, urging missionaries to come and help the Wanika of Kenya 'to move to joy in the Holy Spirit from dread of evil spirits'.[28] Both missionaries and Western adventurers associated these fears with fatalism, or the Africans' alleged tendency to simply accept an atypical natural phenomenon as indicative of their destiny. This perceived fear stood in stark contrast with the athletic and sporting ideals that missionaries brought with them to colonial Africa.

However, in hunting, African and European notions of the divine's relationship to the natural world interacted in intriguing ways. This is particularly true with regard to predatory animals. Many missionaries reported that Africans believed that certain animals were possessed by a spirit or, perhaps, that a spirit might send an animal with a certain message – an omen. There were obviously

---

26 Duff MacDonald, *Africana; or, the Heart of Heathen Africa* (2 vols, London, 1882), vol. 2, pp. 73, 82–83; see also H.W. Garbutt, 'Native Witchcraft and Superstition in South Africa', *Journal of the Royal Anthropological Institute of Great Britain and Ireland*, 39 (1909), pp. 530–558. Here, of course, I am speaking in very general terms. Exceptions and qualifications notwithstanding, I do think that my characterisation captures a general disposition among many Protestant missionaries between the late nineteenth century and the mid-twentieth century.

27 'Cow Witchcraft', *Central Africa*, April 1931, p. 99. The quote comes from Jesus' words in the Gospel of Luke 12:32.

28 Ludwig Krapf, *Travels and Missionary Labours in East Africa* (London, 1860), pp. 246–247.

variations across Africa. For example, the Ngoni of present-day Malawi, among other ethnic groups, believed that the spirit of a great hunter might inhabit a lion or leopard. Others, in Kenya and Uganda, believed that once such an animal attacked people, it was really an evil spirit that animated it. The animal itself was not so much the problem as the evil spirit within it, which was often translated into English as 'demon' or 'devil'.[29] Therefore, when missionaries killed a predatory animal, they were committing an act of metaphysical violence, but, it should be noted, at the behest of Africans.[30] In killing 'demons' in this way, missionaries combined European sport with their ultimate goal: the religious conversion of non-Christians. Taken together, Western technology and Western religion might work to dispel a perceived fear in African hearts.[31]

Commonly found in sub-Saharan Africa were various relationships between hunting and spirit mediums. The workings of this relationship could be seen among hippopotamus hunters in Bunyoro. In this case, it was maintained that the ghost of a slain hippopotamus might come back to haunt, plague or pester the hunter who had killed it.[32] The Nyoro dealt with this threat by consulting an expert of the *mmandwa* (or *emandwa*) cult. The latter initiated the hunter into the cult. Following this, the initiate made a shrine to the spirit of the slain animal. It was not uncommon for the association between particular individuals and certain species of animal to have more sinister connotations. Take, for example, the Watts who, while in the Kiangi district of Tanganyika, reported that local people feared that some of their peers were wont to turn into preying animals at night in order to terrorise villagers. Watt encountered one woman who had been tied to a tree and left, having been accused of becoming a hyena. Watt decided to free her.[33] But missionaries, sometimes out of necessity, were drawn into a curious blending of Western and African hunting rituals and techniques. In Uganda, Ruth Fisher told of a group of men who solicited her husband's help in killing a homicidal leopard. Her husband, rifle in hand, led the villagers to hunt it, but only after participating in a ceremony performed by the chief for the protection and success of the expedition. The missionary was successful in

---

29 Frank Carpenter, *Uganda to the Cape* (New York, 1928), p. 95.

30 Such activities were not limited to Africans. See John Patterson, *The Man Eaters of Tsavo* (London, 1907), for the quintessential adventure-danger travelogue in which European religion plays little – if no – role in the killing of 'demonic' lions.

31 Ibid.; L.N. Meredith, 'In Tiger-Country of Central Provinces', *The CMS Outlook*, 65/776 (November 1938), pp. 241–242; 'Cow Witchcraft'; Watts, *Heart of Savagedom*, p. 60. See also Carpenter, *Uganda to the Cape*, in which a missionary reported that there were some 35 different 'devils', and one in every leopard.

32 J.H.M Beattie, 'A Note on the Connexion between Spirit Mediumship and Hunting in Bunyoro, with Special Reference to Possession by Animal Ghosts', *Man*, 63 (1963), pp. 188–189.

33 Watt, *Heart of Savagedom*, p. 60. Connections between witches and such animals as hyenas were common in southern Africa as well. See Garbutt, 'Native Witchcraft'

dramatic fashion, as he shot the leopard as it pounced at him from its lair.[34] Such blending was not unique to Africa, since L.N. Meredith wrote from India of the local priests' 'extraordinary' abilities to find elusive and dangerous animals:

> If a tiger is harassing a village the Baiga priest is called; he walks round the village saying mantras; then at some spot he will knock nails into a tree up to the head; thereby the tiger's jaws are closed! Curiously enough the claw marks of the tiger can be seen on the tree in his effort to tear out the nails! Be it as it may, these men possess uncanny powers over the tiger, and sportsmen pay for their knowledge, or invariably fail to shoot a tiger.[35]

What these examples indicate is that hunting, though rightly often considered to be a display of imperialistic might and masculinity, could also constitute an ambiguous forum in which missionaries interacted with Africans, including 'pagan' spirit mediums, in the pursuit of game. Edward Steinhart has argued that colonial Kenya's professional safari hunting guide was, in reality, a 'hybrid' figure – a white hunter who had adapted his European techniques from closely observing African hunters. What the cases discussed above foreground is that such exchanges were not merely tactical, but could take a ritualistic and metaphysical character as well.[36]

But hunting was not the only way in which missionaries simultaneously engaged in adventurous sport and metaphysical combat. Equally intriguing were instances in which missionaries intentionally scaled mountains accompanied by nascent converts. These African converts to Christianity remained convinced that they must not ascend the mountains because they would experience retribution, which they frequently attributed to ancestral spirits, as only their traditional priest was able to approach the summit for specific, sacrificial purposes. With converts in tow, some missionaries reached the summit, went to the sacrificial place, and then descended – all with the purpose of demonstrating that no such retribution followed. Here, once more, European adventure and missionary vocation combined with the view to dispelling fear of a particular natural object from the hearts of African converts.[37] But what effect did all of this have upon African converts to Christianity? How did missionaries' efforts translate? How were they perceived?

---

34 Fisher, *Borders of Pigmy Land*, pp. 56–58.

35 Meredith, 'Tiger-Country', p. 242.

36 Steinhart, *Black Poachers, White Hunters*, p. 109. See also Karl Weule, *Native Life in East Africa*, trans. A. Werner (New York, 1909), pp. 199–202.

37 'Mlinga the Spirit Mountain', *Central Africa*, November 1927, pp. 229–230.

Conversion, 'Divinicide' and African Agency

Unquestionably, the colonial period was a time of significant changes in African political, cultural and social structures. Religious conversion was part of these changes. The literature on African conversion (be it to Christianity or Islam) is rich and voluminous.[38] Though an extended discussion of the phenomenon of conversion is beyond the scope of this chapter, my use of the term reflects Robin Horton's Weberian 'intellectualist' model of conversion. The adjustments in African worldviews that the religious dimensions of missionary big game hunting were expected to usher in corresponded to a shift in emphasis from patterns of causality based primarily on lesser, more intimate, spirits to ones predominated by a singular God and/or the purely natural functions of the world. It should be noted that such a conception of conversion was not defined by a dramatic, instantaneous recognition of personal sin, followed by a shift in personal piety and spirituality, producing a sense of spiritual rejuvenation and a change in religious adherence. Such was the general evangelical ideal, stemming from the trans-Atlantic revivals of the eighteenth and nineteenth centuries. But, in the African context, adjustments often happened gradually, indicating a spectrum across which Africans moved, rather than a sharp line to be crossed at once (from, say, 'enchanted' to 'disenchanted', if such a thing were possible). Take, for example, the following exchange reported by a missionary in Nyasaland (Malawi):

> Regarding this form of superstition [of taking heed of "ill-omened" snakes], I once asked a native that strongly professed his belief in it whether he would turn back on meeting a snake in his way if I gave a letter for Zomba, and told him that it had to go on quickly. He said, "No; he would not turn back when a white man sent him." Why? "Because", he said, "you would laugh at me, and ask why I had not brought the snake home for a specimen!"[39]

Here the young man does not disavow a belief in omens, but merely makes his observance of them contingent. While an admonishment – however minor – from a missionary may have been the occasion for this man's stated adjustment

---

38 Matthew Engelke, 'Discontinuity and the Discourse of Conversion', *Journal of Religion in Africa*, 34/1–2 (2004), pp. 82–109; Robin Horton, 'African Conversion', *Africa*, 41/2 (1971), pp. 85–108; Lamin Sanneh, *Translating the Message: The Missionary Impact on Culture* (Maryknoll, 1989), pp. 211–238; Andrew Walls, *The Missionary Movement in Christian History: Studies in the Transmission of the Faith* (Maryknoll, 1996), pp. 43–54; John D.Y. Peel, 'Conversion and Tradition in Two African Societies: Ijebu and Buganda', *Past and Present*, 77 (1977), pp. 108–141; Tomas S. Drønen, *Communication and Conversion in Northern Cameroon: The Dii People and Norwegian Missionaries, 1934–1960* (Leiden, 2009).

39 MacDonald, *Africana*, p. 83.

in his behaviour (though not, apparently, in his fundamental belief structure), previous examples indicate that, quite often, it was missionaries who acted at the Africans' behest and within African worldviews in order to help solve imminent threats to their well-being or dilemmas in their spirituality.

Central to the relationship and tensions between missionary and African views was the combat between forces of good and redemption and those that were malevolent and destructive. But one ought not to view missionary and African systems of thought as completely mutually exclusive. Christianity itself allows for a range of beings (saints, angels, Satan, and demons) that interact with the believer and his or her world.[40] As Birgit Meyer has argued with respect to southeastern Ghana, 'the Devil was the link between the missionaries' and the Ewe's worldview'.[41] In killing demonic predators or supposedly disproving the potency of potentially malevolent spirits, missionaries hoped to demonstrate the superior power of the Christian God to overcome what they thought was a dilemma in African spirituality: their fearful disposition to the relationship between the divine and the natural world. In overcoming evil powers and dangerous animals through hunting and adventure, missionaries tried to reorient the Africans' understanding of the relationship of the divine to the natural world. In this sense, from the missionaries' perspective, conversion was intimately bound up with an altered disposition to nature. Missionaries sought to dispel the fear of the malevolent forces at work in nature, which – they were convinced – formed the basis of many Africans' worldviews. Instead, missionaries depicted nature as being under the providential guidance of a single supreme God, who worked through it benevolently to a redemptive end. As Meyer argues, the experience of African Christians within mission and independent churches attests to the persistence of a belief in the intimate relationship between the evil forces and the physical world, as indicated by exorcisms performed to expel demonic spirits from individuals.

Interestingly, in the sources that form the basis of this chapter, evangelical missionaries, despite personally believing in the reality of Satan, almost completely ignored this malevolent spiritual dimension of their own theology when discussing matters of the spirit in the natural world, particularly with regard to hunting. Many tended to relegate the existence of Satan to the realm of 'spiritual warfare'. Take, for example, Joe Church's Bible study entitled 'The Devil as a Man-eating Lion'. Even though his Bible readings began with 1 Peter 5:8 (quoted at the outset of this chapter), they ended with Proverbs 22:13: 'The slothful man saith, "There is a lion without, I shall be slain in the streets."'

---

40 Robin Horton, *Patterns of Thought in Africa and the West: Essays on Magic, Religion and Science* (Cambridge, 1993), chapter 9.

41 Birgit Meyer, '"If You Are a Devil, You Are a Witch and, If You Are a Witch, You Are a Devil." The Integration of "Pagan" Ideas into the Conceptual Universe of Ewe Christians in Southeastern Ghana', *Journal of Religion in Africa*, 22/2 (1992), p. 106.

The 'devil' here, therefore, is really 'slothfulness'.[42] One ought not to assume, however, that his hearers would have appraised the message metaphorically. The overall purpose in killing demon-possessed animals was undoubtedly to display the superior power of the Christian God in overcoming such malevolent forces as the Africans considered to be active in the world. But it is further evident that what quite often happened is that such malevolent spiritual forces were rarely killed outright. H. Mwangi, a Kenyan evangelist, expressed a hybridised form of this understanding, when he wrote to Anglican missionaries in Uganda that 'Satan is not so dangerous when he comes as a roaring lion as when he attacks spiritually'.[43] As was the case with the young man from Nyasaland, Mwangi is close to missionary understandings of 'Satan' as primarily involved in individual spiritual warfare in an immaterial, psychological, and/or relational sense, but he extends this meaning to incorporate the possibility that Satan can also dwell in a material lion.

The analysis presented so far offers a corrective to Bilinda Straight's approach. In 'Killing God',[44] she recounts what she describes as an exceptional moment of colonial metaphysical violence: the instance in which Charles Scudder, an evangelical British missionary to the Samburu, in northern Kenya, fired a shotgun into a cave at which Samburu women left offerings to *N'gai*, the deity that dwelled within. After Scudder's crass actions, devotion at the cave abruptly ended. Straight finds in this act simply another instance of Christianity's proclivity for 'divinicide' – or the killing of non-Christian peoples' metaphysical beings. Divinicide ostensibly levels the metaphysical playing field, which missionaries then populate with Christian theology. What is missing in Straight's account is what this chapter aims to provide: an explication of missionary assumptions and beliefs regarding the relation of the natural to the divine.

Moreover, Straight does not address the role that Africans themselves played in the destruction of certain metaphysical beings or, to put it differently, the fact that missionaries were engaging in metaphysical violence that combined elements of African and European worldviews. For example, as has been seen, it was not uncommon for missionaries to participate in protection ceremonies led by non-Christian chiefs or priests before hunting a murderous lion or leopard.[45] In hunting, missionaries depended heavily upon Africans for migration patterns, tracking expertise and geographical knowledge. The Africans' hidden agency in European hunting parallels their hidden agency as catechists and Bible women

---

42 From Joe Church's *Scofield Reference Bible*, 'Subjects for Addresses', note 34, JEC 14/4–9, box 16, Joe Church Papers, Henry Martyn Centre, Cambridge.

43 H. Mwangi, 'Satan's Attacks on the Kenya Revival', January 1950, JEC 5/4/51, Henry Martyn Centre, Cambridge.

44 Bilinda Straight, 'Killing God: Extraordinary Moments in the Colonial Mission Encounter', *Current Anthropology*, 49/5 (2008), pp. 837–860.

45 Fisher, *Borders of Pigmy Land*, pp. 56–58; Meredith, 'Tiger-Country', pp. 241–242.

in missionary evangelism. When hunting and evangelism were combined, as was the case with some missionary hunting, one must continuously consider the playing-out of African agency in the metaphysical exchanges that took place. Indeed, metaphysical violence was not a colonial or missionary introduction, and the overcoming of evil and malevolent spirits was an abiding concern for many African converts and non-converts alike.

Conclusion

Missionary guns were not strong enough to completely level the spiritual playing field. So, far from being 'killed', local deities or demons were often reinterpreted, or ascribed new meanings and powers. This was a process of mutual interpenetration between worldviews. Such integration was complex, and it is poorly served by simplistic or mono-causal explanations of the way that Christianity and Christian monotheism spread.

African converts incorporated the missionaries' adventurous sportsmanship into their own, distinct Christian spiritualities. One intriguing example from Ghana will have to suffice here. The author is a woman convert, Afua Kuma, and her recorded prayers offer a glimpse into the effects of the missionaries' gun-wielding evangelism.

> Jesus you are the Elephant Hunter, Fearless One!
> You have killed the evil spirit and cut off its head!
> ...
> Okokodurufo: the strong-hearted One,
> Whose works are indeed stout-hearted:
> You stand at the mouth of the big gun
> While your body absorbs the bullets
> Aimed at your followers.
> ...
> Among powerful rifles, you are the elephant gun.
> Jesus enables the hunter to kill the elephant.[46]

Through a complex maze of poetic symbolism and metaphor, Kuma emotively evokes the presence and power of Jesus. This is not simply the result of a brash imperial levelling of the metaphysical playing field. For Kuma, Jesus has a thoroughly spiritual agency within a thoroughly African worldview. What could be seen as the flaunting of imperialistic masculinity (hunting with European

[46] Afua Kuma, *Jesus of the Deep Forest: Prayers and Praises of Afua Kuma*, trans. Jon Kirby (Accra, 1980), pp. 7–8.

guns) thus becomes subject to the concerns, spirituality and worldview of an African convert.

Missionary big game hunting in colonial Africa was a part of the deeper missionary objective of cultivating what they believed to be a proper conception of the relationship between the natural and the divine. This chapter broadly defines this shift in disposition as moving from one of fear to one of benevolent, redemptive providentialism. This gave missionary hunting a peculiar significance vis-à-vis other forms of colonial hunting. Evangelicals encountered God through the natural world.[47] Missionaries believed nature could be the means through which God made known his benevolent and redemptive purposes for humanity. Whether it was two young doctors believing that God spoke to them through an adventure travelogue to establish a new mission, or missionaries thanking God for a well-placed shot that killed a charging lion, they conceptualised God's benevolent work in the world as leading ultimately to redemption – the salvation of all peoples. Many missionaries hunted with a desire to cultivate a similar expectation among African converts. For both missionaries and their converts, hunting was a way in which God could truly become present – even in the form of an elephant gun.

[47] Simon Schama, *Landscape and Memory* (New York, 1995), pp. 502–513.

Chapter 4

Cockney Sportsmen? Recreational Shooting in London and Beyond, 1800–1870

Matthew Cragoe

> Ye fowlers! manly strength your toils require;
> Defiance of the summer's burning sun
> And winter's keenest blast of hail or storm,
> Of ice or driving snow; nor must the marsh
> That quivers wide deter you, nor the brake
> That seems impervious, in whose thorny depth
> You struggle long, and lose the cheerful day,
> 'Till bursting through, again the sylvan scene
> Tranquil and smooth re opens to your view.
>
> – John Vincent, *Fowling: A Poem in Five Books* (1808)

In the mid-nineteenth century, English field sports were widely held to be an important source of individual character-building and national and imperial well-being.[1] When John Henry Walsh first published his *Manual of British Rural Sports* in 1856, he was at pains to highlight the link between field sports and the recent campaign in the Crimea: '[T]he perils of the steeple chase, the hunting field or ... the modern cricket field', he suggested, had been key factors in preparing for the rigours of war the men who fought so valiantly at Alma, Inkermann, Balaclava and Sebastopol.[2] Not only did such pastimes improve the moral health of the people, keeping them out of public houses and betting shops, but the 'vigour, courage, and power of endurance' they imparted were what enabled the people of Britain to 'withstand, as a nation, the encroachments of their neighbours'.[3] Writing a decade after Walsh, Robert Blakey claimed similar

---

[1] I would like to thank Brenda Assael, and the members of the Modern British History seminar at the Institute of Historical Research for their comments on an earlier draft of this paper. J.A. Mangan and Callum McKenzie, 'Martial Conditioning, Military Exemplars and Moral Certainties: Imperial Hunting as Preparation for War', *International Journal of the History of Sport*, 25/9 (2008), p. 1141.

[2] John H. Walsh, *Manual of British Rural Sports* (London, 1856), pp. 8–9. The book had gone through 17 editions by 1888.

[3] Ibid., pp. xv–xvi.

virtues for the gun. Shooting, he argued, offered 'well regulated enjoyment' and was connected with 'all that is manly, energetic, and healthful'. Its virtues did not stop there, however. In the context of world affairs, he claimed, it played a crucial role 'in sustaining and strengthening that invincible courage, and skilful use of warlike weapons, now rendered necessary for the maintenance of our national existence, and the consolidation of our independence and power'.[4] Hardiness and endurance were two of the greatest qualities with which shooting was supposed to equip its practitioners. At its finest, shooting pitted the hunter against nature itself, forcing him to encounter and overcome all that the environment could throw at him in pursuit of his quarry – 'keeping the mind and body upon full stretch', as Blakey put it in reference to deer stalking in the Highlands, or as the lines of John Vincent's hymn to the 'manly' toils of fowling quoted in the preamble to this chapter imply.[5]

Not all forms of shooting were equally admired, however. For much of the nineteenth century, those who did not trek across the countryside in pursuit of the high-status species covered by the game laws were derided as 'cockney sportsmen'. The scope of this phrase, and what it connoted when applied to the recreational shooting enjoyed by ordinary Londoners during the first half of the century, is the subject of this chapter. As will be seen, the description, ubiquitous in the first half of the century, lost its currency in the second, as shooting underwent a series of fundamental changes in the mid-Victorian period.

The term 'cockney' had long carried the connotation of a townsman who did not understand, and was ignorant of, country life.[6] The *Oxford English Dictionary* quotes an example from 1640, in which a London cockney innocently asks 'whether Hay-cocks were better meat broyl'd or roasted?', and it could still be found used in this way during the nineteenth century. Alfred, Lord Tennyson, for example, once berated a woman who had attended a reading of his poem *Maud* for not knowing what kind of bird would sit on a high garden wall and call 'Maud, Maud, Maud'. Her guess – 'a nightingale?' – brought a snort of derision from the poet laureate: 'What a cockney you are!' he is reported to have exclaimed. 'Nightingales don't say Maud. Rooks do, or something like it. Caw, caw, caw, caw, caw.'[7]

The twinning of 'cockney' with 'sportsman' appears to have been a more distinctively nineteenth-century phenomenon. What triggered the association is not immediately clear; by 1800, however, the cockney sportsman was entering upon half a century of comedic dominance – a stock character on the

---

4 Robert Blakey, *Shooting: a Manual of Practical Information* (London, 1865; 1st edn, 1854), p. 1.

5 Ibid., p. 161.

6 Gareth Stedman Jones, 'The "Cockney" and the Nation', in D. Feldman and G. Stedman Jones (eds), *Metropolis London: Histories and Representations since 1800* (London, 1989), p. 281.

7 Quoted in Jerome H. Buckley, *The Victorian Temper: A Study in Literary Culture* (New York, 1964; 1st edn, 1951), pp. 135–136.

London stage and in humorous literature. In the first part of this chapter, it will be suggested that, whatever its origins, the joke upon which the 'cockney sportsman' turned had as much to do with his transgressions of accepted codes of masculinity and class bound up in the notion of 'sportsmanship' as with the boundary between countryside and town. In the following section, two groups often derided as cockney sportsmen will be examined: the casual urban shooter who pursued small birds and the crack shots who honed their skills in pigeon-shooting contests hosted by such establishments as the Red House in Battersea. The chapter ends by offering some reflections on the decline of the cockney sportsman as a figure of fun in later Victorian England, and the extent to which this reflected the challenges facing recreational shooting by the century's end.

The Cockney Sportsman

During the first half of the nineteenth century, the figure of the cockney sportsman secured an established place in the standard repertoire of British comedy. He – and it was invariably a 'he' – was to be found on the stage, in the circus ring, in the humour pages of newspapers and periodicals, and in satirical prints. Much of the laughter generated by the cockney sportsman character revolved around either slapstick humour or plays upon the cockney's traditional ignorance of the countryside. However, as this section will also argue, the prints, in particular, suggest that the humour also contained a more complex and multilayered character and played upon the cultural distance between the urban 'swell', out for a day's recreational hunting in the countryside, and his gentlemanly rural counterpart.

Throughout the early nineteenth century, newspapers teemed with stories of the cockney sportsman, and the commencement of the shooting season on 1 September also provided the opening for the annual round of jokes about the capital's sportsmen. It was a nation-wide indulgence. In 1801, for example, the *Aberdeen Journal* reported

> Yesterday the Cockney Sportsmen sallied out to their annual amusement, and, according to custom, made a great deal of sport. We have not heard of any thing being killed, although several thumbs and fingers have been found in the roads. A goose received a desperate wound on Stockwell Commons, and the sparrows were much alarmed about the three-mile stone. Upon the north side of the Metropolis the firing was tremendous, and several bull-dogs and mastiffs were wounded, not being staunch to the heel. Happily we do not hear of any two-legged creature, man

> or bird, being killed by any accident. Indeed the linnets, canaries, and parrots, had all been carefully removed from the windows.[8]

This kind of example could be replicated many times over, as could the idea that poachers around London were busy killing game that they could sell to cockney sportsmen to provide them with a suitably impressive 'bag' on their return to the capital.[9]

The cockney sportsman became the butt of endless apocryphal jokes and stories. One favourite of the mid-century was the poem published in 1846 that recounted the story of a man from Cheapside ordered into the countryside for his health.[10] The man took up residence at the house of a farmer near Sudbury, in Essex, who, thinking to cheer him up by offering him some sport, lent him a gun and his highly prized hunting dog, Ponto. Out they duly sallied, and all went well until, to the horror of the cockney, Ponto suddenly froze. Thinking the dog had suffered some kind of seizure, and terrified that he might expire, the cockney picked him up and carried him several miles back to the farmer only to learn that he was simply 'pointing'. The poem proved highly popular and was endlessly repeated during the next two decades.[11]

Other tales portrayed the cockney as the victim of the cunning countryman. A classic was the story of the cockney sportsman who was invited to shoot as many ducks on a village pond as he liked in return for 2s 6d; only when he had paid his money and started shooting did he discover his error, as the real owner of the ducks, alerted by the sound of gunfire, appeared on the scene.[12] This was a close relative of another apparently unremitting source of amusement to the early Victorians: the story of the cockney who, having gone out hunting, succeeded only in wounding a valuable domestic animal – a chicken, a cow and, on one occasion, two donkeys on Hampstead Heath – and was then obliged to pay substantial compensation to the farmer.[13]

Another location in which the cockney sportsman flourished was the stage. With punning names such as 'Ned Neverkill' or 'Simon Snapshot',[14] the cockney

---

8 *Aberdeen Journal*, 9 September 1801.

9 'First of September: Gazette Extraordinary!!', *Morning Post*, 2 September 1806; *Morning Post*, 1 September 1801 (for poachers); 'A Trip to the Lakes', *Lancaster Gazette*, 15 December 1827.

10 *Essex Standard*, 8 January 1847.

11 This same story was told by P.E. Dove at the Literary and Scientific Association in Arbroath, as part of a lecture on 'The Wild Sports of Scotland', to illustrate the ignorance of the cockney sportsman. *Dundee Courier*, 3 January 1856. It also turned up as a free standing joke in 'Varieties', *Cheshire Observer*, 2 June 1860, and the *Lancaster Gazette*, 28 June 1862.

12 *Trewman's Exeter Flying Post*, 27 November 1856.

13 *Essex Standard*, 8 September 1832; see also 'Sporting Intelligence Extraordinary', *Morning Post*, 26 October 1805, and *Leicester Chronicle*, 8 October 1842.

14 *Hampshire Advertiser*, 2 September 1837; *Era*, 10 October 1869.

sportsman became established as a standard comic character in the theatre, appearing alongside Harlequin, Panteloon, Clown and Columbine at the Sans Pareil Theatre in the Strand in 1814, for example.[15] Theatrical performances and pantomimes in the capital continued to feature the figure of the cockney sportsman for the rest of the century,[16] and in 1841 there was even a ballet at the Queen's Theatre entitled 'The Sailor's Return, or the Cockney Sportsman'.[17]

Perhaps the most enduring arena for the cockney sportsman, however, was the circus ring.[18] In 1833, Batty's Equestrian Circus advertised 'An entire new Comic Equestrian Scene, called the Cockney Sportsman', and stuck with it for many years. In 1837, Ducrow's Royal Arena of Arts followed suit, offering the public a 'comic extravaganza' entitled 'The Cockney Sportsman or the 1st of September', involving four horses and an appearance by Ducrow himself.[19] By the 1850s, the cockney sportsman was standard circus fare. William Cooke's Circus Royal greeted the 1854 season with 'a new Hippodramatic Sporting Extravaganza' on the theme, while, three years later, a Liverpudlian audience at Hengler's Cirque Variete were kept 'in roars' by the cockney sportsman jokes played out before them.[20] How much the genre evolved it is impossible to say, since no accurate reports of the cockney sportsman in the ring have survived. However, the nature of the characters involved in the drama suggests that the same basic joke was repeated again and again. In the 1870s, for example, Cooke's Royal Circus production of 'The Cockney Sportsman or the Londoner out for a day's recreation' in Aberdeen featured three stock characters: the squire, Simon Pure, a countryman, and a cockney sportsman, Young Popham, 'a Pawnbroker's Son, of London', a cast very reminiscent of those common during the 1830s.[21]

The cockney sportsman thus had an enduring presence in the consciousness of Victorians as a stock figure of fun, an easy reference point for a certain kind

---

15 See *Morning Chronicle*, 10 January 1814; *Morning Post*, 12 November 1828, Royal Olympic Theatre, Newcastle and Wych-sttreet, Strand, new comic pantomime: 'Tithe Sheaf or the Cockney Sportsman'; *Morning Post*, 27 December 1831, review of 'Hop o' my Thumb and his Brothers', set in Wales; *Morning Chronicle*, 27 December 1860, Drury Lane, Christmas pantomime *Peter Wilkins*.

16 *Standard*, 3 September 1839, review of 'The First of September'; *Standard*, 30 March 1848, 'Harvest Home'; *Era*, 1 February 1852, 'Annie Tyrrell or Attree Copse'; *Pall Mall Gazette*, 14 September 1868, 'Blow for Blow'.

17 *Era*, 21 February 1841

18 Brenda Assael, *The Circus and Victorian Society* (Charlottesville and London, 2005); Marius Kwint, 'The Legitimization of the Circus in Late Georgian England', *Past and Present*, 174 (2002), pp. 72–115.

19 *Caledonian Mercury*, 2 November 1837. Ryan's Circus offered a performance with a similar title at Bristol in 1837; *Bristol Mercury*, 16 September 1837.

20 *Bristol Mercury*, 21 January 1854 (Cooke's); *Liverpool Mercury*, 2 November 1857; *Era*, 22 November 1857 (Hengler's). Macarte's and Clarke's Magic Ring, at Newcastle also ran an equestrian feature on the Cockney Sportsman in the 1850s. *Newcastle Courant*, 15 January 1858.

21 *Aberdeen Weekly Journal*, 2 July 1877.

of bumbling, slapstick incompetence. Yet it is arguable that there was something more to the stereotype than this. Another medium in which the cockney sportsman enjoyed a considerable vogue, especially in the first three decades of the century, was the satirical print. In 1800, for example, two of London's foremost printmakers, James Gillray and Charles Williams, each produced four-piece sets on the theme of the cockney sportsman. In the next 30 years, their example was followed by many others, and all the leading printmakers – including Thomas Hood, Isaac Cruickshank and Robert Seymour – seem to have turned their hand to the subject at some point. In these prints, a more complex layering of meanings inherent in the cockney sportsman character becomes visible. The humour derives only in part from the townsman's misunderstanding of the countryside; it also plays on his ignorance of the codes that animated the kind of ideal sportsman described by the authors of field sport manuals encountered in the introduction.

The set produced by Gillray provides an instructive introduction to the genre. The series follows two friends who go out hunting in the countryside to the north of London. They are well-heeled, expensively dressed individuals: one a morbidly obese character whose real interest in the outing, as emerges later in the series, is eating; the other a raw-boned, would-be dandy with a big blue cravat and effete townsman's boots – a genuine cockney 'swell' of the type commonly encountered in the first half of the nineteenth century.[22] They have their counterparts in the two absurd dogs that accompany them: a small poodle-like creature whose fur has been cut to resemble a lion, and a round-faced mongrel-breed cur. They are decidedly town dogs, just as their masters are town men.

The first picture in Gillray's series depicts the cockney sportsmen 'marking game', and the humour resides in a play upon this term. The dandy has spotted or 'marked' his game: a cluster of decidedly non-game birds feasting on the carcass of a dead horse. However he is holding the gun all wrong and when he accidently squeezes the trigger in his excitement at spotting the birds, he discharges the shot into the huge, upended posterior of his friend, bowled over by his round-faced dog who has raced in undisciplined excitement through the bars of the stile, and thereby 'marking' him. The remaining prints in the set continue the theme: 'Shooting Flying' shows the dandy firing his gun indiscriminately at a flock while jumping mid-air; 'Recharging' depicts the obese friend tucking in to a huge picnic while the dandy reloads his fowling piece; and the last in the set, 'Finding a Hare', has the party creeping up on a very tame-looking domestic rabbit.

---

22 These prints suggest a slight revision to the fascinating chronology of characteristics associated with the 'cockney' by Stedman Jones, '"Cockney" and the Nation', pp. 284–288.

Figure 4.1 James Gillray, 'Cockney Sportsmen marking game', 1800
The first plate of four in Gillray's famous series. Courtesy of Matthew Cragoe.

The traditional 'cockney' joke – of the townsman failing to understand the countryside – undoubtedly ran throughout the prints. Their clothes and accoutrements were wrong; they failed to identify their quarry correctly. However, the humour also resided in a number of other factors. The prints mocked their social pretensions, their obvious assumption that what they were engaged in was actually 'hunting' of the type described in the formal literature of shooting. First, the location was not some wild desolate heathland, where skill and endurance are required to hunt a quarry down; this was the extra-urban scrubland just north of London – so close, indeed, that St Paul's can be seen looming in the background. Second, the quarry they hunt are not the noble high-status species pursued by the traditional sportsman, but random birds, such as crows, and domesticated animals such as the rabbit. There was a clear implication that what you hunted in some way defined you. Just as those who hunted game species arrogated to themselves some of the qualities of skilfulness, intelligence and cunning with which they endowed their prey – the idea of such anthropomorphising was, after all, to render the hunt a competition between equals – so cockney sportsmen hunting non-game species were tarred with the low status of their quarry. Third, the pictures invert all the tropes of character-building bruited in the formal literature. Far from breeding physical hardiness, a sense of endurance and self-control, these 'swell' hunters are by turns effeminate, gluttonous and prone to over-excitement, firing wildly at anything. What is more, they are terrible shots: there would be little hope for England if the defence of her possessions rested on the skill of these sportsmen.

The jokes running through Gillray's prints were very similar to those in the sets produced by Williams and later artists. The vein of humour they tapped, however, was richer than the old joke about the cockney townsman being bewildered by the ways of the countryside, or the slapstick fun of someone getting a round of shotgun pellets embedded in their backside. It also mined a subtly anthropomorphic seam of ideas about class and 'manliness' that were associated with hunting. In the next section, attention turns to the groups whom contemporaries chose to describe as 'cockney sportsmen', and to the question of what resemblance the comedic image bore to the metropolitan reality.

The Sporting Cockney

Despite the remarkable growth of interest in the history of sport and leisure in the last 25 years, remarkably little has been written about recreational shooting. In part, this may stem from the association of shooting with the aristocracy, and the associated world of game licences and game laws.[23] However, the game laws, in Britain as in other European countries, applied only to relatively small lists of 'high-value' species; most animals and birds were unprotected and were fair game for anyone that chose to take them. Since there was no law requiring that guns be licensed until 1870, shooting flourished.

There were certainly large numbers of guns in circulation – one estimate sets the figure at 150,000 on the eve of licensing. Guns were readily available to the would-be purchaser. Alongside high-end specialists like Mortimer in the Strand and Barnett in Oxford Street, and later Purdey's, in first Oxford Street and then South Audley Street, many ironmongers sold basic firearms for between five and ten shillings apiece in the early nineteenth century.[24] Beyond this there was doubtless a flourishing trade in second-hand weapons, often apparently purchased through a pawnbroker,[25] and, below this, the world of the homemade firearm. Three sportsmen arrested in 1826 for causing damage to fruit trees in a market garden near Rotherhithe, for example, were armed with an extraordinary

---

23 Hugh Cunningham, *Leisure in the Industrial Revolution* (London, 1980), pp. 17–19; Alastair J. Durie, 'Game Shooting: an Elite Sport, c. 1870–1980', *Sport in History*, 28/3 (2008), pp. 431–449.

24 John Lowerson, *Sport and the English Middle Classes, 1870–1914* (Manchester, 1993), pp. 230–232.

25 'The Cockney Sportsman', *The Oddfellow*, 12 October 1839.

selection of weapons. One was carrying a rusty old blunderbuss; another, a short weapon comprising a pistol barrel 'fixed very clumsily into the stock of an old musket'; and the third, 'a thing which he called his fowling piece, the barrel of which, about a yard and a half in length, appeared to have been once an old gas pipe[,] ... fixed to the stock by means of pieces of iron hoops, over which were pieces of list'. With guns such as these, they were probably more danger to themselves than anything else.[26]

The men involved in this case were two tailors and a shoemaker, and other court cases reported in newspapers before the various police and magistrates courts from time to time between 1820 and 1860 offer an insight into the people who engaged in shooting in and around London.

Table 4.1 Cases involving firearms heard at police and Magistrates' courts in London, *c.*1820–1860

Date	Name	Occupation	Location	Charge
1826		Tailor Tailor Shoemaker	Rotherhithe	Damage to property
1833	Lush	Greengrocer and tea-dealer	Strutton Ground, Westminster	Shooting pigeons
1834	Bishop	Tailor	Notting Hill	Discharged gun through a window
1836	Possey Wilson	Thermometer maker Tailor	Highgate	Sporting
1837	Harrison Collins	Engineers	Millwall	Assault
1845	Greenfield	Butcher	Camberwell	Willful damage
1845	Philip Ernst	Dyer	Clapton	Attempted murder
1846	Westbrooke Balcombe	Shoemaker Watchmaker	Camden	Assault
1858		Tradesman	Walworth	Shooting pigeon
1858	Noble	Paper hanger		Shooting pigeon

While this table is necessarily impressionistic, it does suggest that shooting was a recreation less of the labourers and more of the artisan and the shopkeeper, something that may be associated with the expense of acquiring a weapon, shot and powder.

26 *Sheffield Independent*, 9 December 1826.

I've got a visper for you, Sir; I don't vish to be impertite,
but next time you shoots a bird vot I've brought to my
call, I'll shoot you into a clay pit, that's all!!

Figure 4.2 Robert Seymour, 'The Bird Catcher'
The clothes worn by the bird-catcher indicate a social and economic marginality that differentiates him sharply from the cockney sportsman. Courtesy of Matthew Cragoe.

Certainly, both cartoonists and commentators drew a sharp distinction between the cockney sportsman and the bird-catchers who netted and trapped live birds for sale. As Gareth Stedman Jones makes clear, the mid-century 'cockney' was not conceived to be a member of the 'dangerous classes' in the way that the contemporary costermonger was.[27] An illustration by Robert Seymour of a confrontation between a bird-catcher and a cockney sportsman who has just killed a bird successfully lured by the former sets out the differences graphically. The cockney sportsman is dressed in perfectly ordinary clothes; the bird-catcher, however, merits the description provided by the *Preston Guardian*'s London correspondent, who complained that every Sunday morning 'swarms of low-looking men, most of them as vile in their language as there are blackguardly in appearance, invade almost every suburban field with their nets and other apparatus', stripping the fields of song-birds.[28]

Sunday morning was a very bad time to be a bird in or near London, as both bird-catchers and cockney sportsman sallied forth to pursue their mutually incompatible missions. The two tailors and the shoemaker picked up shooting across market gardens near Rotherhithe explained that they went out on Sunday morning 'to amuse themselves the only spare day they had in the week', and to kill 'a few "pee wees" to make a pie for their Sunday dinner'.[29] They were not alone. In 1846, the *Era* carried a report that the inhabitants of Plaistow and East Ham had been much annoyed by cockney sportsmen coming out on Sundays to shoot in the marshes.[30] The 'sportsmen' were said to generally belong to the working classes of Shoreditch and Whitechapel, and to muster in some force – 30 to 40.

Not all cockney sportsmen were so careful about where they loosed off their fowling pieces. Sporting in the rural areas beyond London was one thing, but a much more annoying, and potentially dangerous, form of hunting involved shooting in built-up areas. A number of cases came before the authorities in which people were accused of risking life and property by shooting in town. Robert Seymour's cartoon, 'A Lark', which captures well the casual disregard of the cockney sportsman hunting in the suburbs, was played out in many real-life examples.

In 1834, for example, a tailor named James Bishop was bound over for 12 months after discharging his gun through the window of a house in Notting Hill while the family were inside getting ready for church. Bishop's explanation in court, that 'he was aiming at a bird, when on discharging the gun it gave a sudden twist and the contents perforated the Complainant's window', might seem to carry its own moral. The magistrate ordered that his gun be detained at the station house.[31]

27 Stedman Jones, '"Cockney" and the Nation', p. 288.

28 *Preston Guardian*, 20 November 1869.

29 *Sheffield Independent*, 9 December 1826.

30 *Era*, 25 October 1845.

31 *Morning Post*, 25 February 1834.

Figure 4.3 Robert Seymour, 'A Lark'
This cartoon draws attention to the dangers posed by cockney sportsmen letting off fowling pieces in residential areas. Courtesy of Matthew Cragoe.

In 1833, a greengrocer and tea-dealer, of Strutton Ground Westminster, named J.L. Lush, was charged with shooting four pigeons belonging to his neighbours John Sampson and John Langlet.[32] Apparently, he was so 'particularly addicted to shooting' that he was 'in the habit of firing at lighted candles at night in his yard'. In the day time, by contrast, his chief amusement was shooting at pigeons that flew over his yard, 'to the great terror and annoyance of the fancy in that line.' He was fined 40s for each bird and their value – a total of £9 1s 6d.

In 1845, Lambeth St Magistrates granted the Vicar of Camberwell, Revd John Gore Storie, a warrant against a local butcher, Mr Greenfield, who regularly shot blackbirds, thrushes and sparrows in the vicinity of the vicarage and was both causing immense damage to the hedges (estimated at 30 shillings) and putting the inhabitants in fear of going out into the garden.[33] Despite having promised Storie to cease and desist on several previous occasions when challenged, he persisted in the activity.

Finally, in 1854, Ilford Petty Sessions heard the case of a man employed in West Ham who was charged with having let off a gun in a thoroughfare on a Sunday morning.[34] Mr Pluxton, a farmer of Cann Hall Lane, received several shots in the face. He 'complained of the great increase in the number of Cockney sportsmen, and expressed a hope that some means might be adopted to put a stop to the practice, which was now becoming an intolerable nuisance'. The Bench agreed, and having said that they would do all in their power to put down the practice, fined the defendant 20s and costs.

As these examples hint, the problem of cockney suburban shooting moved further out of the centre of town as the century wore on. George Cruickshank's famous cartoon, 'London Going out of Town', published in 1829, picks up nicely the extent to which, as the bricks and mortar of the town advanced, the countryside and all the species that inhabited it retreated. The real-life urban sportsman had to travel further and further to indulge his passion as the century progressed, and many could not, or would not, wait until they got into deep countryside. Thus while the classic 'cockney sportsman' story of the 1820s involved incidents in Westminster or the Mile End Road, by the 1840s, the newspapers were covering incidents in Camberwell. By the 1850s, the front line had moved further still, and now it was the residents of West Ham, Lewisham and Peckham Rye whose turn it was to be

32 *Morning Post*, 17 April 1833.

33 *Morning Post*, 26 February 1845.

34 *Essex Standard*, 13 January 1854.

> not only annoyed but considerably alarmed at the Sunday visitation of cockney sportsmen with guns, who fire at everything, perfectly regardless of consequences, and caring nothing as to who or what came within range of their shot.[35]

One important aspect of the cases reviewed in this section was the very clear distinction magistrates drew between recreational shooting and poaching. The fear expressed in the Chartist newspaper, *The Charter*, in 1839 – that if a rural police force were introduced, officers would 'rake up innumerable trumpery cases, and fill the goals with petty offenders, whose crimes would consist of cutting a walking-stick, pulling a turnip, or going out with a gun to shoot hedge-sparrows' – did not materialise in practice.[36] Magistrates were careful to differentiate between the amateur hunter out for some recreational shooting and the dark figure of the poacher. As Professor Edward Christian of Gray's Inn argued in a treatise on the Game Laws published in 1817, a Justice of the Peace had to satisfy himself of the intent of the defendant from the circumstances stated by the witnesses 'that the party was in pursuit of Game and not of sparrows, larks or fieldfares'.[37] The views of any witnesses called by the prosecution as to the intentions of the defendant were immaterial: 'the circumstances proved ought always to be such as would lead a jury to believe that such was the intent.' The cockney sportsman was undoubtedly a nuisance, subverting as he did the suburb's role as a place of class exclusivity and physical safety,[38] but he was not automatically a criminal.

The 'cockney sportsman' tag was used by contemporary commentators to cover the activities of another class of gunmen in nineteenth-century London: those who shot at sparrows or pigeons released from traps. This pastime, as will be seen, was rather different from the recreational hunting described above, and it involved men of a higher social background. What earned them pejorative inclusion within the ranks of the cockney sportsmen is the subject of the remainder of this section.

The origins of competitive pigeon shooting are somewhat obscure: the *Sporting Magazine* referred to it as 'established' in 1793, though it must be said that very little notice of it was taken in the press at this time.[39] The heyday of pigeon shooting undoubtedly came later, in the first half of the nineteenth century, when the sport came under the patronage of the Red House club at

---

35 'Police Intelligence', *Morning Chronicle*, 22 July 1856.

36 Letter of 'Reformator' on 'The Rural Police Scheme', *The Charter*, 14 July 1839.

37 Edward Christian, *A Treatise on the Game Laws* (London, 1817), pp. 157–158.

38 Anne Witchard, '"A Fatal Freshness": Mid-Victorian Suburbophobia', in L. Phillips and A. Witchard (eds), *London Gothic: Place, Space and the Gothic Imagination* (London, 2010), pp. 29–34; Simon Joyce, *Capital Offenses: Geographies of Class and Crime in Victorian London* (Charlottesville, 2003), pp. 39–40, 53–57.

39 Robert Blakey, *Shooting: a Manual of Practical Information* (London, 1854), p. 145.

Battersea and was developed there into an ever-more competitive and refined activity.[40] For much of the century, the only set of public rules regarding pigeon shooting were those used in Battersea; any disputes regarding the rules were referred to *Bell's Life in London*, and they gave their opinions with regard to what the practice was at the Red House.[41]

The Red House at Battersea occupied a prominent position on the south bank of the Thames, directly opposite the Chelsea Hospital. Close by were Astley's amphitheatre and the famous Vauxhall gardens. Mentioned in every good tourist guide to the beauties and pleasures of London from the second half of the eighteenth century, the Red House was, according to one, a favourite haunt of London's 'Sunday citizens'.[42] Its tea gardens were particularly popular in the summer time, the natural destination for Londoners enjoying the famously pretty walk along the river-bank at this point.[43] The Red House backed onto the Battersea Fields, some 300 acres of open and common land that had a reputation for lawlessness, gang fights and duelling: the famous encounter between the Duke of Wellington and the Earl of Winchilsea came off here in 1829 at the height of the struggle over Catholic Emancipation.[44]

The blessings of Battersea were not, therefore, unmixed, and the Red House itself was more than a tea garden. Like other public houses on the margins of London – the Old Hats on the Uxbridge Road for example – the Red House was a centre for sports of various kinds. Given its position on the water, it was unsurprising that it was a centre for rowing competitions; however, it also hosted quoits, pedestrianism and pugilistic encounters throughout the early nineteenth century.[45] The common denominator was, of course, gambling, and it was into

---

[40] Cf. Mark A. Kellett, 'The Power of Princely Patronage: Pigeon-Shooting in Victorian Britain', *International Journal of the History of Sport,* 11/1 (1994), pp. 63–85, who suggests that pigeon shooting did not really become popular until the period after 1860; and John Martin, 'Pigeon Shooting' in T. Collins, J. Martin and W. Vamplew (eds), *Encyclopedia of Traditional British Rural Sports* (London, 2005), p. 207.

[41] 'Marksman', *The Dead Shot* (New York, 1863), p. 237.

[42] John Fisher Murray, *A Picturesque Tour of the River Thames in its Western Course* (London, 1845), pp. 10–11.

[43] John Hassell, *Picturesque Rides and Walks: With Excursions by Water* (2vols, London 1818), vol. 2, p. 91; Parliamentary Papers (hereafter PP), 1833, XV, *Report from the Select Committee on Public Walks*; p. 361, Q 209–211: evidence of Mr James Bailey.

[44] Peter Cunningham, in his *Hand Book of London: Past & Present* (London, 1850; 1st edn, 1849), p. 39; Fisher Murray, *Picturesque Tour,* pp. 10–11; PP, 1837–1838, XV, *Report from Select Committee on Metropolis Police Offices,* pp. 419–420, Q 515: evidence of Mr Thomas Bicknell, Superintendent of Police.

[45] Old Bailey Proceedings Online (www.oldbaileyonline.org), January 1831, trial of Richard Curtis (t18310106-114). Curtis recalled having been involved in an hour and a half long fist-fight at the Red House; *Bell's Life in London*, 31 October 1847, p. 6, and 14 September 1851, p. 1.

this ready-made groove that pigeon shooting slotted when the sport became popular at the turn of century.

The sport of pigeon shooting was popularised both through *Bell's* and through the columns of the newspapers that reported the results for a nation-wide audience. In its most rudimentary form, pigeon shooting was a very simple sport.[46] Each competitor would shoot at an agreed number of birds, released one by one. The pigeons were released from a shallow box, about a foot long and 8 inches wide, set into the earth, with a sliding lid; the lid was removed by a man sitting right by the gunner pulling on a piece of string on the gunner's command of 'pull'. The gunner stood 21 yards from the trap and was not allowed to raise his gun to his shoulder until the bird was on the wing; the bird itself must fall within 100 yards of the box or be deemed a lost shot. The trap was then filled with a new bird, and the next competitor took his shot; apparently it was normal just before putting a bird into the trap to pull out a few feathers from the tail 'to make them lively, and thus go off keenly'.[47] Each shooter must be back at his mark and ready to shoot within five minutes of the last shot. The hits and misses were recorded by an arbiter, who also decided points such as whether a bird could be deemed to have 'risen' sufficiently to permit a shot, and whether the bird had dropped within the specified ground. Naturally enough, the competitor with the most hits at the end of the competition was the winner.

However, rules at the Red House quickly became more sophisticated, as experienced marksmen sought to stretch themselves. The Red House was attracting an elite sporting clientele, very different from the Sunday-morning recreational hunters after a pee-wee or two to put in a pie for dinner. Men such as Hon. George Anson, Captains Ross and Bentinck, Messrs Biddulph, Osbaldeston, and George Shoobridge were the crack shots of their day, and they developed pigeon shooting in new and interesting ways to make it more challenging.[48] As before, the 'ground', the area within which a bird must drop in order to be counted measured 100 yards, but under the new rules developed at the Red House, candidates were faced with not one but five traps, and on their command of 'pull', two would be opened at random.[49] Moreover, the Red House gunners favoured a particular breed of pigeon, the Blue Rock, reputed to be 'swift and strong' on the wing and difficult to kill 'by reason of their very thick feathers'.[50] In the provinces, the older form of single-trap shooting remained popular.

---

46 The following account is from Blakey, *Shooting* (1854 edn), p. 145.

47 'Stonehenge' [John Henry Walsh], *The Shot-Gun and Sporting Rifle* (London, 1859), p. 12.

48 'Craven' [John William Carleton] (ed.), *The Sporting Review* (April 1847), p. 284.

49 'Marksman', *Dead Shot*, pp. 230–233. Which traps were pulled was determined by dice rolled behind the gunman's back; Walsh, *British Rural Sports*, p. 44.

50 'Pigeon Shooting', *The Sporting Magazine* (August 1861), p. 109; 'Stonehenge', *Shot-Gun*, p. 11.

Although pigeon shooting in the fashionable Red House attracted the social elite, the sport had the power to attract good shots from across the social spectrum.[51] This was particularly the case after the creation of the Volunteer Forces in 1859: the prospect of participating in shooting competitions, often in a highly subsidised environment, appears to have been an important factor persuading men from a range of backgrounds to enlist.[52]

Pigeon shooting had its critics, however. Unlike the classic field sports hymned by John Henry Walsh and Robert Blakey, the tendency of shooting from traps was to encourage recourse to public houses, since publicans were the principal promoters of the sport, and the wagering of large sums on the outcome of matches. As a vehicle for gambling, pigeon shooting was endlessly flexible. A survey by *Bell's Life* of the significant pigeon-shooting contests which had taken place around Britain in 1839 reveals that, alongside regular matches shot for money wagers, cups, rifles and even hogs, contests were arranged in which special conditions were introduced.[53] Some were simple modifications, such as the match for a stake of £50 at Hornsey Wood Tavern, where the gunners shot at 12 birds each from 25 rather than 21 yards; others were far more complicated, such as the bet at the Red House where a Mr Alexander wagered that a Mr Wells could not kill a single bird within 20 minutes standing 50 yards from the trap, the bird to fall within 50 yards of the trap. Ward relieved Alexander of 20 sovereigns by bringing down a bird with his third shot.

If the gambling and association with public houses represented one way in which pigeon shooting fell short of the idealised behaviours associated with field sports, the lack of true sporting ability or instinct formed another focus for criticism. Robert Blakey, for example, argued that 'the liberal mind' experienced a 'repugnance at the idea of first confining, and then liberating from confinement, hundreds of domestic animals doomed to instant death, with a very slender probability in their favour, when a moderate shot will bring down fourteen or fifteen, and some nineteen out of twenty'.[54] Lt Col. Peter Hawker, in his much-reprinted volume of *Instructions to Young Sportsmen*, hit a similar note when he compared pigeon shooting to badger-baiting. He also went on to criticise the level of skill required.

---

[51] Lowerson, *Sport and the English Middle Classes*, p. 37; Mike Huggins, *The Victorians and Sport* (London, 2004), pp. 3, 25, 115.

[52] H. Cunningham, *The Volunteer Force: A Social and Political History, 1859–1908* (London, 1975); Lorna Jackson, 'Patriotism or Pleasure? The Nineteenth Century Volunteer Force as a Vehicle for Rural Working-Class Male Sport', *The Sports Historian*, 19/1 (1999), pp. 125–139.

[53] *Bell's Life*, 12 January 1840.

[54] Blakey, *Shooting* (1854 edn), p. 147

> So little is the art of pigeon shooting the criterion of a good shot that many of the very best performers at this are scarcely third rate shots at other birds, and some of them perfect cockneys in every other kind of shooting.[55]

John Henry Walsh agreed, remarking that he saw no difference between pigeon-trap shooting and the 'turnip butchery' of battue shooting, where partridge were driven onto the guns by beaters.[56] The comment of the *London Magazine*, in August 1826, that moorland shooting was beyond the 'mere Cockney sportsman or Battersea-shooter' shows both how synonymous the terms had become and the identity of the yardstick against which they were being measured.[57]

There was, however, a small group of writers who were prepared to give pigeon shooting its due, and to acknowledge its place in the sporting firmament. In 1847, John William Carleton, the editor of the *Sporting Review*, acknowledged that pigeon shooting provided good practice for the novice gunner, while enabling the experienced gunner to keep his hand in.[58] 'Marksman' went even further in 1863.[59] While acknowledging the general superiority of field shooting, he maintained that pigeon shooting was 'unquestionably, the finest practice for the aspirant to excellence, in the use of the gun at flying objects, of any that is used'. The sport required skill, steadiness, and 'perfect coolness' to stand and shoot in a public arena like a pigeon-shooting match.

> I have seen many persons who are dead shots in the field, completely eclipsed by very inferior sportsmen at a pigeon-match; and this entirely because the nerves of the one were so much quieter and under better control than the other.

The vast sums being wagered on these matches undoubtedly unsettled the equilibrium of some competitors. For a third champion, pigeon shooting had a more patriotic justification: it was, he said, 'an Englishman's sport, since it is a free one, in which a man may exercise his gun, although of the unprivileged class'.[60]

For the most part, however, pigeon shooting was considered a decidedly inferior pastime to country shooting, demanding neither the level of skill nor the levels of character that field sports were held to involve. As such, it offended

55 Peter Hawker, *Instructions to Young Sportsmen in All that Relates to Guns and Shooting* (London, 1830; 1st edn, 1814), pp. 219–220.

56 Walsh, *British Rural Sports*, p. 70.

57 John Scott and John Taylor, in *London Magazine*, 5 (August 1826), p. 498.

58 'Craven', *Sporting Review*, p. 283.

59 'Marksman', *Dead Shot*, pp. 238–239.

60 Blakey, *Shooting* (1854 edn), p. 148.

mid-Victorian sensibilities in a slightly different way than did the approach to shooting adopted by the townsmen who stalked the fringes of London on a Sunday morning. What bound both groups together in the moral consciousness of the period, however, was a perception that neither measured up to the standards of skill, character and behaviour exhibited by sporting gentlemen in the countryside.

The End of the Cockney Sportsman

In a variety of forms, therefore, recreational shooting was common in London during the first half of the nineteenth century, and the cockney sportsman was a recognisably real figure for contemporaries. Yet in the second half of the century the fame of the cockney sportsman dwindled, just as a series of measures were brought in which changed the nature of the sport that his real-life counterparts pursued. In this final section these changes will be reviewed.

At one level, the demise of the 'cockney sportsman' as a stock figure of fun reflected a simple change in public taste. The mid-nineteenth century saw a considerable revolution in attitudes to the proper treatment of man and beast alike, both of which ultimately impacted on the cockney sportsman. In 1866, the great mid-Victorian translator and publisher, Henry George Bohn, reprinted a selection of Seymour's sketches. The collection became a landmark, but not in the way Bohn had hoped. Far from being acknowledged as a high point of comic invention, the collection became a feature in the landscape against which later Victorians could measure how far their civilisation had evolved. Initial reviews set the tone. The *Publishers' Circular and Booksellers' Record* said sniffily that the content would not 'now be thought up to the requirements of our more enlightened public'; it was repetitive and seemed to be devoted to satirising people's petty misfortunes.[61] The cartoons, meanwhile, exhibited none of the arts that 'would in these days have made the fortune of a new magazine'. Nearly 20 years later, the incredulity at how such things could have been deemed amusing still prevailed: 'It is extraordinary how fond our jolly fathers were of these jokes', remarked the *Daily News* in 1883, '… That someone should blunder, and someone be hurt, shot, kicked, or half-drowned – these were the favourite jests of our ancestors.'[62]

This reaction, as Henry Miller has recently argued, exemplified an important trend in the Victorians' self-fashioning. By 1850, the 'scurrilous, grotesque,

61 *Publishers' Circular and Booksellers' Record*, 29 (1866), pp. 113–114; 'Two Green Leaves', *The Graphic*, 26 March 1870; *The Graphic* 12 September 1874. *Lloyd's Weekly Newspaper*, 13 January 1867, has a more favourable review.

62 *Daily News*, 27 October 1883.

vulgar, and immoral' underpinnings of Georgian caricature, captured in the Gillray and Williams cartoons, had been superseded by something more 'sentimental and amiable'.[63] 'The central tenets were that humour should be good natured and sympathetic,' writes Miller, 'should laugh with people not at them, and should gently point out foibles rather than expose them.'

The cockney sportsmen who appeared in the pages of mid-Victorian novels, often drawn by Leech, were altogether more sympathetic than their forebears. An excellent example was the character created by Robert Surtees: John Jorrocks, grocer and fox-hunter extraordinaire. The *Pall Mall Gazette*, for example, was keen to insist upon the difference between this sportsman of cockney extraction and the classic stereotype of the cockney sportsman. 'It would be the greatest of mistakes to confound Jorrocks with that sort of cockney whose sporting misadventures were the subject of Seymour's caricatures', it suggested.

> He is certainly as arrant a cockney as was ever born within the sound of Bow bells, but then he has the sporting instinct as truly as Peter Beckford or Assheton Smith. He hunts because he likes it, and his natural gifts favour his inclination.[64]

Similarly, Leech's 'simple-minded, sport-loving, philistine paterfamilias, Mr. Briggs',[65] who made his debut in *Punch* in 1849, had a different appeal for mid-Victorians than did his Georgian forebears. In 1865, *Blackwood's* remarked that

> Briggs is a totally different character from the Cockney sportsman who was the butt of Gillray or Seymour. It is impossible not to feel sympathy and respect for the perseverance and resolution with which he pursues his object, or affection for the good humour with which he meets repeated disappointment.[66]

If the change in public taste regarding the style of humour forms one part of the explanation of the cockney sportsman's dwindling purchase on the comedic sensibilities of later Victorians, attention must also be paid to two further factors that assaulted the constituent parts of his identity – as a 'cockney' and as a 'sportsman'. The cultural construction of the 'cockney', as Gareth Stedman Jones has demonstrated, was highly elastic, undergoing a variety of transformations during the nineteenth and twentieth centuries.[67] One key aspect of this was that the social status of the cockney underwent a protracted slide during the

---

63 Henry J. Miller, 'John Leech and the Shaping of the Victorian Cartoon: The Context of Respectability', *Victorian Periodicals Review*, 42/3 (2009), pp. 268–269.

64 'Jorrocks and Sponge', *Pall Mall Gazette*, 30 July 1874.

65 *Punch*, 18 July 1891, p. 3.

66 *Blackwood's Edinburgh Magazine*, 97/596 (1865), p. 466.

67 Stedman Jones, '"Cockney" and the Nation', pp. 280–309.

nineteenth century: the prosperous characters represented in the pictures of Gillray and Williams, who could look a wealthy tea-merchant like John Jorrocks in the eye, belong to the first half of the nineteenth century; by the century's end, the cockney has been reconfigured as flasher, brasher, and more definitively proletarian figure, replete with his own wit and his own stoicism. The character no longer fit the part that had been written for it.

At the same time, assumptions as to what constituted acceptable sporting behaviour were being contested. Evangelical sensibilities had long fuelled the drive for the gentler treatment of animals. The Society for the Prevention of Cruelty to Animals was formed as early as 1824, and, from an early date, the small birds that formed a staple of the cockney sportsman's bag became a target for evangelical sentimentality.[68] Even the hard heart of a London magistrate could be moved by the fate of small birds, as two Sunday-morning sportsmen detained near Highgate discovered in 1836.[69] The constable, who arrested the pair for 'sporting', handed the magistrate a bag containing the dead birds he had seized.

> Mr Rawlinson – Hello! What have we here? – Eight cock sparrows.
>
> Constable – They are not all sparrows, your Worship; most of them are poor harmless robins.
>
> Mr Rawlinson – What a shame to shoot such birds. I'm sure none but a cockney would do such a thing. In the country robins are held in the greatest reverence, on account of their tameness and beautiful notes.

The prisoners promised not to re-offend and were discharged, but their weapons were detained at the police house.

The killing of small birds was condemned on grounds of cruelty, but also because they performed such a useful service to the gardener, eating small insects and the like. By the mid-century, more and more voices were being heard against the casual slaughter of what one correspondent of the *Morning Post* called 'the melodious little inhabitants of our hedges and trees', with the clergy very much to the fore.[70] In 1862, for example, a pamphlet entitled *Bird Murder; or, Good Words for Poor Birds – A Tract for the Times – By a Country Clergyman* appeared, which drew attention to the fact that many vestries, especially in Midland

---

68 Brian Harrison, 'Animals and the State in Nineteenth-Century England', *English Historical Review*, 88/349 (1973), pp. 786–820; Chien-Hui Li, 'A Union of Christianity, Humanity, and Philanthropy: The Christian Tradition and the Prevention of Cruelty to Animals in Nineteenth-Century England', *Society and Animals*, 8/3 (2000), pp. 265–285.

69 *London Dispatch*, 23 October 1836.

70 Letter of 'Humanitas', *Morning Post*, 13 January 1826.

parishes, still paid bounties on the heads of small birds. By the end of the decade, pigeon shooting, too, had begun to attract public condemnation based on the cruelty of the slaughter involved, rather than its supposed limitations in relation to the building of character and gunmanship.[71]

If the tide of sentiment was running against the cockney sportsman, what decisively clipped his wings was the passage of legislation. The Parliament of 1868–1874, over which Gladstone's first administration presided, is more usually celebrated for the distinctive programme of legislation that saw the Church in Ireland disestablished; a new, comprehensive system of elementary education introduced; and the protection of the secret ballot afforded to all those voting in parliamentary elections.[72] However this same Parliament also passed a clutch of measures that seriously impacted on recreational hunting: the Sea Birds Protection Act (1869), the Gun Licence Act (1870), and the Small Birds Protection Act (1872).

The first of these was a Private Member's Bill aimed at the 'sport' of those who hired boats and took them a little way out to sea in order to shoot seagulls on the nest. The second was a government measure, introduced by Gladstone's Chancellor of the Exchequer, Robert Lowe. The new gun licence required that anyone carrying a gun outside their dwelling place or its immediate curtilage must have a licence, which cost 10s. The landowners managed to secure an amendment which recognised that farmers needed to shoot vermin on their farms, and so allowed them to extend the definition of 'curtilage' to include their own land. 'A farm-house', as one newspaper put it, 'would be as incomplete without a gun as a draught-horse or a sheep-dog.'[73]

The landowners' sectional interests aside, the Bill seems to have been largely uncontroversial. The MP for Leicestershire, Mr P.A. Taylor, argued in the Commons 'that the Government proposal could be regarded as nothing short of an attempt to disarm the people, and that it seemed to be prompted by a desire to bring our laws and customs into harmony with those of the most despotic continental countries',[74] but this was not picked up in any subsequent discussion of the Bill. For the most part, people seem simply to have paid the new licence fee without demur. Official returns suggest that around 125,000 licences were sold across the UK during the first year of the Act's operation, and that this

---

71 Kellett, 'Power of Princely Patronage'; Anthony Taylor, '"Pig-Sticking Princes": Royal Hunting, Moral Outrage, and the Republican Opposition to Animal Abuse in Nineteenth- and Early Twentieth-Century Britain', *History*, 89/293 (2004), pp. 30–48.

72 Jonathan Parry, *Democracy and Religion* (Cambridge, 1983) is the standard account.

73 *Newcastle Courant etc*, 16 June 1871.

74 *Newcastle Courant etc*, 1 July 1870.

rose gradually thereafter.[75] Prosecutions at Petty Sessions of those caught out shooting without a licence became a common occurrence.[76]

Nevertheless, there was a very limited extent to which Taylor was right. The Act does seem to have effectively disarmed the cockney sportsman – the local Londoner who wished to indulge in a little recreational game shooting. The *Morning Post* reported on 10 December 1870 that the large flights of skylarks and fieldfares driven by the cold weather onto suburban commons and fields had been unmolested by cockney sportsmen 'in consequence of the powers now possessed by the police and others to demand an inspection of licences by all persons carrying guns'.[77] In the course of the winter, various other reports of this 'excellent and happy result' of the Gun Licence Act were heard from different parts of the county,[78] and there were plenty who welcomed the Act as a measure that promised 'to abate a great suburban nuisance and danger'.[79]

If the Gun Licence Act represented one check to the cockney sportsman's activities, further legislation was to limit their range still further. In 1872, Parliament passed another Private Member's Bill imposing a close season for wild birds, the Wild Birds Protection Act. Henceforth any person taking a wild bird between 15 March and 1 August was liable to a fine of 5s for each bird killed, wounded, taken, or exposed for sale. A total of 78 species of bird were protected, including the cuckoo, curlew, gold finch, hedge-sparrow, robin, swallow, titmouse, wagtail and wren. There was no longer any reason, as the *Bristol Mercury* remarked, 'why country pedestrians and farmers, as well as birds, should not be protected from the danger caused by cockney-like sportsmen'.[80] One Magistrate pronounced the Act the 'death warrant of Cruelty'.[81]

These Acts did not, in one fell swoop, do away with all popular hunting. The Wild Birds Protection Act was a first instalment only; a Select Committee was appointed in the following session of Parliament to consider its extension and the legislation was significantly amended in 1880. Gun licensing, meanwhile, remained only the mildest impediment to the ownership of a firearm. Indeed, the number of gun licences issued each year grew, as Alastair Durie records, from 100,000 in 1871 to 150,126 in 1881, 238,026 in 1904–1905 and 300,000 in 1914.[82]

75 PP, 1874, XV, *Seventeenth Report of the Commissioners of Her Majesty's Inland Revenue on the Inland Revenue, for the year ended 31 March 1874*, p. 689. Since farmers were not required to license their guns, the official figure of licences sold underestimates the number of guns in circulation.

76 PP, 1873, XIII, Report from the Select Committee on Game Laws, p. 497, QQ. 11347–8.

77 'Protection of small birds', *Morning Post*, 10 December 1870.

78 Letter from 'Starling, Blackbird & Co' of Narbeth, *Standard*, 4 February 1871; *Hampshire Advertiser*, 28 December 1870.

79 *Sheffield & Rotherham Independent*, 25 June 1870.

80 *Bristol Mercury*, 17 August 1872.

81 Letter of 'A Country Magistrate', *Birmingham Daily Post*, 21 March 1873.

82 Durie, 'Game Shooting', p. 434.

Yet legislative interventions certainly affected the ability of the urban sportsman to prosecute his hobby with as much freedom as once had been the case, and they had their counterpart in changes affecting the rural sporting scene. Shooting became a less socially exclusive activity. In 1881, Gladstone's second administration secured the passage through Parliament of the Ground Game Act, which gave tenant farmers the long sought right of killing rabbits and hares on their own property.[83] On large estates, meanwhile, shooting had become an industry.[84] The introduction of battue shooting led to colossal numbers of birds being killed; as John Henry Walsh remarked, however, this was 'butchery' not sport.[85] Raymond Carr cites the example of the Holkham estate, where, in 1790, 3,000 birds were shot in a year; in 1880, the same number was dispatched in a single day.[86]

* * *

In this new version of rural 'sport', where the 'sportsman' did little other than stand and shoot birds driven into his field of fire by beaters, it was hard to sustain the older notion of field sports, in general, and shooting, in particular, as the particular sources of those individual qualities on which the future of the nation might rest. By the third quarter of the nineteenth century, organised games like rugby and football were now assigned that role, instilling what John Tosh has described as 'the character-building qualities of courage, self-control, stoical endurance, and the subordination of the ego to the team'.[87] The hunter as hero had retreated to the wilder margins of the Empire. Sportsmen and the reading public alike rallied to tales from a new frontier of human endurance and courage, captured in stories of the lone (white) gunman pitted against the ferocity of a man-eating tiger, or marshalling all his resources to pursue and subdue his quarry deep in the wilderness.[88]

---

83 John Fisher, 'Property Rights in Pheasants: Landlords, Farmers and the Game Laws, 1860–80', *Rural History*, 11 (2000), pp. 165–180; F.M.L. Thompson, 'Landowners and the Rural Community', in G. Mingay (ed.), *The Victorian Countryside* (2 vols, London, 1981), vol. 2, pp. 459–460.

84 Lowerson, *Sport and the English Middle Classes*, pp. 37–38.

85 Walsh, *British Rural Sports*, p. 70.

86 Raymond Carr, 'Country Sports', in Mingay, *Victorian Countryside*, p. 483; for a case study over the *longue durée*, Alastair J. Durie, '"Unconscious Benefactors": Grouse-Shooting in Scotland, 1780–1914', *International Journal of the History of Sport*, 15/3 (1998), pp. 57–73.

87 John Tosh, *A Man's Place: Masculinity and the Middle-class Home in Victorian England* (Yale, 2007), p. 189.

88 Joseph Sramek, '"Face Him like a Briton": Tiger Hunting, Imperialism, and British Masculinity in Colonial India, 1800–1875', *Victorian Studies*, 48/4 (2006), pp. 659–680; Callum McKenzie, 'The British Big-Game Hunting Tradition, Masculinity and Fraternalism with Particular Reference to "The Shikar Club"', *The Sports Historian*, 20/1 (2000), pp. 70–96.

In a world thus configured, the cockney sportsman lost much of his appeal. Although circus audiences continued to enjoy the slapstick fun to be had out of a townsman – and a Londoner at that – being outwitted by a countryman, little of the subtler social commentary embodied in the cockney sportsman survived. Without an idealised country gentleman hunter to play off, the old dichotomies that lay at the heart of the joke made no sense. Since it was no longer only the high-born who could hunt the high-status species, the parallel joke drawing an equivalence between the cockney and low-bred quarry no longer stuck; and once the character-building qualities claimed for hunting – endurance, coolness, manliness – ceased to be convincing, jokes about the cockney's rashness, sloth and greed lost their point. In any case the cockney himself was now required to fill a different set of roles, those more pertinent to a society facing up to the challenges of class conflict rather than the division between rural and urban society. The cockney sportsman accordingly left the stage, and the gunmen of London were left to pursue what was left of their sport, underided.

PART II
Resisting Guns: Edged Weapons and the Politics of Indigenous Honour

Chapter 5

'They Disdain Firearms': The Relationship between Guns and the Ngoni of Eastern Zambia to the Early Twentieth Century[1]

Giacomo Macola

The development of increasingly efficient firearms and their contribution to Western imperialism are master themes of nineteenth-century historiography.[2] Conversely, the survival of earlier tactical and weapons traditions has received much less scholarly attention, partly, no doubt, because of the Eurocentric and technologically deterministic nature of much contemporary military history.[3] To be sure, continuity in military hardware on the peripheries of European imperial expansion can sometimes be explained away as the simple consequence of commercial isolation and general lack of economic opportunities. Yet wilful resistance to foreign military technology was also frequent and important. Such instances of technological conservatism – this chapter contends – are best interpreted in terms of the local cultural structures and social dynamics that underpinned them. After all, as Marshall Sahlins famously demonstrated, it is on the basis of 'existing understandings of the cultural order' that people 'organize their projects and give significance to their objects'.[4]

The first section of the chapter discusses some salient episodes in the pre-colonial political history of the Jere Ngoni of present-day eastern Zambia. My aim is to isolate the key features of their relationship with firearms. It will be argued that, despite having repeatedly experienced the potential of the new weapons of destruction and having had ample opportunities to gain access to them through trade, Ngoni fighters consistently rejected their adoption for war purposes. An explanation for this seemingly aberrant behaviour is offered in the second part of

1 The research on which this chapter is based was sponsored by the Netherlands Organisation for Scientific Research (Nederlandse Organisatie voor Wetenschappelijk Onderzoek).

2 See, e.g., Daniel R. Headrick, *The Tools of Empire: Technology and European Imperialism in the Nineteenth Century* (Oxford, 1981) and *Power over Peoples: Technology, Environments, and Western Imperialism, 1400 to the Present* (Princeton, 2010).

3 Jeremy Black, *Rethinking Military History* (London and New York, 2004).

4 Marshall Sahlins, *Islands of History* (Chicago, 1985), p. vii.

Map 5.1 Eastern Zambia and Malawi in the late nineteenth century
Adapted from John McCracken, *Politics and Christianity in Malawi, 1875–1940: The Impact of the Livingstonia Mission in the Northern Province* (Cambridge, 1977). Drawn by Judith Weik and Jack Hogan.

the chapter, which focuses on Ngoni pre-colonial social institutions and notions of honour and masculinity as reflected in the ethnographic record.

The third section of the chapter examines the chronology and modalities of the subjugation of the Ngoni at the close of the nineteenth century. Its main purpose is to illustrate the endurance and ultimately devastating effects of what a colonial historian felicitously termed the Ngoni's 'cult of cold steel'.[5] The chapter ends with a discussion of the impact of conquest on Ngoni militarism. It will be submitted that, far from spelling the immediate end of Ngoni heroic honour culture, colonialism provided an opportunity for its recasting. In this respect, my approach heeds Thomas Spear's call not to overstate colonial ability to invent and manipulate African identities and institutions.[6] However, insofar as firearms are concerned, the effects of the British take-over *were* revolutionary. Construed as a 'martial race' partly on account of their reliance on edged weapons and devotion to the ideal of close combat, the Ngoni were recruited in substantial numbers into colonial paramilitary police forces, one of whose distinguished traits was, paradoxically, the right to bear firearms. Under the new circumstances, then, the gun became everything it had not been in the pre-colonial context, replacing the assegai as a central symbol of bellicose masculinity and major vehicle for individual improvement.

Ngoni Experiences of Firearms in the Second Half of the Nineteenth Century

Early in the second half of the nineteenth century – following a four-decade-long process of migration, military conquest, demographic expansion and political segmentation – a number of Ngoni groups acquired a dominant position over large swathes of the territory corresponding to present-day Zambia and Malawi. Paramount among these were the Jere Ngoni, the descendants of Zwangendaba Jere, who had left KwaZulu-Natal after the defeat of his ally, Zwide of the Ndwandwe, at the hands of Shaka's Zulu in the late 1810s and had eventually led his growing (and growingly heterogeneous) followers as far north as Ufipa, in south-western Tanzania.[7] Upon Zwangendaba's death in the 1840s, his 'snowball state' (to use John Barnes' famous definition[8]) had fragmented into several autonomous sections, the most important of which were headed by two

5 Lewis H. Gann, *A History of Northern Rhodesia: Early Days to 1953* (London, 1964), p. 89.

6 Thomas Spear, 'Neo-traditionalism and the Limits of Invention in British Colonial Africa', *Journal of African History*, 44/1 (2003), pp. 3–27.

7 Norman Etherington, *The Great Treks: The Transformation of Southern Africa, 1815–1854* (London and New York, 2001), pp. 114–121, 275–277.

8 John A. Barnes, *Politics in a Changing Society: A Political History of the Fort Jameson Ngoni* (London, 1954).

of his sons: Mpezeni (b. *c.*1830[9]) and M'Mbelwa (b. *c.*1840[10]). By the 1860s, the former and his partisans were temporarily settled among the Nsenga of present-day Petauke district of eastern Zambia and Maravia district of western Mozambique and would shortly thereafter shift the heartland of their polity eastward, overrunning the Chewa of *Mkanda* and others, near contemporary Chipata.[11] In the meantime, the Ngoni of young M'Mbelwa had succeeded in imposing their sway over the Tumbuka and related peoples of Mzimba district, northern Malawi.

Both of these migrant groups – notwithstanding the recasting of their southern identities that the incorporation of substantial numbers of captives during, and after the completion of, their treks was bringing about – claimed origins in present-day South Africa, spoke northern Nguni dialects (which, however, were being rapidly marginalised to the advantage of the local languages spoken by captive wives) and were deliberately organised and trained for war, being based on the principle of age-set regiments and the (theoretical) military mobilisation of all able-bodied males. The Jere Ngoni of M'Mbelwa I and Mpezeni I have commonly been portrayed as 'formidable' military machines that attached 'little importance' to external trade, the cattle, captives and, to a lesser extent, grains necessary to energise their political economies being obtained mainly by means of 'frequent raids' against their neighbours.[12] As already noted by Leroy Vail,[13] modern scholarly depictions owe much to earlier missionary characterisations: for Walter A. Elmslie – a missionary in the northern Ngoni's heartland since 1885 – 'cattle-lifting was a constant occupation in the dry season' for the 'proud warriors of Ngoniland', whose 'hordes' kept surrounding districts in a permanent 'state of terror and distress'.[14] Without wishing entirely to deny all historicity to these descriptions (for the Ngoni did maintain a degree of contested military ascendancy until the arrival of European forces in their respective areas at the close of the nineteenth century), this section of the chapter contends that both

---

9 Carl Wiese, *Expedition in East-Central Africa, 1888–1891: A Report*, ed. H.W. Langworthy (Norman, 1983), p. 160, fn. 41.

10 T. Jack Thompson, 'The Origins, Migration, and Settlement of the Northern Ngoni', *Society of Malawi Journal*, 34/1 (1981), p. 16.

11 Inheritable political titles and names, such as *Mkanda*, are italicized, unless they refer to one specific individual incumbent.

12 The first and the third quotes are to be found in John McCracken, *Politics and Christianity in Malawi, 1875–1940: The Impact of the Livingstonia Mission in the Northern Province* (Cambridge, 1977), p. 8; the second in William E. Rau, 'Mpezeni's Ngoni of Eastern Zambia, 1870–1920', unpublished PhD thesis, UCLA, 1974, p. 171.

13 Leroy Vail, 'The Making of the "Dead North": A Study of the Ngoni Rule in Northern Malawi, *c.* 1855–1907', in J.B. Peires (ed.), *Before and after Shaka: Papers in Nguni History* (Grahamstown, 1981), p. 230.

14 Walter A. Elmslie, *Among the Wild Ngoni* (Edinburgh, 1899), pp. 50, 78, 89.

the Ngoni's military might and their putative economic insularity have been hitherto overestimated.[15]

After settling athwart the contemporary boundary between eastern Zambia and western Mozambique in *c.*1860, the Ngoni of Mpezeni I had several chances to experience the power of firearms. In 1863, what is very likely to have been one of their raiding parties failed to storm the stockade of Chinsamba's, a Mang'anja settlement on the Linthipe River, to the east of present-day Lilongwe. David Livingstone, who visited the village on the day after the attack, discovered that: 'Chinsamba had many Abisa or Babisa in his stockade, and it was chiefly by the help of their muskets that he had repulsed the Mazitu.'[16] During the same decade, Mpezeni also came up against the guns that Chikunda mercenaries were occasionally putting at the disposal of harassed Chewa and Nsenga communities.[17] Indeed, the alliance between the Nsenga of *Mburuma* and the Chikunda of the Afro-Portuguese warlord, Chikwasha, may have been one of the factors behind Mpezeni's decision to shift the heartland of his kingdom to Chipata between the late 1860s and the early 1870s.[18]

Mpezeni himself enlisted the services of heavily armed Chikunda elephant hunters against the Chewa chief Mkanda, who may initially have invited the Ngoni into his own territory, but who would eventually be overrun by the new arrivals – and their hired guns – in about 1880.[19] The deceased Mkanda's heir and some of his subjects took refuge with the important Chewa leader Mwase Kasungu, whom the Zambian Ngoni, just like their northern Malawian counterparts, would always fail to subdue on account of his impressive defences and arsenal. In 1889–1890, Mwase's troops could apparently muster 'more than three thousand firearms', mainly muzzle-loading muskets.[20] Guns – as is made clear by the German-born ivory trader and diarist Carl Wiese – were by

15 While drawing on insights from the literature on the Ngoni of Malawi, this chapter's main focus is on the Zambian Ngoni, or Ngoni of *Mpezeni*. A fuller treatment of the former groups will be provided in Giacomo Macola, *The Gun in Central Africa: A Social History to the Early Twentieth Century*, forthcoming.

16 David and Charles Livingstone, *Narrative of an Expedition to the Zambezi and its Tributaries* (London, 1865), p. 502; and Andrew D. Roberts, 'Firearms in North-Eastern Zambia before 1900', *Transafrican Journal of History*, 1/2 (1971), p. 5, where '1861' should read '1863'. 'Mazitu' was one of the local names given to Ngoni invaders.

17 See, for instance, Albert J. Williams-Myers, 'The Nsenga of Central Africa: Political and Economic Aspects of Clan History, 1700 to the Late Nineteenth Century', unpublished PhD thesis, UCLA, 1978, pp. 293–294. The Chikunda were the former armed slaves of Mozambican estate-holders whom manumission turned into professional traders, hunters and state-builders from *c.*1850; Allen F. and Barbara S. Isaacman, *Slavery and Beyond: The Making of Men and Chikunda Ethnic Identities in the Unstable World of South-Central Africa, 1750–1920* (Portsmouth, NH, 2004).

18 Williams-Myers, 'Nsenga of Central Africa', p. 266.

19 Wiese, *Expedition*, p. 275.

20 Ibid., p. 168.

this stage a central feature of Mwase Kasungu's capital. While 'almost all' of its numerous inhabitants had access to them, local metalworkers – including Chibisa, the king's brother – found the technology of even Wiese's 'modern weapons' intelligible and accessible.[21] This being the case, it is scarcely surprising that Kasamba Malopa, the important *induna* of Mtenguleni, one of the two capitals of Ungoni, Mpezeni's heartland, should have shown little willingness to confront so well-armed an enemy.[22]

To be sure, Mwase Kasungu's position was unique. No other Chewa chief could equal his resources and military preparedness vis-à-vis the Ngoni. Ngoni raids, indeed, forced the southern Chewa and their leaders, the most notable of whom was the then holder of the ancient *Undi* title, to seek refuge on mountainous retreats, where they eked out 'a miserable existence ... suffering almost every year from famine'.[23] Yet, even in Maravia – Edouard Foà noted with reference to the year 1891 – Mpezeni's people were 'frequently defeated', as they were only armed with 'bows, shields, assegais and knobkerries', whereas Undi's people had guns and powder.[24] At this point, it is important to stress that muzzle-loaders were not necessarily inherently superior to Ngoni shields and *armes blanches*. They became so only when associated with such strongly fortified positions as both Mwase and Undi evidently made use of in the last decades of the nineteenth century. This is aptly borne out by the fate that befell a tactically naïve party of 'Arab' traders on the upper Bua River in 1887. Having been lured out into the open by a youthful Ngoni advance regiment, the traders, who believed 'these to be their only enemies', fired 'their weapons almost all at the same time' and began to give chase to the retreating *impi*. They soon fell into the hands of two more regiments, who lay in ambush. 'The Arabs were then attacked ... from the rear. Not having time to reload, they began to be massacred by club and *assegai*.'[25]

Victories such as this may have contributed to the Zambian Ngoni's choice not to deploy firearms for military purposes. The use of the word 'choice' is appropriate, for what we are confronted with is a deliberate decision, rather than the enforced consequence of lack of economic opportunities. Indeed, there

---

21 Ibid. pp. 244, 252, 254.

22 Ibid., p. 168.

23 Alfred Sharpe, 'A Journey through the Country Lying between the Shire and Loangwa Rivers', *Proceedings of the Royal Geographical Society*, 12/3 (1890), p. 156; Wiese, *Expedition*, pp. 113, 117, 131–132.

24 Edouard Foà, *Du Cap au Lac Nyassa* (Paris, 1901; 1st edn, 1897), p. 280. Foà's original passage reads as follows: '*les gens de Mpéseni ... sont pourtant encore souvent battus, n'étant armés que d'arcs, de boucliers, de sagaies et de casse-tête, tandis que leurs adversaries, les Agoas ou gens d'Oundi, ont des fusils et de la poudre.*'

25 Wiese, *Expedition*, p. 179.

are enough indications that, *pace* the common historiographical stereotype, the Ngoni of Mpezeni were not entirely excluded from long-distance trading networks. The aforementioned Wiese's first visit to Mpezeni dated to 1885–1886. The establishment of a semi-permanent trading base in Mtenguleni followed suit.[26] By then, other merchants were active in the area. These included both Zambezian entrepreneurs – such as the Portuguese Joaquim Augusto do Rego, whose presence in Ungoni was reported in 1889, and the Afro-Portuguese Francisco Jose Pacheco, active among the Nsenga and neighbouring people since 1884[27] – and coastal traders, who, according to Wiese's probably exaggerated description, 'continually visit[ed]' Mpezeni, in spite of the dangers posed by some of the same king's less disciplined regiments.[28] During Wiese's stay in Ungoni in 1889–1890, at least one Zanzibari caravan made Dingeni its temporary headquarters while waiting to 'buy slaves and ivory'.[29] While there is little doubt that Mpezeni sought to monopolise the trade in ivory, of which he sometimes 'had a great deal',[30] the evidence relating to his involvement in the slave trade is more contradictory. The Arab-Swahili of Dingeni did undoubtedly acquire some slaves.[31] Yet they were also reported to have 'sold slave women to Mpezeni', who, Wiese remarked, '[bought] people but never [sold] them'.[32]

The primary Ngoni import was cloth, which by the late 1880s was becoming common both among women, 'especially the rich ones', and men, some of whom were 'beginning to dress as Portuguese subjects'.[33] King Mpezeni himself had 'many suits and clothes of good quality'.[34] But firearms also featured prominently among Mpezeni's possessions. Wiese, in fact, described him as an enthusiastic 'collector' of weapons 'of modern workmanship'. These, however, were 'never use[d]'.[35] What was true of Mpezeni was also apparently true of his people as a whole. 'The Ngoni' – the usual Wiese reported –

---

26 Harry W. Langworthy, 'Introduction: Carl Wiese and Zambezia', in Wiese, *Expedition*, pp. 4–5.

27 Wiese, *Expedition*, pp. 139, 195.

28 Ibid., pp. 155, 186.

29 Ibid., p. 185.

30 Ibid., pp. 191, 165.

31 Ibid., pp. 185–186.

32 Ibid., p. 191. Alfred Sharpe, who visited Mpezeni in 1890, wrote that the Jere Ngoni king did 'a brisk trade in slaves, selling to the Arabs'. 'A Journey from Lake Nyassa to the Great Loangwa and Upper Zambezi Rivers', *Proceedings of the Royal Geographical Society*, 12/12 (1890), p. 745.

33 Wiese, *Expedition*, p. 152.

34 Ibid., p. 160.

35 Ibid., p. 195. Mpezeni's arsenal, Wiese went on, included 'an abundance of different revolvers, Snider, Spencer, Remington, Lefaucheux, and other carbines, as well as huge amounts of different percussion arms, some presented to him, others taken in wars'.

> have remained very conservative regarding their armament. Although nowadays many Ngoni already possess firearms, some two or three thousand in my computation, they do not use them in war to the fullest extent. They use, as in the early times, wooden clubs, *assegais* (they carry two or three of them), and shields, which are elliptically shaped and made of oxhide.[36]

Hunting, too, was carried out without the aid of firearms, as the Ngoni were 'not experienced enough' in their use.[37] The military propensities of the Jere Ngoni of present-day Chipata remained unchanged over the course of the next few years, for early in 1896 Lt Col. R.G. Warton, of the North Charterland Exploration Company (NCEC), could still write that Mpezeni's forces were 'unaccustomed to handle' even the 'obsolete guns' to which they had access, 'preferring assegais and bows and arrows'.[38] Warton's first meeting with Mpezeni confirmed the validity of the information he had initially received. The Ngoni – he wrote in August of the same year – 'do not rely very much on firearms in case of war, almost invariably using their assegais and knob-kerries'.[39]

Explaining Technological Conservatism

The few regional specialists who have pondered over the Jere Ngoni's reluctance to adopt firearms for military purposes have tended to accept anthropologist Margaret Read's old statement to the effect that the 'antique guns' put in circulation 'by the Portuguese and Arabs were no match for the Ngoni skill with the long throwing-spear or the short stabbing-assegai, and the bullets were such

---

36 Ibid., pp. 153–154. See also Carl Wiese, 'Beiträge zur geschicte der Zulu im norden des Zambesi, namentlich der Angoni', *Zeitschrift für Ethnologie*, 32 (1900), p. 196, where Wiese speaks explicitly of 'avoidance' (*vermeidung*) of firearms on the part of Mpezeni's Ngoni. 'Even today' – he elaborated in the same article – 'the spear, shield and club are their only weapons, just as in the time of their fathers before they crossed the Zambezi. Although they amass firearms from enemies during battles, these would only be used to fire salute shots at festivities.' My thanks to Judith Weik for translating the above passage.

37 Wiese, *Expedition*, p. 157.

38 R.G. Warton to P. Forbes, 2 April 1896, in *North Charterland Concession Inquiry: Report to the Governor of Northern Rhodesia by the Commissioner, Mr. Justice Maugham, July 1932* (London, 1932), appendix 28.

39 R.G. Warton to Secretary (NCEC), 6 August 1896, in 'Copy by E.H. Lane Poole of correspondence relating to the North Charterland Exploration Company (East Loangwa District), 1896. Also relating to the Angoni Rising', (hereafter Poole Papers), Archives of the Livingstone Museum, no ref., Livingstone, Zambia. I am indebted to Jack Hogan for making copies of this important source available to me.

that the stout cow-hide shields of the Ngoni could withstand them'.[40] In light of the significant military setbacks described above, this minimalist, utilitarian, explanation is either wide of the mark or, at best, able only to tell a limited part of the story.[41] I contend that a closer investigation of Ngoni socio-cultural institutions is needed if we are to make sense of the technological conservatism described in the previous section of this chapter.

Ngoni armies were made up of age-set regiments open to all able-bodied youths and men. While the ultimate origins of this principle of social and military organisation are to be found in developments that had affected KwaZulu-Natal, the homeland of the leaders of the initial wave of migrations, between the eighteenth and the nineteenth centuries,[42] the achievement-oriented aspects already inherent in the Nguni archetype were enhanced by the experience of migration, during which the need forcefully asserted itself to incorporate – and secure the allegiance of – large numbers of foreign captives. Thus, successful participation in the Ngoni military system – and in the state of semi-permanent warfare of which it was both a consequence and a cause – was made to serve the twin purpose of bringing about a common (though by no means fixed or irreversible) identity and favouring social promotion for scores of captives, which soon formed the majority of the members of all of the Ngoni communities in east-central Africa.[43] In the 1930s, the northern Ngoni ethno-historian Chibambo described the process in the following terms:

> In [war], ordinary people and slaves came together ... The slaves who showed their courage and strength in war quickly received their freedom and many also had villages of their own because of the people they had captured in war; others obtained the standing of men in authority.[44]

To be sure, Chibambo's account is romanticised and papers over internal tensions; yet there is no need to reject its essential contours. Indeed, making specific reference to the Ngoni of Mpezeni, the eyewitness Wiese wrote that

---

40 Margaret Read, 'Tradition and Prestige among the Ngoni', *Africa*, 9/4 (1936), p. 461, fn. 1; Barnes, *Politics in a Changing Society*, p. 59; and Vail, 'Making of the "Dead North"', p. 249.

41 This, of course, is not to deny that the large shields and short stabbing spears of the Ngoni *did* pose a formidable challenge to people armed only with bow and arrow. See, for instance, Livingstone, *Narrative of an Expedition*, pp. 556–557, and Donald Fraser, *Winning a Primitive People* (London, 1922; 1st edn, 1914), p. 29.

42 John Wright, 'Turbulent Times: Political Transformations in the North and East, 1760s–1830s', in B.K. Mbenga et al. (eds), *The Cambridge History of South Africa. Volume I: From Early Times to 1885* (Cambridge, 2010), pp. 221–223.

43 Barnes, *Politics in a Changing Society*, pp. 36–37, 40; Andrew D. Roberts, *A History of Zambia* (London, 1976), p. 119.

44 Quoted in Margaret Read, *The Ngoni of Nyasaland* (London, 1970; 1st edn, 1956), p. 29.

prisoners of war were not only used 'on agricultural work and other services', but were also 'adopted later into Ngoni customs, language, and dress and sometimes reach[ed] important positions'. He knew 'many who now have influence and wealth'.[45]

Among the Ngoni of both Zambia and Malawi, promotion within the regimental structure went hand-in-hand with the possibility of partaking of the patronage networks through which raided captives and cattle were distributed among deserving members of specific army units.[46] In the late nineteenth century, Ng'onomo was the living embodiment of the opportunities for individual achievement thrown open by a system in which military skills were valued as much as – if not more than – birth. Originally a Thonga captive from Delagoa Bay, Mozambique, Ng'onomo, having 'proved himself a warrior braver and more successful than most', had been remunerated by M'Mbelwa 'with wives, slaves and cattle', and he had eventually risen to the position of overall leader of the northern Ngoni army.[47] It was perhaps with a view to regulating their rewards mechanisms that the Ngoni of both M'Mbelwa and Mpezeni developed formal means to honour gallantry. A kind of 'decoration, or a military order', consisting of 'enormous ox horns', was in use among both groups,[48] while M'Mbelwa also bestowed the title of 'Master of the Stockade' on the first warrior to have scaled the palisade of a besieged village and slain one of its inhabitants. When brought before the presence of M'Mbelwa, the hero of the day carried 'in his hand the bow or gun of the man he [had] killed', danced, and was given a bullock 'as a signal token of [the king's] princely admiration'.[49]

Military bravery, then, was the key to unlock the potential for self-advancement inherent in Ngoni institutions. But notions of 'military bravery' – an important component of John Iliffe's category of 'heroic honour' – are far from universal and always depend on a given social group's approved codes of behaviour.[50] Insofar as the Ngoni military is concerned, the practice of hand-to-hand fighting with spear, knobkerrie and shield was precisely one such prescriptive, hegemonic norm. The 'right to command respect' among the Ngoni – and to reap the social and material advantages of such a respect – was forged in the heat of close combat, which – the Maseko Ngoni told Last in 1886 – did not involve the use of missile weapons. Rather, 'on coming to close quarters, [Ngoni fighters] strike their opponents' legs, and when they have brought them down,

45 Wiese, *Expedition*, p. 152.

46 Barnes, *Politics in a Changing Society*, pp. 30–32, 37.

47 Fraser, *Winning a Primitive People*, p. 51.

48 Wiese, *Expedition*, p. 155, and Chibambo, in Read, *Ngoni of Nyasaland*, pp. 36–37.

49 Fraser, *Winning a Primitive People*, pp. 37–39.

50 John Iliffe, *Honour in African History* (Cambridge, 2005), pp. 4–6.

then spear them'.[51] No Ngoni voice is on record as having openly described the gun as the 'coward's weapon'.[52] Yet it is clear that not all forms of warfare were construed by the Ngoni as equally commendable and respectable. It was because they tainted dominant notions of honour that firearms, their potential military usefulness notwithstanding, were regarded with 'disdain' (*dédain*) by both Mpezeni's Ngoni and the Maseko Ngoni.[53] Hence, the most that could be done with them was to leave them in the hands of independent foreign mercenaries (see above), who had no stake in infra-Ngoni competition for upward mobility and who could therefore afford to ignore the demanding requirements of Ngoni heroic behaviour. In pre-colonial Ungoni, firearms never became central symbols of masculinity. Rather, it was the stabbing spear and the knobkerrie that most readily epitomised the qualities of the ideal Ngoni fighter. *Imigubo* war songs celebrated the spear's distant origins – 'Do you hear? The Ngoni come from the south-east' – and the wealth it had made possible to accumulate: 'No Paramount can be poor because of the spear/Then why are you running away?'[54]

The struggle for status and self-improvement – one that expressed itself in a high degree of personal investment in the core values of Ngoni militarism – was principally the affair of the youths, from whose ranks the majority of captives were taken. Both coeval observers and later scholars have stressed the significance of inter-generational cleavages among the various Ngoni groups and the pressures brought to bear on Ngoni leaders by their younger followers. In particular, while senior politico-military chiefs, having already attained a high position in society, were not averse to regulating – if not abolishing altogether – warfare and raiding with a view not to antagonising encroaching Europeans, younger warriors belonging to newly instituted regiments resisted this tendency on account of their eagerness 'to obtain wealth, or wives and slaves'.[55] Unlike their seniors, junior soldiers 'felt that unless they established their social credentials in war – "washed their spears in blood" – the system into which they were trying to gain a substantial hold would have no meaning or significance'.[56] The result – as will be further seen in the next section – was that a good many of the raids witnessed by early European observers on the outskirts of the major

---

51 J.T. Last, 'A Journey from Blantyre to Angoni-Land and Back', *Proceedings of the Royal Geographical Society*, 9/3 (1887), p. 186. The Maseko were a separate Ngoni group settled in central Malawi.

52 These were the words of Mangwanana Mchunu, of the uVe, a regiment inaugurated by the Zulu king Cetshwayo in *c.*1875. Quoted in Ian Knight, *The Anatomy of the Zulu Army from Shaka to Cetshwayo, 1818–1879* (London, 1995), pp. 215, 268.

53 Foà, *Du Cap au Lac Nyassa*, p. 298, and, by the same author, *La traversée de l'Afrique du Zambèze au Congo français* (Paris, 1900), pp. 72–73.

54 Read, *Ngoni of Nyasaland*, p. 46.

55 Elmslie, *Among the Wild Ngoni*, p. 98.

56 Rau, 'Mpezeni's Ngoni', p. 260.

Figure 5.1 Members of a youthful Ngoni regiment
Edouard Foà, *Résultats scientifiques des voyages en Afrique* (Paris, 1908).

Ngoni kingdoms of both Zambia and Malawi represented local – as opposed to centrally organised and sanctioned – initiatives, and that it was most probably the youths, rather than their elders, who acted as a check on military innovation, insisting that these same wars be fought with obsolete, but prestige-enhancing, edged weapons and shields.

The openness of the Ngoni social system and its relationships with specific notions of heroic behaviour and masculinity illuminate – I think – the root cause of resistance to firearms, particularly (and paradoxically) among such groups as, in more rigidly stratified societies, are commonly viewed as the natural supporters of 'progress' and 'change'. Since their ability to rise to position of greater power and influence depended on the extent to which they succeeded in proclaiming their 'right to respect' by excelling in warfare and socially prescribed fighting methods, young warriors of captive origins cannot have been keen to drop their hard-won military skills with a view to taking up a foreign technology – firearms – which did not offer comparable prospects of material rewards and which might indeed have hampered their progress towards full adulthood and social acceptance.

The Subjugation of the Ngoni

The history of Ngoni–European relations during the decade that preceded the war of January 1898 has been told in considerable detail before.[57] For our purposes, a chronological summary suffices that highlights the motives of the protagonists and the structural causes of the conflict. Between the summer of 1889 and the spring of 1891, a Portuguese expedition resided in Mtenguleni, one of Mpezeni's two capitals in Ungoni. Its unofficial leader, the trader Carl Wiese, to whose writings reference has repeatedly been made above, became very close to the Ngoni king, whom he had already visited twice in the recent past. Indeed, on 14 April 1891, a few days before the departure of the last members of the expedition, Mpezeni was persuaded to sign a concession that granted Weise – rather than Portugal – extensive mining and other rights in the heartland of the kingdom.[58]

For the next few years, notwithstanding the Wiese concession and the widespread belief that Ngoniland was rich in gold, Mpezeni's Ngoni were left undisturbed by Europeans. The Anglo-Portuguese treaty of June 1891 had placed the bulk of Mpezeni's country in British territory. But in the early 1890s Cecil Rhodes' chartered company, the British South Africa Company (BSAC), to which the administration of the British sphere to the north of the Zambezi had been delegated, was in no position to even attempt to penetrate Ungoni. Meanwhile, the infantry forces of the British Central Africa Protectorate (BCAP) (Nyasaland/Malawi), which would become known as the Central African Rifles (CAR) from the mid-1890s, were being kept fully occupied by prolonged Yao resistance in the southern part of the country.

---

[57] The best accounts are provided by Barnes, *Politics in a Changing Society*, chapter 3; Rau, 'Mpezeni's Ngoni', chapter 7; and Langworthy, 'Introduction'.

[58] Wiese, *Expedition*, pp. 368–369.

European actions only began seriously to affect Ngoni politics late in 1895, when the armed forces of the Protectorate stormed Mwase Kasungu's fortified town in the context of their anti-Yao operations.[59] Mwase, it will be remembered, was one of the few Chewa leaders whom Mpezeni's Ngoni had always failed to subjugate. Understandably, the Ngoni king regarded his old enemy's demise as a 'threatening' development.[60] A further unmistakable sign that the tide of European occupation was fast approaching Ungoni was the inauguration, in mid-1896, of the first British South Africa Company's station in the future Eastern Province of Zambia. Fort Jameson – as the BSAC post was known until 1898 – was sited among the Chewa of *Chinunda*, some 50 miles to the north of Luangeni, the village of Mpezeni's senior wife. Though Warringham, the founder of station, and his meagre force of 25 police had been urged to obtain the 'good will and friendship of Mpezeni',[61] the very existence of the Maxim-equipped post and the fact that it fell well within the Jere Ngoni's traditional raiding territory are likely to have compounded their anxieties about future intercourse with the encroaching *vishanzi* ('people from the sea'[62]).

In mid-1896, the short-lived Rhodesia Concessions Company was superseded as the representative of European business interests in eastern Zambia by the North Charterland Exploration Company, which had obtained a large land grant and related mineral rights from the BSAC on the strength of Wiese's 1891 treaty with Mpezeni.[63] Relations between the Ngoni hierarchy and the NCEC's party, initially led by Warton and including a sizeable private police force raised by Wiese in Nyasaland, did not start on a hostile footing. Employing Mpezeni's friend Wiese as go-between and deemphasising the British nature of its expedition, the NCEC was granted permission by Mpezeni to build a base (known as Fort Young in 1897–1898) in Luangeni and to carry out a series of prospecting expeditions in surrounding areas.[64] Mpezeni clearly hoped to use the NCEC as a bulwark against the feared administration of the BCAP, which had already given abundant proofs of its powers, and the BSAC, which the Ngoni king probably regarded as a mere extension of the Protectorate.

Yet tensions were not long in coming to the surface between a more and more easily identifiable war party led by Mpezeni's son and heir, Nsingo, the army commander whose compound was located in the immediate proximity of Fort

[59] A detailed account of the campaign is to be found in 'The Mwasi Kazungu Campaign: Extract from Mr. A.J. Swann's Report to Commissioner Johnston', *British Central Africa Gazette* (hereafter *BCAG*), 3/4 (15 February 1896).

[60] R.G. Warton to P.W. Forbes, 16 July 1986, Poole Papers.

[61] W. Honey to F.C. Warringham, 30 March 1896, Poole Papers.

[62] Rau, 'Mpezeni's Ngoni', p. 247.

[63] Langworthy, 'Introduction', p. 24.

[64] Warton to Secretary, 6 August 1896, and P.W. Forbes to Secretary (BSAC), 9 October 1896, both in Poole Papers; Langworthy, 'Introduction', pp. 31–32.

Young, and NCEC employees, under acting manager Major G.R. Deare between September 1896 and mid-1897. As early as October 1896, Forbes, the Blantyre-based BSAC Administrator, had reported that the NCEC prospectors were 'living ... at the pleasure' of Mpezeni, since 'a large portion of the people [were] very averse to whites settling in their country'.[65] Deare, too, became increasingly aware of what he euphemistically called 'signs ... of disaffection' towards him, his fellow Europeans and their African employees.[66] Such hostility was closely related to Nsingo's growing awareness that the presence of Europeans and their armed retainers in Ungoni and surrounding areas would eventually spell the end of such raiding system as had hitherto enabled the warriors of the age groups he represented to prove their military prowess in battle and, in so doing, enhance their standing in Ngoni 'meritocracy'.[67] Having already attained a high socio-economic status on the strength of their past military performance and display of heroic honour, politico-military leaders of Mpezeni's generation could at least contemplate the possibility of compromising with the white intruders and yielding to the latter's opposition to continuing warfare. Conversely, from the point of view of younger warriors, the NCEC and the other Europeans posed an unacceptable threat to a way of life and system of social relationship whose benefits they had not yet fully reaped.[68]

From the beginning of 1897, Mpezeni struggled to keep the lid on pressures emanating from the war party, and the occurrence of unsanctioned raids on the periphery of Ungoni was reported on several occasions.[69] The traveller Hugo Genthe witnessed the immediate aftermath of one such foray near Fort Jameson in May 1897:

> On my arrival [at Chuaula's] I found the male population all under arms, and the women crying. A raiding party of Mpeseni's people had attacked them suddenly that morning. Ten women were killed in the gardens and twenty-two were taken away as prisoners. An old man and one of Chuaula's children had been very

65 Forbes to Secretary, 9 October 1896.

66 Deare's account was published as 'Eighteen Months with the Last of the Slave Raiders', *Weekend Advertiser* (supplement to the *Natal Advertiser* [Durban]), 6 April–11 May 1929. In the typescript copy of the text that I consulted (Archives of the Livingstone Museum, LM2/4/93/8), the quoted passage is to be found on page 31.

67 I borrow this useful expression from Roberts, *History of Zambia*, p. 119.

68 First advanced by Barnes, *Politics in a Changing Society*, this reading of inter-generational conflict in Ungoni on the eve of the war of 1898 was accepted and elaborated on by Rau, 'Mpezeni's Ngoni', pp. 258–262, and Langworthy, 'Introduction', p. 29.

69 'Local news', *BCAG*, 3/22 (1 December 1896) and 4/1 (1 March 1897); 'Extracts from Mr. A.J. Swann's Report on the Marimba District for the Year 1896' *BCAG*, 4/7 (15 April 1897).

> severely wounded. Their entrails hung out frightfully torn wounds, inflicted most likely by barbed spears.[70]

Genthe was a perceptive observer and grasped the social logic behind the militancy of young Ngoni or Ngonised spearmen: 'to understand their motives one must know that the young Angonis are not allowed to participate at the big games unless they already have "wetted their spears", that is, killed somebody.'[71]

The assertiveness of Nsingo's followers, evidenced by increasing levels of violence on the outskirts of the kingdom, led the BSAC, which could now count on the promised support of the CAR,[72] to change its hitherto prudent approach. By the end of 1897 – when 'hardly a month [went] by without several raids being reported'[73] – Warringham, the Company's official in Fort Jameson, and his superior in Blantyre, Acting Administrator Daly, had reached the conclusion that the Ngoni had to be 'smashed' and 'dealt with' once and for all, 'if in the future the BSA Co. wishes to collect labour in these districts for any purpose whatsoever'.[74] However, the playing out of inter-generational tensions in Ungoni continued to the end, for it is said that, late in 1897, when Wiese and three other European employees of the NCEC were besieged in Fort Young by Nsingo's soldiers – a development that prompted the BCAP administration to dispatch a large-scale military expedition across the border – Mpezeni still sent provisions to the beleaguered party.[75] Mpezeni and the old guard probably still hoped that a rapprochement might be effected, and this uncertainty as to the final outcome of the power struggle in Ungoni may account for the fact that, when war with the *vishanzi* finally came about, the Ngoni were actually unprepared for it.

With no national army mobilisation having been ordered, and with the bulk of Ungoni's population distracted by the yearly *nkhwala* celebrations,[76] the Ngoni were slow to react to foreign aggression. The first CAR contingents reached Fort Jameson during the first half of January 1898.[77] Conflicting reports about the fate of Wiese, still hemmed in Fort Young, convinced their commanding officers to advance towards the Ngoni heartland without awaiting the arrival of Lt Col. W.H. Manning – the CAR's Commandant and Acting

70 H. Genthe, 'A Trip to Mpeseni's', *BCAG*, 4/13 (1 August 1897).

71 Ibid.

72 H.H. Johnston to Salisbury, 6 February 1897, in *North Charterland Concession Inquiry*, appendix 35.

73 'Mpezeni', in *BCAG*, 5/1 (15 January 1898).

74 P.H. Selby to Acting Administrator (Salisbury), 16 December 1897, and P.H. Selby to Acting Administrator (Salisbury), 31 December 1897, both in Poole Papers.

75 Deare, 'Eighteen Months', pp. 41–42.

76 Langworthy, 'Introduction', p. 39.

77 W.H. Manning to Salisbury, 13 January 1898, encl. in FO to CO, 1 March 1898, National Archives of Zambia (NAZ), Lusaka, NW/HC4/2/1, vol. 3.

British Commissioner in the Protectorate – and his additional troops.[78] The invading force consisted of about 350 African and Sikh regular soldiers and 200 porters under the general command of Captain Brake, who was assisted by three fellow European officers and Warringham, of the BSAC. Accompanying the troops were two Maxim machine guns, two 7-pounder RML mountain guns, and over 20,000 rounds of rifle ammunition.[79] Brake's column left Fort Jameson on 18 January 1898, reaching Fort Partridge, the abandoned former post of the Rhodesia Concessions Company, some 25 miles to the north of Luangeni, on the evening of the same day. No contact with the Ngoni was made during the march, despite the progression of the British force being hampered by heavy rain and despite it having to pass through several potential ambush positions that the Ngoni might well have taken advantage of.[80]

Once the CAR entered central Ungoni on 19 January, the default position of such regiments as could be mustered at short notice was to attack in the open, seeking to make direct contact with the enemy. The first significant encounter of the war took place to the north of Luangeni/Fort Young, when a 200-strong *impi* schooled in the principles of honour-enhancing close combat was 'seen advancing towards our left front [...]. As the enemy closed down, the left wing of the advanced guard fronted left, and, when the enemy were within thirty yards, poured in a steady volley which checked them, and, on the appearance of the flanking party in their rear, they drew off slowly, with a loss of some half-a-dozen men.'[81] A later account explained that the Ngoni casualties consisted of the 'bravest' of their numbers, who had 'endeavoured to charge our line, but were shot before they got near'.[82]

The key set-piece battles of the war took place around Fort Young in the morning and afternoon of 20 January. On both occasions, the Ngoni made very limited use of firearms. In the morning, the British faced a party of '600 or 700' warriors under the leadership of Mlonyeni, another son of Mpezeni. Mlonyeni's forces were apparently bent on retrieving the 1,000 head of cattle that the British had captured on the previous day. They advanced 'dancing and brandishing their weapons within 300 yards'. The CAR soldiers themselves marched forward with fixed bayonets. 'Fire was opened by the centre companies at about 100 yards distance, two volleys being poured in, and the advance resumed. This silent steady advance impressed the Angoni almost more than the volleys had, and, throwing

---

78 'Mpezeni'.

79 H.E. Brake to A. Sharpe, 20 January 1898, encl. in W.H. Manning to Salisbury, 17 February 1898, National Archives of the UK (NAUK), Kew, London, FO2/147.

80 Brake to Sharpe, 20 January 1898.

81 Ibid.

82 'The Mpezeni Campaign', *BCAG*, 5/3 (26 February 1898).

their spears, they slowly gave way.'[83] The CAR soldiers followed the retreating Ngoni, who were also shelled by the mountain guns, and burnt a number of villages where resistance was encountered. 'The total loss to the enemy, from prisoners' accounts, was about twenty killed and several wounded.'[84]

The afternoon of 20 January witnessed what Rau has described as the 'most deliberate and organized Ngoni attack' against the invaders.[85] Led by Nsingo, 'three long lines' of soldiers deployed in the Zulu-inspired horn-shaped formation strove to encircle the British force, which had returned to Fort Young after the morning battle. The Ngoni pincer movement, however, clashed on a wall of fire, for 'the 7-pounder guns and Maxims were brought into action and the troops fired steady volleys at the advancing masses. The Angoni came on bravely, but as their losses increased, the lines halted and then began to retire slowly ... The order was at once given to charge and the Angoni broke and were pursued for some distance.'[86] More villages – including Nsingo's – were razed to the ground. Brake himself was impressed by the 'considerable courage' showed by the 'Angoni Zulus' in 'advancing to within 50 yards of our men'. It was their technological conservatism, however, that was to blame for their defeat, for Nsingo's spearmen 'could not face the volley firing. Guns amongst them are apparently rare and their fire most inaccurate and their spears never reached the firing line. In no case were they able to charge home.'[87] 'Every officer' involved in the expedition, in fact, recognised that if the Ngoni had 'been armed with guns ... operations would have been vastly more difficult'. Without firearms, however, the Ngoni simply 'expose[d] themselves in their endeavour to charge down on our troops'.[88] Nsingo's routed army may have numbered between 3,000 and 5,000 men, reservists included.[89] Conservative estimates put the number of Ngoni deaths at 50, while no casualties were registered on the British side.[90] In reality, Ngoni losses are likely to have been higher, since 'many dead and wounded [were] carried off' by their retreating comrades.[91]

During the following two weeks, the invading troops – numbering almost 1,000 by 28 January and representing the 'largest and most efficient' force ever collected by Europeans in the region – were split into four mopping-up columns led by white officers. 'Several smart skirmishes' were fought 'in which some of

83 H.E. Brake to A. Sharpe, 21 January 1898, encl. in Manning to Salisbury, 17 February 1898.

84 Brake to Sharpe, 21 January 1898; Rau, 'Mpezeni's Ngoni', pp. 271–272.

85 Rau, 'Mpezeni's Ngoni', p. 272.

86 'The Mpezeni Campaign'.

87 Brake to Sharpe, 21 January 1898; Rau, 'Mpezeni's Ngoni', p. 274.

88 W.H. Manning to Salisbury, 20 May 1898, NAUK, FO2/148.

89 'The Mpezeni Campaign'.

90 'Mpezeni War', *BCAG*, 5/2 (5 February 1898); H.L. Daly to Acting Administrator (Salisbury), 4 February 1898, Poole Papers.

91 'The Mpezeni Campaign'.

the Angoni showed that the old Zulu spirit still lingers amongst them, as they charged fearlessly up to small parties of our men' armed with rifles.[92] Punitive expeditions and search and destroy operations – in the course of which all of the largest villages and fields in Ungoni were burnt down, dozens of additional Ngoni spearmen killed and thousands of cattle requisitioned – lasted until the capture of Nsingo, who was summarily executed on 4 February.[93] Mpezeni surrendered a few days later, as the fires that had destroyed his once 'thickly populated', 'splendid pastoral country' were still smouldering.[94] The old king, who, after all, had sought to avoid a direct clash with the Europeans, was spared his son's fate, being instead briefly interned in Mchinji, in Nyasaland, from April 1898.[95] By this time, the Ngoni were reported to be 'thoroughly demoralized and convinced of the uselessness of attempting anything against troops armed with rifles and machine guns'.[96] Rifles, machine guns and artillery fire had brought to an end the independent history of Mpezeni's Ngoni. However, as will be shown below, the story of Ngoni militarism was not yet over.

The Remaking of Ngoni Militarism

The violent incorporation of Mpezeni's Ngoni into North-Eastern Rhodesia spelled the ultimate end of their war-based society and of the raiding system that had kept inter-regimental advancement mechanisms oiled up. Yet important elements of the pre-colonial pattern of social relationships and honour culture endured even in the changed circumstances of the early twentieth century.

Labour migration towards Southern Rhodesia – which began in earnest immediately after the defeat of 1898 – was a vehicle for the survival of significant traits of the old honour culture.[97] Lust for adventure, resilience in the face of adversity, the cultural belittlement of agricultural labour, and the readiness to put one's life at stake with a view to improving one's social standing might all have been responsible for the rapidity with which 'the idea of going to work' in the Southern Rhodesian gold mines 'caught on with the Angoni' of eastern

---

92 Ibid.

93 Detailed accounts of the 'pacification' of Ungoni between January and February 1898 are to be found in H.E. Brake to A. Sharpe, 29 January 1898, encl. in Manning to Salisbury, 17 February 1898, and in Manning to Salisbury, 17 February 1898.

94 The quoted description of Ungoni is Warton's; Warton to Secretary, 6 August 1896.

95 H.E. Brake to W.H. Manning, 6 May 1898, encl. in W.H. Manning to Salisbury, 16 June 1898, NAUK, FO2/148.

96 H.L. Daly to Acting Administrator (Salisbury), 17 February 1898, Poole Papers.

97 The relationship between twentieth-century labour migrancy and pre-colonial notions of honour in southern Africa is touched upon in Iliffe, *Honour*, chapter 16.

Zambia.[98] Of course, more obvious and, indeed, better-known structural determinants were also at play, for the poverty in which the Ngoni heartland was plunged by the war of 1898 was compounded in 1900 by the imposition of a hut tax, the device with which regional colonial regimes invariably sought to extract labour from their newly acquired African subjects.[99] But compulsion and what historians of an earlier generation used to consider the systemic alliance between capital and the colonial state clearly do not tell the whole story of Ngoni labour migration.

The centrality of African motives and agency emerges with particular clarity when one examines the history and patterns of recruitment into colonial paramilitary police forces, the institutions that offered the most immediate opportunities for the twentieth-century remaking of Ngoni militarism.[100] Before the war of 1898, the *vishanzi* regarded Mpezeni and his followers with a mixture of dread, revulsion and admiration. Early in 1896, Warton, the NCEC's manager, explained that Mpezeni's 'people (Zulus) have entirely driven out the original native population, and he is a terror to the surrounding villages for many miles. He is a slave-raider and exercises the powers over life and death.'[101] Mpezeni – he reported after, and despite, his first positive meeting with the king – was 'an absolute savage. He thinks no more of killing a man than of killing an ox, perhaps not so much.'[102] Underlying these images of martial might and unrestrained brutality was the historical link between the Jere Ngoni and the Zulu of South Africa, whose military ethos and tactics, of course, had made an everlasting impression on British imperial policy-makers and commentators since the 1870s. The connection was made explicit by Harry Johnston, the first Consul-General and High Commissioner in British Central Africa: 'Mpeseni is a very powerful Chief, of Zulu or Matabele origin ... He can put together at least 30,000 warriors into the field, armed chiefly with spears and assegais, and attacking impetuously and with great bravery in the Zulu style.'[103] Martial representations such as these, of course, could be used for alternative – but not necessarily conflicting – purposes: either to convey a sense of African inherent savagery or rather of the worthiness of a likely future opponent. Their sway, moreover, was generally reinforced by the war of 1898, during which – as has been seen above – some at least of the European protagonists interpreted the

---

98 'Report of an interview with Mr. Hayes ...', 15 January 1900, encl. in Secretary to J.H. Hayes, 18 January 1900, NAZ, NER/A3/12/1.

99 Rau, 'Mpezeni's Ngoni', pp. 347–350.

100 My reading of Ngoni experiences in colonial armed forces owes much to Iliffe, *Honour*, p. 234, and Isaacmans, *Slavery and Beyond*, pp. 296–297.

101 Warton to Forbes, 2 April 1896.

102 Warton to Secretary, 6 August 1896.

103 H.H. Johnston, 'Memorandum', 14 July 1896, in *North Charterland Concession Inquiry*, appendix 31.

naked courage displayed by many Ngoni fighters in the face of the enemy's crashing technological superiority as a demonstration of lingering 'Zuluness'.

Images of Ngoni 'pluckiness' (a favourite word in the vocabulary of 'martial race' ideologues) help explaining the decision to recruit some of their number into the newly formed Mashonaland Native Police (MNP), part of Southern Rhodesia's British South Africa Police. But equally – if not more – important were the Jere Ngoni's own predilections and aspirations. Administrator Forbes, who first suggested that an Ngoni contingent be raised for service in Southern Rhodesia, explained that the Ngoni of both Zambia and Malawi were 'willing to go anywhere' and took 'great pride in being made soldiers'.[104] It was on the strength of this assessment that Colin Harding, a Major in the Southern Rhodesia's police and the future Commandant of the Barotse Native Police, was dispatched to Nyasaland and, then, Fort Jameson with a view to enlisting some 300 Lakeside Tonga and Ngoni police.[105] Harding's recruiting mission on behalf of the MNP in the summer of 1898 was a resounding success. As soon as he reached Fort Jameson, between 150 and 200 Ngoni 'young men' were promptly handed over to his charge by local authorities to supplement the levies he had already raised in the BCAP.[106] No coercion was apparently needed at any stage, for the Ngoni joined Harding's incipient force 'readily'.[107] Once in Southern Rhodesia, the Ngoni contingent from North-Eastern Rhodesia is said to have adjusted easily to paramilitary life[108] – so much so that, upon expiry of its original term of service in 1900, some at least of its members 'readily found employment as [mining] compound police'.[109] Four years later, Ngoni from both Nyasaland and North-Eastern Rhodesia still constituted the bulk of the African police in Mashonaland.[110]

As pointed out above, by the time of Harding's visit, the exiled Mpezeni's heartland was a shadow of its former self, having seen most of its large villages

---

104 P.W. Forbes, 'Memorandum', 8 March 1898, encl. in H. Canning to FO, 14 March 1898, NAZ, NW/HC4/2/1, vol. 3.

105 Colonial Fort Jameson (now Chipata), the capital of North-Eastern Rhodesia, is not to be confused with the earlier BSAC post in Chinunda's. The new Fort Jameson was located in the Ngoni heartland.

106 The figure of 200 is given by Colin Harding, *Far Bugles* (London, 1933), pp. 82–83; that of 150 in W.H. Manning to A. Sharpe, 2 November 1898, NAUK, FO2/149.

107 C. Harding, 'Report of a Journey into Northern Rhodesia', n.d. [but late 1898], Archive of the Livingstone Museum, LM 2/3/11/9. I owe this reference to Jack Hogan.

108 Harding, *Far Bugles*, pp. 82–83.

109 Charles van Onselen, 'The Role of Collaborators in the Rhodesian Mining Industry, 1900–1935', *African Affairs*, 72/289 (1973), p. 405.

110 Timothy Stapleton, *African Police and Soldiers in Colonial Zimbabwe, 1923–80* (Rochester, NY, 2011), p. 22.

and fields razed to the ground and suffered 'enormous losses' in cattle.[111] To be sure, it is tempting to interpret the enthusiasm for police service reported by Harding in 1898 as a direct consequence of what the same witness called 'the state of despair and desolation' then prevailing in Ungoni.[112] But – as in the case of labour migration, which kick-started at roughly the same time – an exclusive emphasis on the imperative of survival or what Timothy Stapleton calls 'material motives' and 'concerns' obfuscates the workings of more profound historical processes.[113] The glorified place of warfare in the Ngoni pre-colonial social system, and the determination on the part of aspiring or demobilised warriors to continue to command their 'right to respect' through military pursuits, were the key socio-cultural forces shaping Ngoni choices. Indeed, there are some indications that, in Ngoni eyes, Harding's training camp in Fort Jameson represented the structural equivalent of an old regimental barrack, which chiefs and elders visited regularly to assess progress and which they kept well supplied with provisions.[114] This reading of Ngoni motivations, moreover, is borne out by the fact that Ungoni's gradual economic recovery in the early 1900s did not undermine its young residents' willingness to serve in colonial forces.

For our story, North-Western Rhodesia's Barotse Native Police (BNP) is especially important, since the colonial fixation with the 'trustworthy stranger to police other strangers' meant that the Fort Jameson Ngoni were prevented from dominating the ranks of the North-Eastern Rhodesia Constabulary, the police corps whose area of operations encompassed their home region.[115] Opportunities in the CAR and, to a lesser extent, the BCAP police were also limited, mainly on account of the two forces having been built around nuclei of Lakeside Tonga and Yao volunteers before the final subjugation of the Ngoni.[116] The BNP came into existence once the cost-cutting decision was taken to replace the detachment of white Southern Rhodesian police who had been stationed in Fort Monze, North-Western Rhodesia, since 1898.[117] Officially gazetted at

---

111 R. Codrington to Manager (NCEC), 7 September 1898, NAZ, NER/A3/12/1.

112 Harding, *Far Bugles*, p. 82.

113 Stapleton, *African Police*, pp. 17, 26.

114 Ibid., p. 83.

115 The quoted expression is to be found in David Anderson and David Killingray, 'Consent, Coercion and Colonial Control: Policing the Empire, 1830–1940', in D. Anderson and D. Killingray (eds), *Policing the Empire: Government, Authority and Control, 1830–1940* (Manchester, 1991), p. 7.

116 W.H. Manning, Annual Report on the BCAP, encl. in W.H. Manning to Salisbury, 13 May 1898, NAUK, FO2/147; Risto Marjomaa, 'The Martial Spirit: Yao Soldiers in British Service in Nyasaland (Malawi), 1895–1939', *Journal of African History*, 44/3 (2003), pp. 419–420; John McCracken, 'Coercion and Control in Nyasaland: Aspects of the History of a Colonial Police Force', *Journal of African History*, 27/1 (1986), p. 128.

117 R. Coryndon, 'Notes on Police', encl. in R. Coryndon to A. Milner, 12 July 1900, NAZ, NW/HC4/2/1, vol. 6.

the beginning of 1901, the BNP was mainly employed 'as a defensive force in garrisoning the different stations, patrolling their districts under Officers or white NCOs, executing warrants, as escorts to District Commissioners when visiting the kraals in their District'.[118] What even this dry description makes clear is that the BNP was no different from other colonial police forces, for which the prevention and detection of crime were always subordinate to the defence of the colonial order from internal threats.[119] The paramilitary character of the BNP was a direct consequence of its primary function, and it manifested itself in the heavy emphasis placed in training on arms drill and musketry practice.

Harding, the first Commandant of the BNP, was from the beginning keen on employing some of the same time-expired Ngoni he had earlier recruited for the Mashonaland Native Police.[120] The plan, however, did not materialise at this early stage. By the end of 1902, the BNP consisted of 227 African non-commissioned officers and men under a dozen of white commissioned and NCOs; all of the African recruits hailed from North-Western Rhodesia itself.[121] Since Harding considered this state of affairs dangerous and 'unwise',[122] the BNP ranks were supplemented with 75 Bemba-speakers from North-Eastern Rhodesia in the course of 1903.[123] But aspiring Ngoni military men were not ready to be sidelined and sought to influence recruitment patterns from below. In the early years of the century, Fort Jameson Ngoni were reported to be 'constantly' travelling to Kalomo, the then capital of North-Western Rhodesia and headquarters of the BNP, 'anxious to enlist in the B.N. Police, having walked from N.E. Rhodesia, on their own initiative and unaccompanied by any white man'.[124]

Ngoni pressures bore the desired fruit and, in the summer of 1904, Major Carden was sent to Fort Jameson on recruiting duties. He returned to Kalomo with 124 volunteers, but, according to Harding, he could have easily enlisted

118 C. Monro, 'Report on the Barotse Native Police', encl. in H. Marshall Home to Imperial Secretary, 8 January 1904, NAZ, NW/HC1/2/10.

119 Anderson and Killingray, 'Consent, Coercion and Colonial Control', p. 6.

120 C. Harding to R. Coryndon, n.d. [but mid-Nov. 1899], encl. in J. Chamberlain to A. Milner, 17 March 1900, NAZ, NW/HC4/2/1, vol. 6; C. Harding to Secretary (BSAC, Salisbury), 15 July 1900, NAZ, NW/A6/1/1.

121 F. Hodson. 'Annual Return of Military and Naval Resources of Colonies and Protectorates', 13 November 1903, encl. in H. Marshall Hole to A. Lawley, 7 December 1903, NAZ, NW/HC1/2/10. Though written at the end of 1903, Hodson's return describes the situation obtaining in 1902.

122 C. Harding to R.T. Coryndon, 17 February 1903, encl. in R.T. Coryndon to Secretary (BSAC), 6 March 1903, NAZ, NW/A2/2/3.

123 R.T. Coryndon to A. Milner, 25 May 1903, encl. in Secretary to the Administrator to Secretary (BSAC), 25 May 1903, NAZ, NW/A2/2/3; W.V. Brelsford, *The Story of the Northern Rhodesia Regiment* (Bromley, 1990; 1st edn, 1954), pp. 20–21; Tim Wright, *The History of the Northern Rhodesia Police* (Bristol, 2001), p. 47.

124 C. Harding to Imperial Secretary, 11 January 1905, encl. in Secretary to the Administrator (NWR) to Imperial Secretary, 14 January 1905, NAZ, NW/HC1/2/15.

as many as 500.[125] Of the 124 men brought back by Carden, as many as 90 (72 per cent) were classified as Ngoni.[126] This large influx made the Ngoni the most heavily represented ethnic group in the BNP's rank and file. By the end of 1904, Ngoni police constituted about 27 per cent of a force that now numbered 336 African NCOs and men. The second largest contingent, numbering 50 men, were Ila from North-Western Rhodesia.[127] Once established, Ngoni dominance of the BNP would prove long-lasting. Late in 1905, only '20 local natives' remained in the corps, the bulk of its members being 'composed of natives from North Eastern Rhodesia'.[128] In 1908–1909, all of the new recruits required by the BNP were once more enlisted in Fort Jameson, while many of their predecessors signed up for a further term of three years.[129] By then, the Ngoni (and Bemba) had also managed to penetrate the North-Eastern Rhodesia Constabulary, where the preference for 'alien native' recruits had seemingly been relaxed at the same time as it was tightened in North-Western Rhodesia.[130]

The extent to which the Jere Ngoni's own motivations and socio-cultural inclinations were responsible for carving out a privileged niche for themselves in the BNP (and, earlier, in the MNP) casts doubts on approaches that foreground European agency alone in the fabrication of imperial taxonomies – including the so-called martial race ideology, which Heather Streets' recent work, for instance, presents as being, in essence, a one-way 'British construction', 'born out of specific recruiting needs' and 'nineteenth-century conceptions of race'.[131] The case of the Ngoni shows that the imperial frontier was not a *tabula rasa* awaiting colonial inscription and that the potential for historical invention is never limitless. Nothing is made out of nothing. The colonial subjects' pre-contact military histories and social structures influenced their terms of engagement with European forces and, therefore, the perceptions, imaginings and policies of the leaders of the latter. Sometimes, though not always, specific social or ethnic groups were construed as 'martial' – and therefore targeted as military

125 C. Harding to Imperial Secretary, 23 November 1904, encl. in R.T. Coryndon to Imperial Secretary, 24 January 1904, NAZ, NW/HC1/2/14.

126 'Nominal roll of natives enlisted for service in Barotse Native Police', 4 October 1904, encl. in Harding to Imperial Secretary, 11 January 1905.

127 'Annual Return of Military and Naval Resources of Colonies and Protectorates', encl. in C. Harding to Imperial Secretary, 18 January 1905, NAZ, NW/HC1/2/15.

128 [F. Hodson], Untitled memorandum on the BNP, 9 October 1905, encl. in F. Hodson to Imperial Secretary, 9 October 1905, NAZ, NW/HC1/2/22.

129 'Annual Report on Barotse Native Police for Year 1908–1909', NAZ, NR/A5/1/2.

130 In 1909, North-Eastern Rhodesia's armed police consisted of 310 Africans; of these, 109 were classified as Bemba and 71 as Ngoni. Acting Adminstrator (NER), 'Annual Return of Military and Naval Resources of North-Eastern Rhodesia', 31 December 1909, encl. in Secretary to the administrator (NER) to Imperial Secretary, 2 August 1910, NAZ, NER/A2/3/1, vol. 3.

131 Heather Streets, *Martial Races: The Military, Race and Masculinity in British Imperial Culture, 1857–1914* (Manchester, 2004), p. 3.

Figure 5.2 'Barotse Native Policeman outside hut, with kit laid out for inspection, Kalomo, 1905'

R. Coryndon's Album. Courtesy of Livingstone Museum, Livingstone, Zambia.

or paramilitary recruits – because they *were* martial. This, of course, is not to suggest the existence of a biological predisposition towards warfare among select groups. But a focus on imperial interests and myth-making should not obscure the fact that the pre-colonial experience and social organisation of some communities had been more deeply shaped by warfare and militarism than those of others. Given their pre-colonial war-centred social system and normative universe, it is not surprising that the Ngoni came to be conceived of as Northern Rhodesia's prime 'martial race'. Nor is it surprising that scores of young Ngoni should have volunteered for service in paramilitary forces that appeared to hold out the promise of perpetuating such notions of honour and masculinity as had informed Ngoni socio-cultural structures in the recent past.

Within this framework of continuity, however, there were also sharp ruptures. In particular – to return to the central theme of this chapter – the experiences of the Ngoni in colonial police forces transformed beyond recognition their relationship with firearms. Because of its paramilitary character, the BNP was equipped with single-shot Martini-Henry breech-loading rifles during the first 10 years of its existence.[132] Firearms symbolised the very essence and repressive potential of the BNP, and proficiency in their use was systematically promoted through regular musketry courses held both at the corps' headquarters and in peripheral commands.[133] In this light, a series of fairly precise parallels can be drawn between the lives and expectations of pre-colonial Ngoni fighters and those of BNP recruits. In pre-colonial times – as has been argued above – warfare had offered young Ngoni spearmen the chance to display their heroism and, by so doing, unlock the potential for self-advancement inherent in their social institutions. Given the comparatively peaceful nature of the occupation of North-Western Rhodesia by the BSAC, opportunities for real warfare while serving in the BNP were, in practice, limited to a few anti-slavery and anti-smuggling patrols. Thus, rather than expressing itself in open battle and in the heat of close combat, inter- and infra-regimental competition for bravery now took the form of shooting contests, with 'Marksman Badges' for proficient riflemen taking the place of the horn-made 'military order' described by Wiese in the late 1880s.[134]

Warfare and raiding, of course, had also enabled junior Ngoni soldiers to ameliorate their socio-economic status through the direct accumulation of both material wealth and dependents. While raiding *sensu stricto* could not be openly endorsed by the BNP, police service still offered ample opportunities for preying

132 A. Cree, 'Regimental miscellanea', in Brelsford, *Story of the Northern Rhodesia Regiment*, p. 117; Hodson, 'Annual Return'.

133 See, e.g., Monro, 'Report on the Barotse Native Police'; Harding to Imperial Secretary, 23 November 1904; 'Report on Barotse Native Police, 1910–1911', NAZ, NR/A5/1/2.

134 The award of 'Marksman Badges' is mentioned repeatedly in regimental orders dating to 1907–1909, NAZ, NW/B2/1/6.

on civilians. In condemning the police's prevailing conduct, Marshall Hole, who served as North-Western Rhodesia's Acting Administrator in 1903, urged the BNP's commanding officers to bear in mind that the 'uncivilized instincts' of local recruits could not be 'eradicated' by a 'few months drill and acquaintance with discipline'.

> On the other hand, the possession of firearms and their familiarity with the use of them, together with the right to wear uniform, are calculated to puff them up with a sense of their own superiority and importance, and to tempt them to adopt a bullying and overbearing demeanour when brought into contact with the unarmed population, unless they are restrained by the presence and supervision of a white Officer or Non-Commissioned Officer.[135]

In emphasising the connection between 'possession of firearms' and 'overbearing demeanour', Marshall Hole – his crass racism notwithstanding – was certainly pointing to a real problem – one which Regimental Order no. 1 of 23 February 1903 had sought partly to tackle by stipulating that '"in future all Police going on leave will go unarmed"'.[136] Yet – as Acting Commandant Monro coyly put it – the tendency 'to presume at times on the natives' while on active service was unaffected by the Regimental Order, and indeed it remained a distinguishing trait of the BNP for several years to come.[137]

Between September 1907 and September 1908, for instance, at least three BNP recruits were either fined or given (short) prison sentences for various sexual related offences, ranging from 'assault with violence and indecent assault' (Mumbwa command) to 'sending to a village and procuring by force a native woman' (Mongu) and 'interfering and sleeping with a native woman bringing grain into camp' (Mumbwa).[138] Nine more policemen were punished for common or violent assault, the victims of police violence being, in two cases, 'native prisoners'.[139] Other crimes against civilians reported during the same period included 'creating a disturbance in a native village', 'taking a basket of meal from a native woman and striking her on the head' and even 'abduction'.[140]

* * *

135 H. Marshall Hole to J. Carden, 19 August 1903, encl. in H. Marshall Hole to A. Lawley, 19 August 1903, NAZ, NW/HC1/2/9.

136 Quoted in J. Carden to H. Marshall Hole, 30 August 1903, encl. in H. Marshall Hole to A. Lawley, 21 September 1903, NAZ, NW/HC1/2/10.

137 Monro, 'Report on the Barotse Native Police'.

138 Regimental orders, 19 September 1907, 16 May and 26 September 1908, NAZ, NW/B2/1/6.

139 Regimental orders, 19 September and 31 December 1907, 15 and 20 April 1908, NAZ, NW/B2/1/6.

140 Regimental orders, 25 January, 25 February and 7 April 1908, NAZ, NW/B2/1/6.

Competition between warriors for upward mobility and the exalted position of the latter vis-à-vis less militarily prepared agriculturalists had been central elements of Ngoni pre-colonial militarism. Both features found a new lease of life within the BNP; yet both were transformed by firearms. For Ngoni recruits – as for their pre-colonial predecessors – military life remained the mainspring of heroic honour, a means of masculine affirmation and the ultimate source of socio-economic advancement. What changed was that such exploits were now predicated on the deployment and effective appropriation of a technology – firearms – that young Ngoni had previously rejected as inadequate to foster their self-improving goals. Within the ranks of the BNP, if not in Ungoni as a whole, the Martini-Henry had become the modern assegai.

Chapter 6

'Hardly a Place for a Nervous Old Gentleman to Take a Stroll': Firearms and the Zulu during the Anglo-Zulu War[1]

Jack Hogan

The classic academic statement on the Zulu use of firearms during the Anglo-Zulu war of 1879 was written by Jeff Guy in 1971.[2] Despite being published over 40 years ago, Guy's 'Note' has remained the most influential work on the subject. In essence, his view was that the 'Zulu failed to adapt their strategy and tactics in any way which might have enabled them to deploy firearms more effectively' than they did during the conflict.[3] Subsequent scholarship built on Guy's insights without altering them substantially. Ian Knight argued that the Zulu failed to take advantage of firearms and that, 'if the Zulu began the war believing that the large quantities of guns in their possession made them the equal of the British, they merely assumed that it would be sufficient to use those arms in support of their existing tactics'.[4] A more recent interpretation tells us that

> by and large, the Zulu did use traditional weapons and tactics, although some British officers noted that, on occasion, Zulu forces opposed them with breech-loading rifles. It appears that while individual Zulu shot well, the Zulu as a whole did not make a concerted effort to adapt their battlefield tactics to the new weapons, except on rare occasions.[5]

Again, one of the editors of a new major work on Zulu identity notes that the Zulu 'did eventually acquire a considerable number of firearms by the 1870s,

[1] I am indebted to Paul la Hausse de Lalouvière for commenting on an early draft of this chapter. Needless to say, I remain solely responsible for any errors or misconceptions that might remain in the article.

[2] Jeff Guy, 'A Note on Firearms in the Zulu Kingdom with Special Reference to the Anglo-Zulu War, 1879', *Journal of African History*, 12/4 (1971), pp. 557–570.

[3] Ibid., p. 561.

[4] Ian Knight, *The Anatomy of the Zulu Army from Shaka to Cetshwayo, 1818–1879* (London, 1995), pp. 213–214.

[5] William Storey, *Guns, Race and Power in Colonial South Africa* (Cambridge, 2008), pp. 272–273.

but they failed to adapt their tactics to the new weapon'.[6] In reality, all these are simple reiterations of Guy's original thesis.

This chapter seeks to make two points. First, it argues that Zulu use of firearms during the Anglo-Zulu war was in specific cases more effective than is commonly assumed. In order to do so, it is first necessary to outline the nature of the sources commonly used by regional specialists and examine the interests that informed them and their flaws. Second, this chapter contends that the Zulu perception and use of firearms, and the choices made about them have not been fully understood in the broader cultural context in which they were necessarily embedded. Drawing almost exclusively on Zulu accounts, the final substantive section of the chapter will draw wider conclusions about Zulu tactics and dispositions on the battlefield.

The Zulu had been exposed to firearms from the 1820s, as attested by the development of an extensive vocabulary reflecting a significant degree of familiarity with the new technology. One often quoted Zulu oral informant listed nine separate words for types of firearm, along with a description for each.[7] To these, I would add two further, more general, words: *isiBhamu* and *isiTunyisa*. The first is clearly onomatopoeic, while the second derived from the Zulu word *Tunyisa*, 'to make smoke'.[8] It is reasonable to suppose that such lexical complexity was the result of the presence of large numbers of various types of firearms in Zululand. Barring *isiBhamu* and *isiTunyisa*, the other terms imply a technical and practical understanding of the construction and purpose of a range of different guns.[9]

Commonly quoted figures suggest there were about 20,000 guns in Zululand on the eve of the war: 500 were breech-loaders, 2,500 modern percussion rifles, 5,000 percussion muskets, and the rest flintlock muskets.[10] The majority of these, it is suggested, were obtained in the 1870s, although the volume of the gun trade had been increasing for at least a decade. But this estimate appears to misunderstand the relevant original source. Portuguese officials did write that 'between 1875 and 1877 20,000 guns, including 500 breech-loaders,

---

6 John Laband, '"Bloodstained Grandeur": Colonial and Imperial Stereotypes of Zulu Warriors and Zulu Warfare', in B. Carton, J. Laband and J. Sithole (eds), *Zulu Identities: Being Zulu, Past and Present* (Pietermaritzburg, 2008), p. 174.

7 Evidence of Bikwayo kaNoziwawa, in John Wright and Colin de B. Webb (eds), *The James Stuart Archive of Recorded Oral Evidence Relating to the History of the Zulu and Neighbouring Peoples* (5 vols, Durban, 1976–2001) (henceforth *JSA*), vol. 1, p. 63.

8 Alfred T. Bryant, *Zulu–English Dictionary* (Pinetown, 1905).

9 For instance, the '*Idhlebe* or elephant gun came from the Boers, not the Portuguese, in the early days of Mpande's reign ... The *umakalana* was a double-barrelled gun and only given to men of position ... The *Ifili* had a long range, and a small bullet and small cap, the same size cap as the *umakalana* gun.' Evidence of Bikwayo kaNoziwawa, in *JSA*, vol. 1, p. 63.

10 Knight, *Anatomy of the Zulu Army*, p. 168.

and 10,000 barrels of powder were imported ... the greater proportion for the Zulus'.[11] But these were *annual* figures, three quarters of which went to the Zulu. When so understood, this would give us total figures for 1875–1877 of approximately 45,000 guns, including 1,125 breech-loaders, and 22,500 barrels of powder. Another account records that between June and July 1878, 400 men were reported to have been sent to Delagoa Bay, to the north of Zululand and under Portuguese control, to collect 2,000 breech-loaders.[12] What these sparse data make clear is that, regardless of how one quantifies it exactly, the trade in firearms with the Zulu was extensive, and that the weapons thus acquired were not all second- and third-rate old muskets.

Figures such as these also suggest that, by the time of the Anglo-Zulu war, the majority of Zulu fighters were equipped with firearms. King Cetshwayo (1872–1879) made sure this was the case, instructing, for instance, the lightly armed members of the uVe *iButho* (regiment; pl. *amaButho*) who had gathered at his kraal immediately before the war to buy guns the next day.[13] When coupled with the relatively complex vocabulary for firearms, widespread ownership would indicate familiarity with such weapons and some degree of skill in their use. The scholarly consensus suggests this was not the case. It is the origins of this consensus that we must first interrogate.

A 'Celibate Man-Destroying Gladiator': Depicting the Zulu

Samuel Martin's study of British perceptions of the Zulu from the early nineteenth century to the end of the Anglo-Zulu war names three sources as the basis for the national press's estimation of the force the British would face in South Africa. These were the published dispatches of imperial officials, which appeared in the Blue Books, a *Précis of Information Concerning the Zulu Country* compiled by the Intelligence Branch of the Quartermaster-General's Department, and, 'most widely quoted at the time', a semi-official pamphlet entitled *The Zulu Army*.[14] Other potential sources existed, such as accounts of the Zulu by traders, adventurers and the like, but even Sir Bartle Frere, the

11 Ibid. Knight lists Guy's 'Note' as his source, but omits the 'annually'. Guy ('Note', p. 560) gives his source as 'C.O. [Colonial Office] 879; 17/208, no. 4, Foreign Office to Colonial Office, Report by Vice-Consul, Mozambique, 22 Sept. 1879, 6.'

12 British Parliamentary Papers (henceforth BPP), C.2242. Enclosure G7, in appendix III, report of H.C. Shepstone and G.M. Rudolph, Utrecht, 4 December 1877, p. 70.

13 Evidence of Mpatshana kaSodondo, in *JSA*, vol. 3, p. 305.

14 Samuel Martin, 'British Images of the Zulu, ca. 1820–1879', unpublished PhD thesis, University of Cambridge, 1982, p. 283.

then High Commissioner for South Africa, had only read one of these and a compendium of documents relating to Natal.[15]

The sources available to the military were no more extensive. Although only one edition of the *Précis* was published in January 1879, several confidential editions were produced for internal military use both before and afterwards.[16] The January 1879 edition, identical in both its confidential and publicly available form, seems to have been a compilation of the first July 1878 edition, an addendum to the latter published in November of the same year, and dispatches received up to January. The July 1879 edition appears to have been similarly concocted. But the information presented in successive volumes is not a simple cumulative compilation. As new information was gleaned from dispatches, and as political circumstances changed, the *Précis* was altered. In January, for instance, we read of the Zulu that 'they are neither more bloodthirsty in disposition nor more powerful in frame than the other Kafir tribes of the Coast region'.[17] This passage disappears from the post-Isandlwana edition of July 1879, where we are told instead that

> every Zulu is a soldier, and as a nation they are brave, fond of fighting, and full of confidence in themselves ... There can be no doubt of the warlike character of the Zulu race. Their present military organization would also show that they are capable of submitting to a severe discipline.[18]

Gone, as well, is the assertion from late 1878 that the Zulu were not only full of confidence, but also had an 'exaggerated idea of their own numbers and prowess'.[19] It appears that the British no longer thought this idea was quite as exaggerated. Similar discursive elisions and recalibrations manifest themselves in successive appraisals of Zulu tactics.

---

15 BPP, C.2222. Enclosure 9, in no. 19, minute by Frere, enclosed in Frere to Hicks Beach, 16 November 1878, p. 45. These sources were Nathanial Isaacs, *Travels and Adventures in Eastern* Africa, (2 vols, London, 1836) and John Chase, *The Natal Papers* (Cape Town, 1843).

16 *Précis of Information Concerning the Zulu Country, with Sketch Map, Prepared in the Intelligence Branch of the Quartermaster-General's Department, Horse Guards, War Office. July 1878* (London, 1878) (henceforth *Précis July 1878*); *Addendum to the Précis of Information concerning the Zulu Country, Prepared in the Intelligence Branch of the Quartermaster-General's Department, Horse Guards, War Office. July 1878. November 1878* (London, 1878) (henceforth *Addendum to Précis November 1878*); *Précis of Information concerning the Zulu Country, with a Map, Prepared in the Intelligence Branch of the Quartermaster-General's Department, Horse Guards, War Office. Corrected to January 1879* (London, 1879) (henceforth *Précis January 1879*); *Précis of Information concerning the Zulu Country, with Sketch Map, Prepared in the Intelligence Branch of the Quartermaster-General's Department, Horse Guards, War Office. Corrected to July 1879* (London, 1879) (henceforth *Précis July 1879*).

17 *Précis January 1879*, p. 32.

18 *Précis July 1878*, p. 33.

19 *Addendum to Précis November 1878*, p. 14.

The accounts of the Zulu Army presented in the *Précis* were 'principally extracted from a Memo published by the direction of Lieutenant-General Lord Chelmsford, for the information of those under his command, Nov. 1878'.[20] The July 1879 edition has a similar attribution, but refers to a version of the memorandum dating to April of that year.[21] The memorandum in question was a pamphlet by Fred B. Fynney. First published as *The Zulu Army* late in 1878, the text was reissued in a new edition in 1879 under the title *The Zulu Army and Zulu Headmen*.[22] Fynney was an interpreter for the Natal government in 1876–1877 and an administrator of native law and special border agent for the Lower Tugela Division in 1878–1879. During the war, he commanded the Border Police force in his district, Colonial Defence District No. VI.[23] In this role, he reported intelligence garnered on the border to military and civilian authorities. In the opinion of Rider Haggard, he was consequently the man who, in virtue of his long service on the Zulu border, 'with the exceptions of the late Sir Theophilus Shepstone and the late Sir Melmoth Osborn, perhaps knew more of that land and people than anyone else of his period'.[24] While on the border during the war, Fynney reported on events in his district, on minor fracas across the border, sent statements collected from Zulus as a means of gathering intelligence, and worked as a translator for the military and civilian authorities. He is reported to have known 'Zulus and the Zulu language well'.[25] Morris records that Fynney 'used his fluent Zulu to dabble in ethnology', and that while he 'had his faults as a border agent ... he had been collecting just such information for years as a hobby'.[26]

While he was certainly one of the best qualified to prepare the pamphlet in advance of the war, Fynney was not involved in combat at any stage, though he may have spoken to those who were, or read such reports as appeared in the press. Writing in November 1878, Fynney pointed out that the 'introduction of firearms' was likely to have wrought 'great changes, both in movements and dress', upon the 'ordinary customs of the Zulu army'.[27] A somewhat different account, which minimised tactical innovation resulting from the adoption of firearms, is to be found in the April 1879 edition of the pamphlet:

---

20 *Précis January 1879*, p. 40. Lord Chelmsford was at the time Lieutenant-General commanding in South Africa.

21 *Précis July 1878*, p. 41.

22 Fred Fynney, *The Zulu Army* (Pietermaritzburg, 1878); Fred Fynney, *The Zulu Army and Zulu Headmen* (Pietermaritzburg, 1879).

23 John Laband, *Historical Dictionary of the Zulu Wars* (Lanham, 2009), p. 102.

24 H. Rider Haggard, *Child of Storm* (New York, 1913), p. ix.

25 BPP, C.2242. Appendix V, enclosure 1, in no. 1, Bulwer to Frere, 24 April 1878, p. 81.

26 Donald R. Morris, *The Washing of the Spears. The Rise and Fall of the Zulu Nation* (London, 1994; 1st edn, 1965), pp. 286, 292.

27 Fynney, *Zulu Army*, p. 6.

> Experience during the present war has shown that ... there has been little or no change [in Zulu military customs], except in the matter of ornaments, the use of which has been almost entirely discarded. Their method of marching, attack formation, &c., remains the same as before the introduction of fire arms among them.[28]

Yet it is hard to reconcile this assertion – one of Guy's key sources[29] – with another statement in the same edition. The Zulu's 'skirmishing' – Fynney maintained – was 'extremely good, and [was] performed even under a heavy fire with the utmost order and regularity'.[30] This line also appears in the July 1879 *Précis*.[31] These internal contradictions are puzzling. There most certainly were first-hand accounts, some of which are quoted below, which make mention of the Zulu's excellence as skirmishers. But these also tended to make the point that the Zulu were using firearms in a way their opponents were not expecting. How, then, can one explain the fact that both Fynney's pamphlets and the series of *Précis* simultaneously recognise and deny changes in Zulu tactics? The key to the mystery – I contend – is to be found in the changing purposes to which depictions of the Zulu were put by the British over the course of the war.

Seeking to justify his aggressive policy towards the Zulu, Sir Bartle Frere infamously wrote in January 1879 that the Zulu king Cetshwayo had endeavoured to revive Zulu 'military power', with the direct intention of 'forming every young man in the land into a celibate man-destroying gladiator'. To this effect, Frere went on, Cetshwayo had organised 'the youth so trained into well disciplined and well armed regiments', obedient to his command. Less well remembered, however, is the fact that Frere also sought to justify his policy by claiming that Cetshwayo had

> obtained fire-arms to an unprecedented extent and in numbers which cause him to be regarded by his own people, as well as by most other native tribes in South Africa as more than a match for any power, native or European, as, in fact, the greatest military power in South Africa.[32]

Frere had previously informed Hicks Beach, the Secretary of State for the Colonies, that, in his view, the Zulu Army was '40,000 to 60,000 strong, well

---

28 Fynney, *Zulu Army and Zulu Headmen*, p. 6.

29 Guy, 'Note', p. 561. Guy appears to have thought that the second, April edition of Fynney's pamphlet was 'issued to all officers at the start of [the] British invasion in 1879'. This is patently impossible.

30 Fynney, *Zulu Army and Zulu Headmen*, p. 5.

31 *Précis July 1878*, p. 42.

32 BPP, C.2252. No. 18, despatch from Frere to Hicks Beach, 24 January 1879, p. 46.

armed, unconquered, insolent; burning to clear out white men'.[33] In the same year Lord Chelmsford made a grim assessment of the defensive situation. 'There are ... but 3 ½ battalions of British infantry and 12 guns to defend a border line of 300 miles, whilst the lowest estimate of the Zulu force is 30,000 drilled fighting men, mostly armed with some kind of firearm.'[34] This is not, of course, necessarily to say that they actually expected the Zulu to put up any serious military obstacle, but that they wished to portray them as a serious threat, and one that needed dealing with.[35] The experience of other colonial wars suggested that they would clear their opponents out in no time.

The British considered their new enemy to be well armed and certainly dangerous, but unlikely to face them in the field. It was soon to become clear to them that this was not at all true. The first major encounter of the war, at Isandlwana, on 22 January 1879, was an imperial catastrophe, resulting in the loss of 52 officers and 739 Colonial and British men, 67 white non-commissioned officers and nearly 500 men of the Natal Native Contingent, an auxiliary force composed of African troops.[36] Paradoxically, the need to justify the disaster suffered at Isandlwana meant that the emphasis on Zulu firepower disappeared from official discourse. Frere's dispatch of 27 January 1879 suggested that the defeat resulted from the British having faced '10 or even 20 times their own force, and [having been] exposed, apparently without any cover or advantage of artificial defence, to the rush of such enormous bodies of active athletes, perfectly reckless of their own losses, and armed with the short stabbing assegai'.[37] This image of the Zulu – of the rush over open ground in the face of heavy fire, an atavistic warrior people in their twilight – has been pervasive and infects the caste of mind behind contemporary work as much as it did the observations of coeval protagonists and policy-makers.

Below, to illustrate the pervasive influence of this set of images and their impact on historical approaches to the Anglo-Zulu war, is a passage from Adrian Greaves' *Forgotten Battles of the Zulu War*. This gives an account of the actions of Lieutenant-Colonel Redvers Buller during the First Battle of Hlobane. Greaves recounts that, on 21 January 1879, as

> the patrol drew near to the foot of Zungwini [a mountain], a number of armed Zulus were spotted at a homestead on the lower slopes and who, when the Boers

33 BPP, C.2222. Enclosure in no. 6, telegram from Frere to Hicks Beach, 5 November 1878, p. 8.

34 BPP, C.2222. Enclosure 2, in no.12, despatch from Lieutenant-General Commanding in South Africa to the Secretary of State for War, p. 18.

35 Adrian Greaves, *Forgotten Battles of the Zulu War* (Barnsley, 2012), pp. xxvi–xxviii.

36 Barring the defence of Rorke's Drift, Isandlwana is the most famous battle of the war. It has been analysed and scrutinised so endlessly and repetitiously that any standard work on the war will provide an account.

37 BPP, C.2252. No. 21, despatch from Frere to Hicks Beach, 27 January 1879, p. 70.

> rode forward to investigate, scattered towards a line of rocks above them then wildly opened fire with their old muskets and rifles. Buller promptly deployed his men to attack and a fire-fight broke out among the boulders, which left one man of the Frontier Light Horse wounded by a thrown spear and at least twelve Zulus dead. The remaining Zulus broke away and fled further up the steep hillside ... Buller ordered his men to ride up the hill to disperse them, hoping them to carry on across the summit of Zungwini to "get a view of Mabamba's *kraal* from above." His men began riding up the difficult and stony hillside but before they reached the crest Zulu numbers had grown and "the hill was too strongly held for us to force it." The massing Zulus had by now taken up their usual "chest and horns" formation and commenced rhythmically beating their spears against their shields. As the Zulu flanks began to descend towards Buller in an attempt to surround his force, Buller halted his men and directed accurate fire against the advancing Zulus. Some warriors were seen to fall under the British fire and the "chest" halted and went to cover among the rocks, a typical Zulu tactic, only for the "horns" to advance rapidly in tolerable order across the open ground. Finding himself totally outflanked by 300 or 400 advancing Zulus on each side, Buller wisely gave the order to withdraw ... The Zulu pursuit continued until Buller rallied and made a determined stand at the White Mfolozi River, finally driving off Zulus.[38]

It is highly instructive to compare this account with the original report on which it appears to be solely based. The report in question was written by Buller himself in the immediate aftermath of the 21 January engagement.

> I have the honour to report that at 4 a.m. yesterday, Mr Piet Uys kindly sent eight of his Dutch Burghers to reconnoitre the top of Zingan Mountain ...
>
> An engagement ensued, during which I reinforced the burghers with 20 dismounted men under Captain Brunker.
>
> Twelve Kafirs were, I know, killed, and I think a few more. One man, F.L.H. [Frontier Light Horse], was wounded with an assegai thrown by a wounded Kafir, and another had a narrow escape. We found four guns and a good many assegais, all of which I had broken, but I did not search the ground thoroughly as I did not think the risk of getting men stabbed by wounded Kafirs worth the result ...
>
> I endeavoured to cross the upper plateau in order to get a view of Mabomba's kraal from above, but the hill was too strongly held for us to force it. With the view of ascertaining the full strength of the enemy, who were coming down to attack us in three columns, I seized a small stony koppie and commenced an

38 Greaves, *Forgotten Battles*, pp. 33–34.

> engagement with the centre column. Our fire soon drove them to cover with a loss of about eight dead (seen, a good many more reported), but meanwhile we were completely outflanked on our right by some 300 Kafirs, who crept round among the stones and kraantzes of the ridge, and our left by some 400 men, boldly moved in tolerable order across the open ground about a mile off.
>
> I accordingly decided to withdraw ... The Kafirs pursued us to the Umfelosi River in force, and about 100 crossed the drifts; but having then secured my retreat I turned on the flats and drove them back. As far as I could see they all returned to the top of Zingan.
>
> We reached camp about 9.30 p.m. ...
>
> I consider that we were engaged with about 1,000 Kafirs, the larger proportion of whom had guns, many very good ones; they appeared under regular command, and in fixed bodies. The most noticeable part of their tactics is that every man after firing a shot or after being fired at drops as if dead, and remains motionless for nearly a minute. In case of a night attack an interval of time should be allowed before a return shot is fired at a flash.[39]

It should be immediately clear from the above that there are several statements in Greaves' account that, while arguably not factually wrong, distort the original sense of the report. Where he reports the earlier engagement, we are simply told the 'Zulus scattered towards a line of rocks above them, opening fire as they did so'. Nowhere, in Buller's report, do we find mention of old muskets, or the melodramatic beating of shields with assegais. Greaves also implies that Buller uses the term 'chest and horns', which he does not. Nor had the Zulus 'fled' in Buller's report. Their movements read more like an attempt to draw the enemy into an ambush and encircle them than a precipitate flight followed by a tactically unimaginative response. It is also striking that no mention is made of Buller's assessment, not only of the quality of the arms in the hands of the Zulus, but also of the manner in which they used them. The notion of the Zulu wildly opening fire is a particularly invidious one, in view of the concluding section of Buller's report. Indeed, the 'typical Zulu tactic' is thought so remarkable by Buller that he not only concludes his report with it, but also recommends what he considered to be an effective counter-measure. How does one account for this gap between the narrative and the source?[40]

---

39 BPP, C. 2260. Enclosure in no.13, report by Buller, 21 January 1879, p. 91. Spelling as in original.

40 In stark contrast to Greaves', most accounts of this skirmish are extremely brief. See, e.g., Huw M. Jones, *The Boiling Cauldron: Utrecht District and the Anglo-Zulu War, 1879* (Bisley, 2006), p. 211, and Morris, *Washing of the Spears*, p. 432. John Laband, *Rope of Sand: The Rise and Fall of the Zulu*

One might argue that the additions are simply inferences from other material, for the Zulu did clash spears on shields and did attack in the 'horns of the bull' formation on other occasions. Whatever the reason for this manipulation – and one must assume it is meant to achieve dramatisation – all it does in reality is to obscure rather than clarify the facts. In Greaves' account, the abaQulusi (the fighters at Hlobane, about whom more will be said below) are recast to conform to a particular conception of proper Zulu battlefield behaviour – a conception rooted in official attempts to account for the Isandlwana disaster.

'I Never Thought Niggers Would Make Such a Stand': Reassessing Zulu Tactics

The essential point being made here is that the closer one examines any account of the Anglo-Zulu war, the less certain the story becomes. In an effort to weld disparate testimonies into some coherent form, scholars often fall back on a meta-narrative of some sort. Of course, historians, whether popular or academic, have not simply relied on a single source. Their arguments are supported by reference to the wider context of the war, the technological deficiencies of the weapons in Zulu hands and the conservatism of the Zulu military system. These views are echoed in many of the sources that scholars have drawn on in making these assessments. The quotation in the title of this paper is taken from one of them. It is from a letter written by an NCO of the 17th Lancers to his brother, which gives his account of the battle of Ulundi, the final engagement of the Anglo-Zulu war. It opens memorably: 'I have great pleasure in informing you I'm not amongst the pegged out, and however much you may be disappointed thereat, I can't help it.' The author then described the Zulu assault on the British square:

> They advanced yelling like madmen, in all about 25,000, the place was black with them, and they kept up a tremendous fire, but fortunately high; still, men and horses were dropping all round, and it was hardly the place for a nervous old gentleman to take a stroll.[41]

This understatement, whilst charming in its own way, is typical of the language of colonial warfare in this period, where the fighting is 'sharp', skirmishes 'brisk', and the enemy 'rather a fine fellow in peace or war, though his mode of warfare is cruel and merciless, like that of all other savages'.[42] Many of the accounts of

---

Kingdom in the Nineteenth Century (Johannesburg, 1995), p. 251, makes the point that the abaQulusi's skill came as a shock to Buller.

41 Anon. NCO, 17th Lancers, in *The North Devon Herald*, 18 September 1879.

42 All quotations are taken from the *Illustrated London News*'s reports on the 'Kaffir War', or Ninth Frontier War (1877–1879), 9 February 1878, 23 March 1878 and 27 April 1878, respectively.

Zulu fire give a similar impression. A Lieutenant of the 20th Hussars remarked that the 'Zulu fire is wild, except in the case of picked marksmen'.[43] Dr Edward Mansell told the press on his return to England that 'one or two of the Zulu leaders made excellent shooting ... Fortunately, however, the mass of the enemy shot far too high.'[44]

However, there are also numerous reports of the Zulu using firearms more effectively than these negative appraisals suggest. Notions of 'effectiveness' are, of course, problematic. They depend on what one sees as the purpose of firearms. To be sure, one cannot escape the tyranny of statistics. The Zulu did not kill large numbers of their enemy with firearms. But references to the Zulu's mode of attack suggest that their tactical integration of firearms reflected a greater familiarity and skill in their use than they have been given credit for. On 22 January 1879, the same day Isandlwana was fought, a Naval Brigade officer at the battle of Nyezane wrote that 'the Zulu had a very strong position, which was chosen with great judgement, as it completely commanded the road'.[45] A Colour-Sergeant of the 99th Regiment elaborated on this:

> I never thought niggers would make such a stand. They came on with an utter disregard of danger. The men that fired did not load the guns. They would fire and run into the bush, and have fresh guns loaded for them, and out again. They fire young cannon balls, slugs, and even gravel. I tell you what it is; our 'school' at Chatham, over one hot whisky, used to laugh about these niggers, but I assure you that fighting with them is terribly earnest work, and not child's play. They were Cetshwayo's picked shots, and came down with the amiable intention of chasing us into the sea. In this, I am happy to say, they failed.[46]

Private Henry Hook, unforgettable to those who have seen the film *Zulu*, describes the Zulu attack on Rorke's Drift as follows. 'During the fight they took advantage of every bit of cover there was ... They neglected nothing.'[47] To this we can add Corporal John Lyons' description: 'The Zulus did not shout, as they generally do; but after extending and forming a half moon, they steadily advanced and kept up a tremendous fire.'[48] At Khambula, on 29 March, Captain Woodgate reported that

---

43 Lieutenant E.R. Courtenay, 20th Hussars, in *The Hereford Times*, 3 May 1879.

44 *The Bristol Mercury and Daily Post*, 29 April 1879.

45 Anon. Officer of the R.N. Brigade, in *The Bristol Observer*, 5 April 1879.

46 Colour-Sergeant J.W. Burnett, 99th Regiment, in *The Dover Express*, 14 March 1879.

47 Private Henry Hook, quoted in Frank Emery, *The Red Soldier: Letters from the Zulu War, 1879* (Johannesburg, 1983; 1st edn, 1977), pp. 127–128.

48 Corporal John Lyons, 2/24th Regiment, in *The Cambrian*, 13 June 1879.

> although individuals showed great bravery, no large formed body could apparently be induced to leave the cover for an assault. Still, creeping up to the crests surrounding the camp on three sides, skilfully availing themselves of existing cover, the Zulus were able to maintain a heavy cross-fire.[49]

At Gingindlovu, on 2 April, Captain Edward Hutton reported that:

> The Zulus continued to advance, still at a run, until they were about 800 yards from us, when they began to open fire. In spite of the excitement of the moment, we could not but admire the perfect manner in which these Zulus skirmished. A small knot of five or six would rise and dart through the long grass, dodging from side to side with heads down, rifles and shield kept low and out of sight. They would then suddenly sink into the long grass, and nothing but puffs of curling smoke would show their whereabouts.[50]

All these instances simply cannot be reconciled with dominant interpretations of the Zulu use of firearms. This is not to deny that the Zulu used traditional tactics, but it is to argue that they frequently demonstrated adaptive skills in their tactical deployment of firearms. Most scholars suggest that it was only when the Zulu had captured a sufficient arsenal of Martini-Henry rifles that they achieved some limited success. However, in the above accounts, it was only at Khambula and Gingindlovu that the Zulu were armed with captured weapons. Most of the other engagements described above predate the acquisition of captured weapons, and although the accounts mentioning them may be somewhat atypical, they do indicate an adaptation of traditional Zulu tactics to incorporate firearms.

This, then, poses the seeming contradiction that the Zulu were in possession of large numbers of firearms, some of a very good standard, but did not deploy them on a larger scale, when the evidence suggests they certainly knew how to do so. It seems clear that the Zulu were making deliberate choices in their use of firearms. On the subject of choice, one further account is suggestive, that of Lieutenant Blaine of the Frontier Light Horse, describing the debacle of the second battle of Hlobane on 28 March. During the advance Blaine recorded that 'the Zulus poured bullets into us from two spots as we went up, and we did the same to them'. During the British withdrawal, however, 'the Zulus were the whole time within a hundred yards behinds us, sometimes even closer. They did not fire much, but were evidently trying to assegai.'[51] This raises the following questions: why were the Zulu trying to assegai when, earlier, they had been firing? Why did the Zulu throughout the war actively seek close combat,

---

49 Captain E. Woodgate, quoted in Emery, *Red Soldier*, p. 174.

50 Captain E. Hutton, quoted in ibid., p. 200.

51 Lieutenant A. Blaine, quoted in ibid., p. 169–170

something which their traditional tactics were designed to achieve, even when this meant facing disproportionate casualties? The answers lie in the crucial cultural importance attributed to concepts of honour and combat, and the role of the assegai in these.

'*Umkontongo we Nkosi!*': The Assegai in Zulu Culture

This section's evidence is drawn largely from the *James Stuart Archive of Recorded Oral Evidence Relating to the History of the Zulu and Neighbouring Peoples*, published volumes of the James Stuart collection held at the Killie Campbell Africana Library in Durban.[52] Having highlighted a tendency to shape evidence to suit a European narrative of the war, this section will, as far as possible, seek to retain passages in their entirety, to allow Stuart's informants to speak for themselves.

Imagery from combat permeated Zulu culture, to the extent that assegais played a part in wedding and doctoring ceremonies.[53] Even out-performing a rival when dancing was described with the phrase '*The one has stabbed the other!*'[54] Perhaps most memorable of all is the phrase given in Bryant's dictionary: '*mus'ukungihlata nga'mkonto munye*', meaning 'don't stab me with a single spear or thrust, do me outright while you are about it', but used colloquially to mean 'don't just half satisfy me with a mere single pinch of snuff'.[55] Imagery of stabbing with the assegai is ubiquitous in the accounts of Stuart's informants. Assegais were never buried with their deceased owners for fear that '*the departed spirits* would or might use them for *stabbing* the living'.[56] If the dead were a threat to the living, the opposite was also true; 'No person who had occasion to pass over the area of ground occupied by the graves [of Zulu kings] might *touch the ground* with (his) stick. Those in charge there would beat him, saying, "What do you mean by stabbing him?" This rule was a great and well-known one.'[57]

Assegais were manufactured by a number of specialised smiths, who enjoyed a position of some status, and were made on the orders of, and delivered to, the king, who would distribute them as he saw fit; 'The king did not give his assegais

52 John Wright and Colin de B. Webb (eds), *The James Stuart Archive* (5 vols, Durban, 1976–2001). For an interesting account of both Stuart and the creation of the published volumes, see John Wright, 'Making the James Stuart Archive', *History in Africa*, 23 (1996), pp. 333–350. Orthography is as presented in the original, where italics represent words or passages given to Stuart in Zulu.

53 Evidence of Ndukwana kaMbengwana, in *JSA*, vol. 4, p. 287; evidence of Ndukwana kaMbengwana, in ibid., p. 320.

54 Evidence of Maxibana kaZeni, in *JSA*, vol. 2, p. 241.

55 Bryant, *Zulu–English Dictionary*, s.v. *um-Konto*.

56 Evidence of Baleni kaSilwana, in, *JSA*, vol. 1, p. 23.

57 Evidence of Ndukwana kaMbengwana, in *JSA*, vol. 4, pp. 290, 296.

to anyone but to selected and brave men who will fight fiercely.'[58] There were many types of assegais, including the *isijula*, with a blade (*ukudhla*) reported to have been between four[59] and nine inches long[60] with a long shank (*umsuka*), the larger *iklwa*, and the *Unhlekwane*, described as being nearly an *iklwa*, with a shank six inches long.[61] There were also *izinhlendhla*, or barbed assegais,[62] and one informant reported that his father fought with 'an *iwaba* assegai, also with *a broad-bladed assegai* (*unhlekwana*)'.[63] Another informant reports the use of poisoned assegais,[64] and brides-to-be would carry an *isinqindi*, 'that is, a knife, double-edged, in reality a kind of assegai with the shaft inserted in a piece of wood to be used as a handle'.[65]

The Zulu had a specific word for those who had stabbed an enemy in battle, *izinxwelera*,[66] but, as a symbol, the assegai transcended its narrow military applications and also epitomised political power and unity, an idea embedded in the Zulu saying '*Umkonto ngo we nkosi!*' or 'The assegai belongs to the king!'[67] This symbolism must derive from its central role, from at least the time of Shaka, as the principal weapon in warfare. But while it certainly had a military purpose, the assegai was not only used in war, and imagery of the assegai in hunting was transposed onto warfare. Speaking of the Zulu who returned from Isandhlwana, Mpatshana kaSodondo recalled:

> There were many *izinxwelera*, perhaps 100 or more ... for their numbers included those who had *stabbed opponents who had already been stabbed by others* (*hlomula'd*); then again those *hlomulaing* became more numerous by reason of the fact that *they had been fighting such formidable opponents, who were like lions,* – for it is the custom among us in lion-hunting that the one who *hlomula's* first, i.e. after the *first to stab,* gets a leg, the second gets *a foreleg,* whilst the last gets *the head.* This custom was observed in regard to Isandlwana because it was recognised that fighting against such a foe and killing some of them was of the same high grade as lion-hunting.[68]

58 Evidence of Mpatshana kaSodondo, in *JSA*, vol. 3, p. 317. See also evidence of Baleni kaSilwana, in *JSA*, vol. 1, p. 41, and of Ndukwana kaMbengwana, in *JSA*, vol. 4, p. 297.

59 Evidence of Nsuze kaMfelafuti, in *JSA*, vol. 5, p. 176.

60 Evidence of Baleni kaSilwana, in *JSA*, vol. 1, pp. 34–35.

61 Evidence of Nsuze kaMfelafuti, in *JSA*, vol. 5, p. 176. On this page is a reproduction of a sketch by Stuart depicting different types of assegai.

62 Evidence of Ngidi kaMcikaziswa, in *JSA*, vol. 5, p. 66.

63 Evidence of Mkehlengana kaZulu, in *JSA*, vol. 3, p. 212.

64 Evidence of Mtshapi kaNoradu, in *JSA*, vol. 4, p. 82.

65 Evidence of Ndukwana kaMbengwana, in *JSA*, vol. 4, p. 267.

66 Bryant, *Zulu–English Dictionary*, s.v. *i-Nxeleha*.

67 Evidence of Ndukwana kaMbengwana, Wright and de B. Webb (eds), *JSA* vol. 4, p. 381.

68 Mpatshana kaSodondo, Wright and de B. Webb (eds), *JSA* vol. 3, pp. 303–304.

Napoléon, Prince Imperial and son of Napoleon III, attached to the British as an observer, was killed on 1 June 1879 in an ambush during a reconnaissance with the 17th Lancers. Accounts of his death suggest that, after firing an initial volley or two, the Zulu managed to catch the prince, who had trouble mounting his horse. Archibald Forbes' telegram for the *Daily News*, sent the day after Napoléon's death, records his injuries as 'No bullet wound; 18 assegai stabs, two piercing body from chest to back, two inside, one destroying right eye.'[69] Bertram Mitford's 1881 account of a journey through Zululand recounts that, when he met the headman charged with the upkeep of the monument to the prince's death, 'the old man was evidently sensible of the *prestige* attaching to himself and his neighbourhood by the possession of such a "lion"'.[70] Evidently, the prince was thought to be prey of sufficient status to warrant his opponents *hlomulaing*. But there is a crucial difference. The stabbing of prey in the hunt did not carry the same risks as the stabbing of a man.

For the Zulu, warfare was hedged about with prohibitions. Ritual preparations before battle were intended to place the Zulu in a liminal state, outside the realms of normal behaviour. The *amaButho* were treated with various medicines. They underwent a vomiting ceremony, and, before battle, could only sleep on the ground or their shields.[71] Following battle, a further series of cleansing rituals took place, which had to be completed before the warrior could return home and re-enter society.[72] If these prohibitions and rituals can be found elsewhere, then knowledge of these can inform us about Zulu choices and actions in war, too. Several of these same practices and prohibitions did apply to actions outside warfare, and these revolved around one thing in particular – killing – and they encompassed the correct way of killing, differences between types of killing, and the consequences of the act.

Most of the major references to killing in the sources revolve around executions, and these were carried out in a grim variety of ways. One Zulu informant gave three: neck-breaking, bludgeoning and throttling.[73] But firearms were also used. In 1876, when the *iNgcugce* female age-set refused to marry the men they had been allocated, Cetshwayo had them pursued and put to death. A participant explained:

> *We finished them off with our guns. We did not use our assegais, for the king had said, "Do not stab them with your assegais, for they will cry out as the assegai penetrates["]*

---

[69] *Daily News*, 20 June 1879.

[70] Bertram Mitford, *Through the Zulu Country: Its Battlefields and Its People* (London, 1883), p. 118. Italics in original.

[71] Evidence of Dunjwa kaMabedhla, in *JSA*, vol. 1, p. 124.

[72] Evidence of Mpatshana kaSodondo, in *JSA*, vol. 3, pp. 302–308.

[73] Evidence of Lugunza kaMpukane, in *JSA*, vol. 1, p. 333.

> *... a person who was stabbed with an assegai would cry out loudly, calling down evil on people.*[74]

The point here is that this evil, known as *umnyama,* and its manifestation, *iqungo* or madness, only applied to killing with an assegai. 'A man is said *to have an inqungo* when he has *stabbed* and has not later on been *doctored* (*setshenxwa'd*) and *treated* (*elatshwa'd*).'[75] This did not apply to any of the other methods of execution listed, including firearms. This suggests that firearms were not understood as falling outside the normal prohibitions surrounding killing because they were an alien technology; other modes of execution did not carry a risk of *iqungo* either. It was only the assegai, not simply a weapon but also a central symbol in Zulu culture, which could bring down *umnyama* on its user. It was precisely this risk which was addressed by the doctoring of an army. As one informant of Stuart's put it: '*iqungo affects those who kill with an assegai,* but not *those who kill with a gun*, for with a gun it is just as if the man had shot a buck, and no ill result follows. This *iqungo* is what is got rid by "*wiping the hoe*".'[76] This rather grisly ritual was explained by the same informant as follows:

> *The wiping of the hoe (ukwesul' isikuba*): done by *those who have killed,* i.e. they will rape a woman; 2 or 3 may rape the same one. They may do this to a woman of any tribe a long way from their own, even though not of that against which they are fighting. This woman may give birth to a child, and such child, it is said, has as *umkangi*, i.e. a mark of a different colour from the rest of the body, and this mark may be on the child's back or in front. This custom is observed in regard to either married women or girls, just what comes, and if they cannot find a woman or girl, they will get a young *umsenge* tree and "*wipe off*" (*sulela*) in that. It is wrong to have connection with one's own people until the "*hoe*" (*isikuba*) has been "*wiped*" (*sulwa'd*).[77]

That this was a necessary prohibition speaks volumes of the power ascribed to the act of stabbing a foe.

It is clear that the manner in which warfare and the rituals that surrounded it were conducted was considered to be important. War was not just an exercise in accomplishing tactical or strategic goals. Central to it was the kill. As the Zulu put it, they 'ate up' the enemy. When stabbing an enemy, the Zulu exaltation

---

74 Evidence of Mtshayankomo kaMagolwana, in *JSA*, vol. 4, p. 133.

75 Nsuze kaMfelafuti, in *JSA*, vol. 5, p. 171.

76 Singcofela kaMtshungu, in *JSA*, vol. 5, p. 344.

77 Ibid. See also Mpatshana kaSodondo, in *JSA*, vol. 3, p. 326, and A.T. Bryant, *The Zulu People as They Were before the White Man Came* (Pietermaritzburg, 1967; 1st edn, 1949), p. 508.

'*Ngadhla!*' was sometimes used, literally meaning 'I have eaten!'[78] This notion, of the total destruction of the enemy, did not only apply to combatants. One praise exemplifies this in a particularly grisly fashion: '*Ngi gwaz' uhlanga; ngi gwaz' isikwebu*', or '*I stabbed the reed; I stabbed the cob*'. This was, Nduna kaManqina explained, the phrase used by the Zulus when the child on a mother's back was stabbed through as well as the mother.[79]

The reforms of the Zulu military attributed to Shaka – the carrying of only one assegai, to force warriors to stab rather than throwing them, and an insistence on ruthlessness – were by the time of the Anglo-Zulu war central to Zulu ideas of the Zulu way of war.[80] To have acted in accordance with these tenets was potentially a source of great prestige.

> The *discussing (xoxaing)* of a *campaign* was done ... for the express purpose of discovering how the campaign had gone – to ascertain who had distinguished themselves (to be thereupon greatly rewarded in cattle by the king) or to learn who had run away or shown the white feather. It thus became essential to *discuss the operations* after a campaign; indeed, it was an inseparable part of military affairs.[81]

Further, during the *xoxaing*,

> the King will have made careful enquiries as to which regiment is entitled to the honour of *being recognised as the one which had stabbed first.* And when this *acknowledging has been done* (for there may have been a lot of *disputing* about this), the *izinxwelera* will go away, cut up their willows and put them on, when they are known as *iziqu.* One always wears these things and keeps the *iziqu* belonging to the campaign in connection with which they were got. If one cuts fresh ones, one asks, 'Have you gone once more and killed others is a fresh battle?' Hence one always keeps the old *iziqu,* though it is permissible to restring them.[82]

The honours one might win in battle were a matter of personal and public pride. Indeed, in Zulu praise-poems the qualities generally praised were courage and

---

78 Evidence of Mtshayankomo kaMagolwana, in *JSA*, vol. 4, p. 144, fn. 135. It appears as '*Ngadla!*' in the footnote.

79 Evidence of Nduna kaManqina, in *JSA*, vol. 5, pp. 3–4.

80 Whether Shaka was truly responsible for this is open to debate; see, for instance, Ngidi kaMcikaziswa, in *JSA*, vol. 5, p. 66.

81 Evidence of Mjobo kaDumela, in *JSA*, vol. 3, p. 141.

82 Evidence of Mpatshana kaSodondo, in *JSA*, vol. 3, p. 305.

ferocity.[83] One of Stuart's informants directly equates courage with close combat: 'Bravery – the stabbing of others – was highly rewarded by Tshaka [*sic*].'[84]

At this stage, one further point must be made. The Zulu who fought in the Anglo-Zulu war were not an undifferentiated mass. As was pointed out in the earlier criticism of Greaves' account of the First Battle of Hlobane, Buller's opponents were abaQulusi. The tactical flexibility brought out so vividly in Buller's report also characterised the engagements of Wood's column with these men. Laband probably exaggerates in calling the abaQulusi 'refugees and renegades cobbled together on the furthest margins of the Zulu kingdom'.[85] Having driven out the inhabitants of the region around Hlobane, Shaka had set up a military homestead, or *iKhanda* (pl. *amaKhanda*), as a means of establishing hegemony over the region. There is some debate over the origins and site of the ebaQulusini *iKhanda*, to which the abaQulusi *iButho* was attached, but it dominated the region.[86] By 1879, and afterwards, the abaQulusi were fiercely loyal to the Zulu royal house and considered themselves 'the Zulu kings' particular followers, and stood by them through all the tribulations that befell the royal house'.[87] The abaQulusi, however, were not enlisted in conventional *amaButho*, but mobilised as, and fought in, their own *iButho*, composed of men from the region alone. They, moreover, were led by Mbilini waMswati, who had 'gained his military apprenticeship in Swaziland', where ambush, hit-and-run tactics and guerrilla warfare were the norm.[88] By way of contrast, barring two small contingents of older *amaButho*, the uDlambedlu and izinGulube, both of which fought at Nyezane, none of the other Zulu regiments who fought during the war had ever engaged the enemy. The most prominent regiments were all formed of men between their mid-twenties and mid-thirties.[89] These would be precisely the men who stood to gain the most from conspicuous bravery in battle. With no battle experience, they would also be the ones most likely to need the mutual reassurance and greater cohesion conferred by close-order massed units. To become *izinxwelera* was to be accepted by the group, and those who were thought to be cowards would be publicly shamed.

> Those who *have fought bravely,* when *in the assembly* and the king has *slaughtered* [cattle] *for* the regiments, will sit in one place and the *cowards* in another, i.e. apart

83 John Illife, *Honour in African History* (Cambridge, 2005), p. 146.

84 Evidence of Madikane kaMlomowetole, in *JSA*, vol. 2, p. 61.

85 John Laband, 'The War-Readiness and Military Effectiveness of the Zulu Forces in the 1879 Anglo-Zulu War', *Natalia,* 39 (2009), p. 45.

86 Jones, *Boiling Cauldron,* pp. 63–65.

87 Laband, *Rope of Sand*, p. 28. See also evidence of Ndukwana kaMbengwana, in, *JSA*, vol. 4, pp. 277–278.

88 Laband, 'War-Readiness', p. 45. See also Jones, *Boiling Cauldron,* p. 6.

89 Laband, 'War-Readiness', p. 41.

> from the heroes. A hero may come up with a dish of cold water in which, maybe, the man has been washing his hands, and dash it all over their meat, saying "*Tell about the campaign.* Let me hear of your exploits in the war. I beat the whole lot of you in what I did by myself." The cowards dare not attack the hero for fear of being killed by the king, so have to grin and bear it. Or a *hero* may give a coward a lump of meat after dipping it in cold water, for the coward to eat, inviting him at the same time to *give his account of the fighting.*[90]

The centrality of the assegai to Zulu warfare and dominant notions of honour placed an imperative on many of the Zulu who fought in the Anglo-Zulu war to close with the enemy, and it is this cultural imperative that accounts for what has been portrayed as an apparently senseless insistence on meeting the enemy at close quarters. For young warriors there was much to be lost or gained through their actions in the field. In contrast to this, the abaQulusi, on the periphery of the kingdom and outside the Zulu Army *sensu stricto*, were not culturally constrained to the same extent, and their mode of warfare did not depend in the same manner on conspicuous individual and corporate bravery. Their apparent lack of 'discipline', perhaps better understood as comparatively lower cohesion as units in the field, led them to suffer the heaviest losses among the *amaButho* in the rout that followed the battle of Khambula.[91] But even among the Zulu *amaButho* themselves, where there might have been a social or cultural imperative to engage the enemy at close quarters, tactics were not mindless or rigid. Yet the image of the Zulu, bravely facing a storm of fire in an attempt to get to grips with the redcoats, has come to dominate our understanding of the war.

Conclusion

This chapter's central contention is that, in some instances at least, the Zulu who took part in the Anglo-Zulu war demonstrated greater ability with, and understanding of, firearms than has generally been assumed. This techno-military competence – as I have argued – has been obscured by a tendency to see the war, and Zulu actions, through the lens of European observers. From this perspective, the argument that the Zulu failed to adapt their tactics, or to make a concerted effort to do so, is a logical one. Laband has termed this 'the intractable problem of the paralysis in Zulu tactical thinking'.[92] But this misses the point. The Zulu were neither blind nor stupid. They could see what was happening on the battlefield and did adapt to changing circumstances, as we have seen. But

90 Evidence of Mpatshana kaSodondo, in *JSA*, vol. 3, p. 306.

91 Laband, *Rope of Sand*, p. 277.

92 Laband, 'War-Readiness', p. 42.

there were limits to Zulu tactical flexibility. Their deployment of firearms was constrained by their cultural approach to warfare – one that foregrounded the roles of close combat and the assegai.

It is this, rather than any failure to adapt tactics, or lack of concerted effort, which explains the apparent contradiction in the Zulu's widespread possession *and* limited use of firearms. They practised with firearms, acknowledging thereby their direct military application, and the need for skill in their use.[93] In contrast, training in the use of an assegai was not a formal process.[94] But firearms never replaced the assegai in terms of the Zulu understanding of their own identity, and so remained subordinate on the battlefield for the same reason. The assegai was a key symbol of Zulu heroic honour and masculinity. It had a long history. Its demise would prove painful – as attested by the following, moving epitaph by Pixley Seme, a founding member of the South African Native National Congress (later the ANC). As Seme told James Stuart in 1922, Cetshwayo's son, Dinuzulu,

> *was a great hunter. He was always organising hunting parties. On the floors of his houses were the skins of all the animals which he had shot, for he did not miss. When he hunted at the Mbekamuzi river, or in the low country (ezansi), where there were predators, he carried a gun, but he did not leave his assegai behind (a large iklwa, as wide as a person's hand), for the heart of a Zulu (inhliziyoka Zulu) places great trust in his assegai. Indeed it is in his assegai that the strength of a Zulu lies. Today the Zulu no longer has his strength, for the assegai has been knocked to the ground by the gun, and it will never rise up again.*[95]

The Anglo-Zulu war was a nasty, short-lived conflict fought for little purpose and at great cost. For many, it remains the archetype of colonial warfare – black bodies and red coats, assegai and bayonet. The war has been a source of endless fascination for scholars, enthusiasts and the general public, particularly in Britain. Cy Endfield's film, *Zulu*, is the most repeated film ever broadcast on television.[96] This highly romanticised construction of the Anglo-Zulu war is the result of two coterminous processes. For the Zulu, the assegai was much more than a weapon. It was the ultimate symbol of honour. For the British, the image of the Zulu warrior, assegai in hand, was a powerful tool with which to burnish their tarnished honour.

93 Evidence of Mpatshana kaSodondo, in *JSA*, vol. 3, p. 328.

94 Ibid., p. 326.

95 Evidence of Seme, Pixley, in *JSA*, vol. 5, p. 273.

96 Greaves, *Forgotten Battles*, p. xxx.

Chapter 7

Steel and Blood: For a Cultural History of Edged Weapons between the Late Nineteenth and the Early Twentieth Centuries

Gianluca Pastori

In late nineteenth- and early twentieth-century military discourse, edged weapons played a role far exceeding their operative importance. To be sure, these weapons' golden age was drawing to a close. Better design, more accurate workmanship and fast-paced technical innovations, not least in the field of ammunitions, were making individual firearms – especially rifles – increasingly effective, cheap and reliable, thereby fostering a parallel evolution in strategy and tactics.[1] However, while, on the one hand, firearms were gradually becoming the 'queens of the battlefield', on the other, there also emerged a more emphatic rhetoric centred on the role of swords, knives and daggers. At the same time, the tactical function attributed to the bayonet was being reconceptualised through a re-reading of the 'lessons' of the revolutionary and Napoleonic wars.[2] In this long-term process, if more effective, long-range fire worked towards depersonalising (and, in a sense, democratising) the act of killing by reducing physical proximity between opponents, military minds seemed, conversely, increasingly concerned with maiming, cutting, piercing and slashing, and with the most direct, emotional and blood-shedding dimensions of warfare. This kind of fascination (which, in different forms, affected both professional soldiers

---

1 Steven T. Ross, *From Flintlock to Rifle: Infantry Tactics, 1740–1866* (London, 1996). A rather 'heretic' history of the early stages of this process is to be found in Giovanni Cerino Badone, *Il bianco dei loro occhi. Storia della potenza di fuoco nelle guerre europee, 1500–1800* (Milan, in press). I am indebted to G.C.B. for allowing me to access different drafts of his work.

2 On French revolutionary tactics, see John A. Lynn, *The Bayonets of the Republic: Motivation and Tactics in the Army of Revolutionary France, 1791–94* (Urbana, 1984). On the Napoleonic wars, see David G. Chandler's classic *The Campaigns of Napoleon* (New York, 1966). On the long-term impact of the French revolution in the military sector, see MacGregor Knox, 'Mass Politics and Nationalism as Military Revolution: The French Revolution and after', in M. Knox and W. Murray (eds), *The Dynamics of Military Revolution, 1300–2050* (Cambridge, 2001), pp. 57–73.

and ordinary people) took two related forms. On the 'negative' side, 'chopping enemies into pieces' became the defining symbol of the worst human brutality, a trademark of the wildest and most irreconcilable of enemies. On the 'positive' side, a steady wall of men and bayonets, such as those depicted in Lady Butler's *The 28th Regiment at Quatre Bras* (1875) or in Robert Gibb's *The Thin Red Line* (1881), came to embody a full corpus of martial virtues: valour, honour, camaraderie and steadfastness. In this cultural environment, until the first battle of the Marne inaugurated the new era of trench warfare, an impetuous charge or a bayonet stand were construed not only as a key test of skill and bravery, but also as a term of reference (maybe *the* term of reference) of human value.

Edged Weapons in Colonial Discourse

At the turn of the nineteenth century, a flourishing of stories, legends and anecdotes about martial bravery, coupled with the emergence of the so-called 'philosophy of the élan vital' in French military circles,[3] helped bring about an almost axiomatic identification between bayonet and military valour. Since the beginning of the century, General Alexander Suvorov's well-known remarks about the irreconcilable divide between 'foolish bullets' and 'wise bayonets' have played an important role in shaping this mindset.[4] At the same time, the gradual emergence of 'Bergsonism' as the main term of reference for French military thinking illustrated a more widespread 'yearning for moral and spiritual regeneration and for the rediscovery of the elementary forces of life' – a yearning that was to gain momentum all over Europe with the coming of the new century.[5]

Nonetheless, until World War I, it was the imperial frontier that provided the main setting for the elaboration of the 'edged weapon discourse' to which this chapter is devoted. This, of course, was partly due to the fact that colonial conquest and imperial policing provided for the bulk of contemporary military experience, at least in quantitative terms.[6] Equally important, however, was the

---

3 Azar Gat, *A History of Military Thought: From the Enlightenment to the Cold War* (Oxford, 2001), pp. 402ff.

4 A late nineteenth-century résumé of Suvorov's thinking is to be found in his biography by Henry S. Spalding, *Suvóroff* (London, 1890). Very similar considerations can be found in William Lyon Blease, *Suvorof* (London, 1920). More recent works, partially redressing traditional stereotypes, are Philip Longworth, *The Art of Victory: The Life and Achievements of Generalissimo Suvorov, 1729–1800* (London, 1965) and the briefer Bruce W. Menning, 'Train Hard, Fight Easy: The Legacy of A.V. Suvorov and His "Art of Victory"', *Air University Review* (November–December 1986), pp. 79–88. Suvorov's *Nauka Pobezhdat was originally published in Russian in 1806.*

5 Gat, *History*, p. 430.

6 For a brief account of Britain's main colonial campaigns, see Jan Henron, *Britain's Forgotten Wars: Colonial Campaigns of the 19th Century* (Stroud, 2002). On France, see Vincent Joly, *Guerres*

fact that the colonies were, in essence, a 'space of opposition', sharply distinguished from the inner space of the Motherland. Stretching beyond the frontiers of 'peaceful' and 'civilised' Europe, overseas territories were remote enough, even in a world that was becoming increasingly small, to permit the existence, among other freakiness and exoticisms, of the most savage and inhuman of enemies. At the same time, along their ill-defined and permeable fringes, empires and colonies supported a constant dialogue between the 'inside' and the 'outside', structuring such mutual perceptions and visions of the 'Other' as a whole range of social and cultural artefacts conveyed to their consumers at home.[7] It was through these channels that the same perceptions and visions became key elements to depict the 'Other' and build the contrast between 'Us' and 'Them'. In these artefacts, 'exoticism' always bordered (and often intersected) the realms of the unruly (or of the un-ruled), with edged and other weapons serving as a marker of difference – one that was made all the more poignant by the tight limitations that regulated their possession, display and use within the peaceful (and pacified) imperial space.

Swords, knives, *tulwars* and daggers, with their unusual shapes and their equally unusual names, soon came to epitomise the 'Other', especially in such emotionally vivid contexts as India's North-West Frontier. Matchlocks (Kipling's 'ten-rupee *jezail*') and short swords were recurrent themes in descriptions of the 'blood-thirsty and treacherous' Pathan, possibly the most stereotypical (and stereotyped) of British enemies. In British imperial discourse, the Pathan was almost invariably presented as a man who 'always carr[ied] arms ... whether grazing cattle, tilling the soil, or driving beasts of burden'.[8] In countless books, pictures and engravings dating back to the first half of the nineteenth century, the Pathan was depicted as 'equally blood-thirsty and treacherous' as the neighbouring Afghan, 'and still more ignorant and barbarous ... Muhammadans of the worst type; intolerant and priest-ridden'.[9] The sheer number of his everyday weapons conjured up a picture of pure and almost primitive violence. The Pathan's weapons were a synecdoche for his nature and aspect, becoming the symbol of a character 'all times liable to be inflamed into conflict by the

d'Afrique: 130 ans de guerres coloniales. L'expérience française (Rennes, 2009). On Italy, see the (heavily biased) Vittorio Giglio and Angelo Ravenni, *Le guerre coloniali d'Italia* (Milan, 1942).

7 On some aspects of this process, see John M. MacKenzie (ed.), *Imperialism and Popular Culture* (Manchester, 1986), and Steve Attridge, *Nationalism, Imperialism and Identity in Late Victorian Culture: Civil and Military Worlds* (Basingstoke and New York, 2003). On the military world, see also John M. MacKenzie (ed.), *Popular Imperialism and the Military: 1850–1950* (Manchester, 1992). For military iconography, see Joan W.M. Hichberger, *Images of the Army: The Military in British Art, 1815–1914* (Manchester, 1988).

8 James Talboys Wheeler, *India and the Frontier States of Afghanistan, Nipal and Burma* (New York, 1899), vol. 2, p. 758.

9 Ibid.

exhortations of fanatical priests' and who acknowledged 'no law, save only the modern injunction that they must not raid in Afghanistan or in the settled British districts'. According to this narrative, the Pathan's main occupation was 'fight[ing] with the tribesmen in Afghan territory, and sally[ing] forth in small bands to plunder the rich villages of the Indian plains … wander[ing] in gangs far and wide in India, pilfering everywhere, and sometimes levying blackmail in lonely villages far away in Madras or Bengal'. British discourse thus depicted an almost animal attitude towards violence: '[t]heir ancestors were wont to harry the countryside, and they obey to this day the overpowering instinct which occasionally impels them to do likewise'; even '[a]mong themselves, they engage in protracted blood-feuds, and quarrel about their women', since 'fighting is the joy of their lives'.[10]

This feral attitude naturally led to an equally feral way of fighting. The Pathan (but the same argument can be applied to almost all of the other 'Others' of colonial discourse, both within and outside India's North-West Frontier) was inherently treacherous and could never be really or fully trusted. Lacking even the basic traits of humanity, the colonial 'Other' defied the rules of war, allowing civilised armies to forsake even the most fundamental principles of the then emerging international law in their intercourses with 'savage tribes'. International laws, British military men maintained,

> apply only to warfare between civilized nations, where both parties understand them and are prepared to carry them out. They do not apply in wars with uncivilized States and tribes, where their place is taken by the discretion of the commander and such rules of justice and humanity as recommend themselves in the particular circumstances of the case.[11]

The Anatomy of Killing

When fighting 'savage tribes', the risk was always of being 'slaughtered', 'butchered', 'cut' or 'chopped into pieces', often by women (*vide* the incumbent destiny of Kipling's *Young British Soldier*) or at their instigation. Women – wrote the writer of *Among the Wild Tribes of the Afghan Frontier* – were 'not exempted' from 'vendetta, or blood-feud, [which] has eaten into the very core of Afghan life'.[12] In some critical situations, women could even act as the men's rallying

10 Lovat Fraser, *India under Curzon and After* (London, 1911), pp. 40–41.

11 *British Manual of Military Law* (1914), quoted in Elbridge Colby, 'How to Fight Savage Tribes', *American Journal of International Law*, 21/2 (1927), p. 280.

12 Theodore L. Pennell, *Among the Wild Tribes of the Afghan Frontier: A Record of Sixteen Years' Close Intercourse with the Natives of the Indian Marches* (London, 1922), pp. 17–18

point, as attested by the (legendary?) Malalai of Miawand. Howard Hensman's narrative of the second Afghan war devotes close attention to highlighting the role of the mother of the former king, Yaqub Khan, portraying her as 'well advanced in years, but still capable, through agents, of doing much mischief' and to become 'chiefly instrumental in raising the *jehad*', despite being a woman in a man-dominated context.[13]

In the undifferentiated wilderness that stretched beyond the imperial borders, there was no real difference between town and countryside. Even in an urban context, sticks, knives and daggers were the weapons of the mob; the same mob that – with the more or less direct support of local authorities – 'treacherously' butchered and cut into pieces two separate British representatives in Kabul: Alexander Burnes, in 1841, and Louis Cavagnari, in 1879.[14]

Mutilation of both corpses and living men (a practice connected with edged weapons both materially and semantically) played a pivotal role in descriptions of the attitudes of frontier peoples towards their enemies. It did not belong to a well-defined – if rejected – ritual system, such as occurred among the people of India's north-eastern frontiers. In the Afghan context, mutilation was construed as the expression of inherent brutality and savagery. 'The violent Afghan 'Other' ... was not restrained by any logic ... Rather, violence was seemingly an expression of Afghan backwardness and lack of restraints.'[15] Significantly, these traits did not emerge in open, hand-to-hand fighting (that the Afghans allegedly normally refused, due to 'their want of organization, their tribal jealousies, and their impatience of regular habits, and of the restraint necessary to render them good soldiers'[16]), but rather when the victims were at their weakest and most vulnerable. In this sense, narratives of the British retreat from Kabul in the winter of 1841–1842, with their wide commercial success, lay the bases for a discourse normally made up of 'bodies ... stripped ... children cut in two ... women as well as men chopped to pieces, many with their throats cut from ear to ear'.[17] Quite significantly, as time went by, descriptions grew more and more truculent, while actual events faded into the remoteness of a national nightmare.

---

13 Howard Hensman, *The Afghan War of 1879–80* (London, 1882), pp. 178, 262.

14 On Burnes' life, see John W. Kaye, *Lives of Indian Officers, Illustrative of the History of the Civil and Military Service of India* (London, 1867), vol. 2, pp. 1–66. A fictional description of his death is to be found in George A. Henty's novel *To Heart and Cabul: A Story of the First Afghan War* (London, 1902), pp. 227ff. For a rather rhetorical portrait of the figure of Louis Napoleon Cavagnari, see Kally Prosono Dey, *The Life and Career of Major Sir Louis Cavagnari, C.S.I., K.C.B, British envoy at Cabul, together with a Brief Account of the Second Afghan War* (Calcutta, 1881).

15 Keith Stansky, '"So These Folks are Aggressive": An Orientalist Reading of "Afghan Warlord"', *Security Dialogue*, 40/1 (2009), p. 88.

16 George A. Henty, *For Name and Fame: Or, through Afghan Passes* (New York, 1896 [?]), pp. iii–iv.

17 Archibald Forbes, *The Afghan Wars of 1839–42 and 1878–80* (London, 1892), p. 107.

As cultural artefacts, the same tropes are manifest in many other colonial representations of the savagery of the 'Other'. They punctuated, for example, narratives of the Italian defeat at Adowa (1896) and its bloody aftermath, especially in such popular exposés as Eduardo Ximenes' richly illustrated diary *Sul campo di Adua* (1897).[18] In Italian narratives, the order of maiming men and corpses was often said to have stemmed directly from Empress Taytu (Betul), whom these representations portrayed as a Valkyrie-like figure, somehow mirroring her Afghan counterparts, raiding battlefields like screaming, mythological Furies. According to the most colourful and imaginative authors, the empress herself personally mutilated wounded or dead soldiers and local auxiliaries on the battlefield. Similar stereotypes (sometimes with explicit reference to sexual and/or religious revenge) are to be found in contemporary narratives of the first phase of the Libyan campaign (1911), especially after the outburst of violence that followed the Arab 'treason' of Sciara Sciat (23–24 October 1911).[19]

Opposed to this view of the edged weapon as inherently barbarous, and intimately linked to the darkest side of human (or unhuman) nature, stood the bayonet, the only edged weapon fully permitted in modern Western armies, besides officers' sabres and such ceremonial paraphernalia as pioneers' axes. Until World War I, when in Allied propaganda it began to be discursively associated with German atrocities in Belgium, the bayonet evoked a wide set of positive, 'white' and 'civilised' values. In the colonial context, it represented a sort of 'steel version' of the volley fire. Its use was predicated on professionalism, self-control and conscious limitation of violence even in the most difficult situations. In last stands, as well as in charges, it was the bayonet that – thanks to the sheer determination that its deployment involved – routed hordes of savage, but morally weaker, opponents. In this sense, the bayonet was an epitome of Western ('European') military value and cold blood. Moreover, in the most desperate cases, it implied acceptance of the supreme sacrifice, thereby highlighting the superior nature of the white soldier/white man displaying its virtue in close combat.

Once again, the colonies were the most convenient place in which to locate this kind of narrative. In the colonial context, the set of negative values attached to the enemy cast an even brighter light on 'Our' heroism. A bayonet charge

---

18 *Sul campo di Adua: Diario di Eduardo Ximens. Marzo–Giugno 1896* (Milan, 1897).

19 On the events in Sciara Sciat and their place in the broader framework of the Libyan campaign, see Francesco Malgeri, *La guerra libica (1911–1912)* (Rome, 1970), pp. 153ff. See also Angelo Del Boca, *Gli italiani in Libia. Vol. 1: Tripoli bel suol d'amore, 1860–1920* (Milan, 1993), pp. 96ff. For a far more emphatic description of the Italian 'genocide' in Libya, see Lino Del Fra, *Sciara Sciat: Genocidio nell'oasi. L'esercito italiano a Tripoli* (Rome, 2011). An almost coeval account of the 'atrocities' perpetrated against Italian soldier by 'Arab tribesmen' is Enzo D'Armesano, *In Libia: Storia della conquista* (Buenos Aires, 1913).

was frequently depicted as an inebriating feast, with its most violent and bodily aspects removed or relegated to the background. It is significant that while 'high' narratives (both official and individual) obliterated or provided adequately sanitised versions of such aspects, piercing, cutting, stomping and the like recurred frequently in rank-and-files' memoirs.[20] Consider, for instance, the words of an officer of the 11th Bersaglieri at the Battle of Henni (26 November 1911):

> At 3.30 I give an order that rouses the Bersaglieri's spirits: "Fix bayonets! Prepare to charge!" Almost immediately a loud resounding shout sweeps over the ground and is heard a mile away, firing those who hear it as with a great blaze of enthusiasm. On from our cover to the enemy's trench we charge, yelling the war-cry "Savoia!" The oasis becomes a mass of Bersaglieri, rushing on in glorious confusion.[21]

A similar tone pervades his description of the men after their successful rush:

> Panting, perspiring, red in the face, my men are beaming with content and laughing, their eyes all aglow with enthusiasm, as they rest in the conquered trench, regardless of the fire of the enemy, whose bullets begin to scream angrily, skimming the parapet.[22]

In this kind of narrative, rush, excitement, valour and glory emerged as the key elements of a 'grammar of the bayonet', somehow opposed to the 'grammar of savagery' built around other edged weapons and their users. This 'grammar of the bayonet', displayed both in popular and military culture and in fictional and non-fictional works, supported a stereotyped and conservative discourse. Paradoxically, the same ways of killing could be conceptualised and presented in opposite terms: either as barbarous practices, relics of a past doomed to be made obsolete by the spread of civilisation, or, conversely, as manifestations of conventional heroism, rooted in valour and discipline. In an age of mechanised

---

20 On rank-and-files' experience of the Libyan war, see Salvatore Bono (ed.), *Morire per questi deserti: Lettere di soldati italiani dal fronte libico, 1911–1912* (Catanzaro, 1992). See also, by the same author, *Tripoli bel suol d'amore: Testimonianze sulla guerra italo-libica* (Rome, 2005). For a useful comparison with a coeval source, see Baccio Bacci, *La guerra libica descritta nelle lettere dei combattenti* (Florence, 1912). An officer's point of view is to be found in Massimiliano Cricco, 'La battaglia di Zanzur dell'8 giugno 1912 nell'inedita testimonianza del tenente Domenico Orsini', in *Atti del X convegno della Società per gli Studi sul Medio Oriente: 'Memorie con-divise. Popoli, stati e nazioni nel Mediterraneo e in Medio Oriente', Milano, Università Bicocca, 9–11 giugno 2011* (Naples, forthcoming).

21 Tullio Irace, *With the Italians in Tripoli: The Authentic History of the Turco-Italian War* (London, 1912), p. 225.

22 Ibid.

warfare, greater numbers and creeping 'commoditisation' of military manpower, an emphasis on the cold steel of bayonets and the cold blood of the men who deployed them served to 're-personalise' killing and, by so doing, support a traditional image of military activity. Going against the grain of historical developments, the 'grammar of the bayonet' made it possible once more to imagine the war as a duel, with its honourable codes and norms, and with its inherent character of 'God's judgement'.

This is one of the reasons why accounts of bayonet charges are often stories of success. In the aforementioned narrative of the Henni's charge, enemies 'discharge a volley or two in frantic haste and then take to their heels', dismayed by 'the mere uproar of our charge'.[23] In a clash of value, the villain is naturally doomed to fail. This was especially true when the bayonet charge represented the last resort to break a stalemate, thus providing the final demonstration of where 'true valour' stood. Another account of an episode of the Libyan campaign provides a self-explanatory example of this kind of pedagogy. During the Battle of Derna (27 February 1912),

> [i]n order to break the obstinate Turkish resistance it was decided to make a counter-attack, which proved the most successful bayonet charge in all the campaign ... With a yell like the cry of some savage beast, the Alpini flung themselves on the enemy ... For a moment the Turks appeared to hold their ground ... Many [attackers] fell wounded and were carried away by the all-compelling force of the stream that swept forward irresistibly. Under the clash of the Italian bayonets the enemy's front line wavered, as if smitten by the rush of air which the charge had driven onward, then broke and scattered in all directions, seized with the customary scare which always overtakes the Arabs when the bayonets flash ... Then the fury of the Alpini became irrepressible. The big, good-natured sons of the mountains, ruddy-faced and sturdy, whose smile is wont to be so kindly and whose glance is as the glance of a child, became for the nonce lost to all pity ... The irresistible shock swept on of its own impetus, and the dark mass of Turkish troops was swallowed up in the grey avalanche of the Alpini that bore all before it.[24]

All of the key elements of the 'grammar of the bayonet' are to be found in this heavily elaborated 'chronicle': the charge as the decisive action; the compact rush of the troops; the cleanness (both material and moral) of the flashing bayonets; the fear of cold steel among 'savages' lacking the moral strength, discipline and ésprit *de corps* necessary to withstand the impact; the charge as an avalanche with a life of its own, routing the enemies by its sheer impact. This bloodless

23 Ibid., p. 226.

24 Ibid., pp. 279–280.

and almost aseptic description strikingly contrasts with the savage onslaught not only of the ruthless Pathan, rushing down form the craggy hills of Afghanistan to 'cut to pieces' the depleted British regiments, but also of the same Arab auxiliaries when charging – in their turn – the Italian forces.[25]

A Shifting Boundary?

Nonetheless, the divide between 'savage enemy' and 'valorous comrade' was extremely thin. The same inhuman opponent, with his practices and his attitudes (and sometimes with his weapons), could be recast as a new friend when joining forces with his erstwhile antagonist against either a common external foe or his fellow (but still unruly) kinsmen. The whole 'martial race' theory – which grew out of the 'Great Mutiny' of 1857–1858 and subsequently shaped the Raj's military system – rested to a considerable extent on this kind of ambiguity. In official discourse, the historical 'loyalty' of the populations of north-western India intermingled with a conscious political and cultural strategy, aimed both at reshaping power structures within the Indian Army and at redrawing the relationship between its social and ethnic components. The Rebellion (and its suppression) 'transformed those troops defending British India into chivalric heroes, while the "mutineers" fighting against British rule became unmanly cowards'.[26] In this case, too, the Afghan stereotype, with its ambiguity and its inextricable mixture of villainy and valour, epitomised, in British eyes, a far wider vision and a general way of thinking about the nature of the 'Other'.

> There are two words which are always on an Afghan's tongue – *izzat* and *sharm*. They denote the idea of honour viewed in its positive and negative aspects, but what that honour consists in even an Afghan would be puzzled to tell you. Sometimes he will consider that he has vindicated his honour by a murder perpetrated with the foulest treachery; at other times it receives an indelible stain if at some public function he is given a seat below some rival chief.[27]

---

25 For a sketch of an Arab 'fierce rush' against Italian positions, see ibid., pp. 145–146. Quite predictably, Arab attitude is explained as a product of their inherent 'substratum of cowardice ... The mind of these people is a strange mixture of treachery and ferocity. They only attack when they are ten to one, as on the fatal October 23 [1911] [the day of 'treason' of Sciara Sciat], or else under the wild impulse of that religious frenzy which fills them with the blood-lust even to their own destruction; an impulse, not of courage, but of epileptic fury, as shown in the battle [of Sidi Mesri] of October 26 [1911].' Ibid., pp. 155–156.

26 Heather Streets, *Martial Races: The Military, Race and Masculinity in British Imperial Culture, 1857–1914* (Manchester, 2004), p. 11.

27 Pennell, *Among the Wild Tribes*, p. 17.

However, the presence of a strong Western ('civilised') guide could turn even this evil to unexpected good. The pedagogical effects of the white man's influence emerges as one core element of the imperial narrative, both in explicitly popular and in supposedly more scientific texts. When dealing with military matters, this narrative depicted European officers (but even more humble soldiers) as both models and mentors for their newly acquired men and comrades, freeing the best traits of their character from the deepest scum and infusing their action with a new, almost transcendent, purpose. Dealing with such idiosyncratic troops required special training and special knowledge, progressively embedded into a body of manuals that taught how to deal with the various 'martial races'. On the one hand, these manuals provided the official view of what these 'races' were, of their main characteristics and of their strengths and weakness. On the other, they helped to reinforce the basic idea of their 'natural' separateness and of the differences existing among them. In so doing, they crystallised a full body of knowledge about their skills and proficiencies, but also, and most importantly, about their ostensible moral features. It was in this field – it was felt – that Western influence could play its most beneficial effects, channelling the recruits' penchant for feral violence towards the pursuit of higher ends. As Victorian novelist George Henty (1832–1902) explained in the fiction *For Name and Fame*:

> when led and organized by English officers there are no better soldiers in the world [than the Afghans], as is proved by the splendid services which have been rendered by the frontier force, which is composed almost entirely of Afghan tribesmen. Their history shows that defeat has little moral effect upon them. Crushed one day, they will rise again the next; scattered, it would seem hopelessly, they are ready to reassemble and renew the conflict at the first summons of their chiefs. Guided by British advice, led by British officers, and, it may be, paid by British gold, Afghanistan is likely to prove an invaluable ally to us when the day comes that Russia believes herself strong enough to move forward towards the goal of all her hopes and efforts for the last fifty years, the conquest of India.[28]

Despite their strong and often virulent imperialistic bias, and despite the openly commercial ambitions of his works, Henty's opinions were not peculiar to his position or linked to their fictional origin. Rather, they elaborated and vulgarised such widespread and long-held pseudo-ethnographic assumptions as formed the background to George MacMunn's *The Martial Races of India*.[29] In the mid-1850s, writing about to the eight-year-old Corps of Guides raised on the Punjab frontier by Captain Henry Lumsden in 1847, Robert Martin –

28 Henty, *For Name*, p. iv.

29 George MacMunn, *The Martial Races of India* (London, [1933]).

former Colonial Treasurer of Hong Kong, founding member of the Statistical Society of London, the Colonial Society and the East India Association, as well as prolific author of contributions on the history of British colonies – argued that:

> Most of the wild and warlike tribes in Upper India are represented in its ranks; the men unite all the requisites of regular troops with the best qualities of guides and spies – thus combining intelligence and sagacity with courage, endurance, soldierly bearing, and a presence of mind which rarely fails in solitary danger and in trying situations.[30]

Following in Martin's footsteps, Colonel George Younghusband observed how Lumsden (whose Promethean figure he depicts in the first pages of his work as the archetype of the 'native troops' officer) consciously raised the first recruits of the Corps in Peshawar, 'then the extreme outpost of the British position in India, situated in the land of men born and bred to the fighting trade, free-lances ready to take service wherever the rewards and spoils of war were to be secured'.[31] And at about the same time, emphasising the sagacity of Lord Curzon's choice to separate the North-West Frontier Province's territories from the settled districts of western Punjab, Lovat Fraser argued that 'the men of the frontier heights are soldiers of fortune, who furnish some of our best fighting material'.[32] Fraser's assessment was consistent with the professional advice of a host of recruiting officers, constantly engaged on the frontier with a view to recruiting the men required to ensure Indian and imperial security.[33] Thirty years earlier, while dark clouds had been gathering on the horizon at the beginning of the second Afghan war, Major Henry Raverty, of the Bombay Native Infantry, had even more outspokenly affirmed that:

> those who have had to deal with Afghan soldiers ... especially in the Panj-àb Guide Corps, the Panj-àb Irregular Force, and the so-called "Balùch" Regiments

[30] Robert M. Martin, *British India: Its History, Topography, Government, Military Defence, Finance, Commerce and Staple Products, with an Exposition of the Social and Religious State of One Hundred Million Subjects of the Crown of England* (London, 1854 [?]), p. 537.

[31] George J. Younghusband, *The Story of the Guides* (London, 1908), p. 5.

[32] Fraser, *India*, p. 40.

[33] For a sketch of the growth of the Indian military establishment until the end of World War I, see *The Army in India and its Evolution: Including an Account of the Establishment of the Royal Air Force in India* (Calcutta, 1924). On the problem of Indian security, see Robert A. Johnson, '"Russian at the Gates of India"? Planning the Defence of India, 1885–1900', *Journal of Military History*, 67/3 (2003), pp. 697–743. On British and imperial security, see John Gooch, 'The Weary Titan: Strategy and Policy in Great Britain, 1890–1918', in W. Murray, M. Knox and A. Bernstein (eds), *The Making of Strategy: Rulers, States and War* (Cambridge, 1994), pp. 278–306.

> of the Bombay Army, which contain a great number, more than half probably, of Afghàns, know what excellent soldiers they make; and, man to man, few nations can surpass them physically.[34]

Lust for prey; extreme powers of endurance; the readiness to face solitary danger; a natural inclination towards the 'best qualities' of spies – all these were the key attributes that made the hills tribesmen 'the best fighting material' for the Indian Army. And, through the mentoring of their officers, this 'raw material' could learn the 'grammar of the bayonet', gradually evolving into 'noble exemplars of the true military tradition',[35] and bridging the gap that divided them from their European comrades. In this perspective, the 'civilising' impact of Western training extended its positive influence even to the way in which 'native' soldiers handled their 'traditional' weapons. In stark contrast with the Afghans, slashing their *tulwar* in a seemingly incoherent way, the Gurkhas used their *kukris* 'in preference to the bayonet' and with sharp, purposeful and decisive gestures, whose effectiveness – adding up to the white man's resoluteness in using his bayonet – forced the (cowardly) enemy to desert the field after having lost the advantage of number.

> William Gale had just re-loaded his rifle when he saw Captain Herbert, who commanded his company, fall to the ground, and three Afghans spring forward to finish him. With a bound Will reached the side of the officer ... In an instant he shot his assailant dead, and then with bayonet stood at bay as the other two Afghans rushed upon him. They had drawn their tulwars and slashed fiercely at him; but he kept them off with his bayonet until a Ghoorka, running up, cut down one of them with his kookerie, a heavy sword-like knife which the Ghoorkas carry, and which they always employ in preference to the bayonet in fighting at close quarters.[36]

Concluding Remarks

At the beginning of World War I, the Boer war (1899–1902), the Russo-Japanese war (1904–1905) and the almost uninterrupted string of violence stretching from the Turco-Italian war in Tripolitania and Cyrenaica to the second Balkan war (1913) had taught European armies hard lessons about the

---

34 Henry G. Raverty, *Notes on Afghanistan and Baluchistan* (Quetta, 1976), p. 163.

35 Douglas M. Peers, '"Those Noble Exemplars of True Military Tradition": Construction of the Indian Army in Mid-Victorian Press', *Modern Asian Studies*, 31/1 (1997), pp. 109–142.

36 Henty, *For Name*, p. 159. A similar scene of cold blood opposed to savage violence is depicted on p. 283ff, when William Gale, walking in the main street of Kandahar, saved the life of Colonel Ripon by opposing, 'sword in one hand and revolver in the other', the three Afghans who had assailed him.

relative advantages of edged weapons and firearms. Nonetheless, the clash of bayonets was still regarded as the climax – both factual and symbolic – of any military encounter. Following a centuries-long tradition, in the first battles of the Great War, Russian soldiers still carried their bayonets always at the ready, in the expectation of a (increasingly less probable) close-quarter fight against German troops and their machine guns. In 1914, French operative manuals, still informed by Bergson's vitalism, recommended the bayonet charge as a given military encounter's decisive action and as the best way to smash the enemy's morale by showing the superior virtue of the French soldier vis-à-vis his antagonists' overreliance on planning and technological superiority. Until 1917, on the Italian front, General Cadorna's tactics revolved around the useless notion of *spallate* ('shoulder pushes'), which resulted in a long series of bloody and unsuccessful head-on attacks against the heavily entrenched Austrian positions on the slopes of the Isonzo River's valley. In an increasingly mechanised and technological warfare, Cadorna's actions simply aimed at depleting the enemy's resources through the massive commitment of infantry forces.[37] The *Strafexpedition* ('Spring Offensive', in Austrian historiography) of 1916 and the Caporetto offensive of the following year, together with the development of trench warfare on the Western front, showed the shape of future things to come, paving the way for such infiltration tactics as would find their first large-scale application during the German *Blitzkrieg*.[38]

But the myth of the bayonet survived even this traumatic experience. After the end of World War II, in such highly politicised conflict as the Korean War (1950–1953), where 'free world' forces faced for the first time their new, radical, 'Communist' enemy, both the American press and official army documents stressed 'the role of the bayonet ... far beyond its intrinsic importance, when the latter is estimated in the very real terms of the battlefield and the thinking of troops about the weapon'.[39] Indeed, in the 'real world', military developments were moving in the opposite direction. 'American troops, left to their own devices, simply threw away the bayonet.' Despite the efforts of some 'occasional strong-minded company or battalion commander', in the average company unit, as of November and December, 1950, the only men possessing bayonets were

---

37 John R. Schindler, *Isonzo: The Forgotten Sacrifice of the Great War* (Westport, 2001).

38 On tactical innovation during World War I and the evolution of infiltration tactics, see Bruce I. Gudmundsson, *Stormtroop Tactics: Innovation in the German Army, 1914–18* (New York and London, 1989). In comparative perspective, see Martin Samuels, *Doctrine and Dogma: German and British Infantry Tactics in the First Word War* (New York, 1992). On the relations between *Blitzkrieg* and the infiltration tactics of World War I, see James S. Corum, *The Roots of Blitzkrieg: Hans von Seeckt and German Military Reform* (Lawrence, 1992).

39 Samuel L.A. Marshall, *Commentary on Infantry Operations and Weapons Usage in Korea: Winter of 1950–51* (Chevy Chase, [1951]), p. 103.

the replacements who had not yet learned that they could heave them and not face a court martial.[40]

40 Ibid., p. 48.

PART III
Controlling Guns: Gun Laws, Race and Citizenship

Chapter 8
The Battle of Dubai: Firearms on Britain's Arabian Frontier, 1906–1915

Simon Ball

The Arabian Frontier

The 'Arabian frontier' was a nexus for the global small arms trade in the years immediately before the First World War. Arms left four great ports in Europe: London, Hamburg, Antwerp and Marseilles. There were alternative land routes, via Russia, and through Turkey to Baghdad, but these routes were sometimes 'not practicable' for the large-scale supply of advanced weapons.[1] Ships passed through the Mediterranean and into the Red Sea, via the Suez Canal. The frontier began at the Indian outpost of Aden. The particular technology of the period meant that it was on leaving Aden that a ship entered a new zone. Ship-to-ship wireless telegraphy was 'generally possible', 'to ships at the head of the Persian Gulf from the entrance to the Gulf of Aden – that is to say, across the whole extent of the Arabian plateau, a distance of over a thousand miles'.[2]

The arms route crossed the Gulf of Aden to Jibuti in French Somaliland. Jibuti was 'a favourite place for the trans-shipment of weapons to Muscat and the Persian Gulf'.[3] Arms were also taken south from Jibuti by railway to the Abyssinian town of Harrar, where they entered British Somaliland.[4] From Jibuti the trade flowed to Muscat in the Gulf of Oman. The weapons were distributed on both shores of the Persian Gulf, to Dubai, Qatar, Kuwait, Bushire, Lingah and Bandar Abbas. Guns for transport to India tended to be landed between Jask and Chahbar on the Gulf of Oman. From there they would be taken by Ghilzai caravan via Helmand to Kandahar for distribution in Afghanistan and

[1] C-in-C East Indies to Admiralty, 20 January 1912, National Archives of the UK (NAUK), London, FO428/10.

[2] The Hon. Arnold Keppel, *Gun-Running and the Indian North-West Frontier* (London, 1911), pp. 131–132.

[3] 'The Traffic in Arms', *Times*, 30 December 1910, p. 3.

[4] Herbert Vivian, 'The English and French in Abyssinia: Some Account of King Menelik's New Railways', *Pall Mall Magazine* (March 1901), pp. 340–353; K.V. Ram, 'British Government, Finance Capitalists and the French Jibuti-Addis Ababa Railway, 1898–1913', *Journal of Imperial and Commonwealth History*, 9/2 (1981), pp. 146–168.

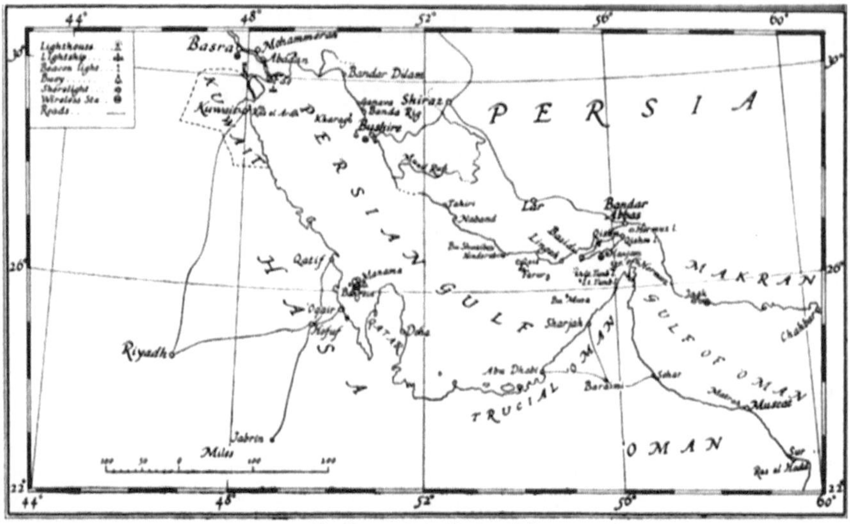

Sketch-map of the Persian Gulf

Map 8.1 Sketch-map of the Persian Gulf
Arnold Wilson, 'A Periplus of the Persian Gulf', *Geographical Journal*, 69/3 (1927), p. 238. Courtesy of John Wiley & Sons Ltd.

the North-West Frontier. Some high-value advanced weapons were shipped to Bombay and then returned by steamer service to Kuwait.

The Arabian small arms trade has attracted periodic attention down the years: its formation had important political and cultural consequences. Beginning in the 1960s, Africanist historians analysed the impact of arms and imperial control on east Africa.[5] Their other main line of enquiry was into the effect of firearms on the Ethiopian *Sonderweg*, between the Battle of Adowa in 1896 and the Abyssinian War of 1935–1936.[6] The extensive literature on the North-West Frontier includes a small number of specialist studies on firearms trafficking.[7]

[5] R.W. Beachey, 'The Arms Trade in East Africa in the Late Nineteenth Century', *Journal of African History*, 3/3 (1962), pp. 451–467; James Cooke, 'Anglo-French Diplomacy and the Contraband Arms Trade in Colonial Africa', *African Studies Review*, 17/1 (1974), pp. 27–41; Gavin White, 'Firearms in Africa: An Introduction', *Journal of African History*, 12/2 (1971), pp. 173–184.

[6] Richard Pankhurst, 'Guns in Ethiopia', *Transition*, 20 (1965), pp. 26–33; R.A. Caulk, 'Firearms and Princely Power in Ethiopia in the Nineteenth Century', *Journal of African History*, 13/4 (1972), pp. 609–630; Harold Marcus, 'The Embargo on Arms Sales to Ethiopia, 1916–1930', *International Journal of African Historical Studies*, 16/2 (1983), pp. 263–279; Jonathan Grant, 'Arms Trade Colonialism: Ethiopia and Djibouti' in his *Rulers, Guns and Money: The Global Arms Trade in the Age of Imperialism* (Cambridge, MA, 2007), pp. 65–77.

[7] Robert M. Burrell, 'Arms and Afghans in the Makrān: An Episode in Anglo-Persian Relations, 1905–1912', *Bulletin of the School of Oriental and African Studies*, 49/1 (1986), pp. 8–24; T.R.

The current chapter is based on the multi-volume confidential print on the arms trade assembled by the Foreign Office between 1906 and 1928. It redirects the focus of study from the presumed crucial 'final destinations' of small arms, Afghanistan/India and Ethiopia, to the 'transit zone' around the 'Arabian frontier'. The arrival of rifles on the frontiers of India was undoubtedly the main mover of political events, but more weapons were actually absorbed within 'the frontier' of Arabia, Persia and Somaliland. At the peak of the trade about 40,000 rifles a year reached the North-West Frontier via the Gulf.[8] This was less than half the annual number that arrived in Muscat. Jibuti saw a volume of approximately 700,000 rifles per year.[9] The chapter begins by highlighting the importance of a skirmish at the then little-noticed town of Dubai.

The Battle of Dubai

On Christmas Eve 1910 the Royal Navy fought a short but deadly skirmish at Dubai.[10] Five British sailors were killed. The commander of the naval landing party, in turn, estimated that he had killed, or wounded, 40 Arab tribesmen. The sailors had landed at Dubai to confiscate contraband weapons. The landing was part of an operation launched in December 1909 by the commander of the East Indies station, Rear-Admiral Edmond Slade. Slade's goal was to prevent the large-scale transit of arms to India via Persia by blockading the sea-lanes around the Straits of Hormuz.

Slade was confident that he could suppress the arms trade in the southern Gulf, but had warned his naval superiors that he would have to land fighting men, and this was bound to lead to violence.[11] 'I have', Slade wrote after a tour of inspection in spring 1909, 'come to the conclusion that ... it must not be expected [it] can be done without some fighting.'[12] Nevertheless, he secured approval to deploy the East Indies Squadron. In his flagship HMS *Hyacinth*, Slade led to the Gulf 'another second-class protected cruiser, HMS *Fox*,

Moreman, 'The Arms Trade and the North-West Frontier Pathan Tribes, 1890–1914', *Journal of Imperial and Commonwealth History*, 22/2 (1994), pp. 187–216; Emrys Chew, 'Militarized Cultures in Collision: The Arms Trade and War in the Indian Ocean during the Nineteenth Century', *RUSI Journal*, 148/5 (2003), pp. 90–96.

8 General Staff, India, 'Report on the Arms Traffic, 1st July 1911 to 30th June 1913', Simla, 1913, NAUK, WO106/6322.

9 Doughty-Wylie to Grey, 12 December 1912, NAUK, FO428/10.

10 Rear-Admiral Slade (at Bushire, about to depart for Dubai) to Admiralty, 25 December 1910, NAUK, FO428/7.

11 Rear-Admiral Edmond Slade (C-in-C East Indies) to the Government of India, 25 July 1910, NAUK, FO428/7.

12 Rear-Admiral Edmond Slade [Bombay] to Admiralty, 7 May 1909, NAUK, FO428/4.

three third-class protected cruisers, *Perseus*, *Philomel*, and *Proserpine* ... three sloops ... a special service vessel ... added specially for blockading work, [and] six steam launches'.[13]

Open violence flared at Dubai as a result of the specific conditions of the blockade. The centre of the Gulf arms trade was Muscat on the Gulf of Oman. Feisal, Sultan of Oman, maintained his insistence on Muscat's trading rights, guaranteed by treaty with Britain and France. Slade could enforce the blockade at sea, but he could not deploy armed parties to seize arms on land. 'The Custom House quay is seldom unencumbered with cases of rifles and ammunition, while every other shop in the bazaar is a rifle shop', an investigator observed in February 1911. 'There is a certain amount of humour in a situation in which a British cruiser is actually at anchor in the harbour, with a dhow loaded to the water-line with rifles and ammunition almost within a cable's length of her.' Searches were not made in 'Muscat territorial waters, since it is suspected that the captured rifles, when returned to the Sultan, find their way back into the possession of the dealers'.[14] Nevertheless, the intensive surveillance of Muscat forced gun-runners to search for alternative routes. 'When', the *Times* reported, 'the authorities found that the arms traffic was being diverted to Debai, preparatory to transhipment to the Persian coast, Admiral Slade was able to do what he could not have done at Muscat. He took his flagship to Debai and landed a party to search for arms.'[15]

Dubai was a prime location for trans-Gulf trade. Dubai was the 'real metropolis of northern Oman', with a growing population that had reached 15,000 in 1900. The town was built around an inlet, or *khor*. It was already clear to knowledgeable observers that Dubai would 'soon outstrip the other towns' of the region.[16] The British had had some hope that the ruling al-Maktum family would not want the arms trade interfering with their flourishing commerce in other goods. The architect of this commercial policy was Sheikh Maktum bin Hasher, ruler of Dubai between 1894 and 1906. In 1904 Dubai became a scheduled port-of-call for Anglo-Indian steamship companies. Sheikh Maktum declared the town a free port.[17] However, when the sheikh died in February 1906 his sons were too young to succeed. Leadership fell to an elderly cousin, Butli

---

13 'British Interests in the Persian Gulf: III – The Operations against Gun-Running', *Times*, 8 July 1911, p. 7.

14 Keppel, *Gun-Running*, pp. 124 and 146.

15 'Leader – Gun-Running in the Persian Gulf', *Times*, 28 December 1910, p. 7. The terms Debai, Dibai and Dubai were interchangeable.

16 S.M. Zwemer, 'Oman and Eastern Arabia', *Bulletin of the American Geographical Society*, 39 (1907), pp. 597–606. Resident for 16 years in Arabia, Zwemer made three journeys along the Pirate Coast as a missionary in 1900 and 1901.

17 Fatma Al-Sayegh, 'Merchants' Role in a Changing Society: The Case of Dubai, 1900–1990', *Middle EasternStudies*, 34/1 (1998), pp. 87–102.

bin Soheil. Although the new sheikh maintained his predecessor's commercial policy, he was 'unreliable and unsatisfactory'.

In addition, Islamic radicalism was on the rise. On one hand the rapidly growing population of Dubai seemed to have little time for Wahhabism. This, at any rate, was view of the Arabian expert, the Revd S.M. Zwemer in 1907.[18] Dubai, however, had become the centre for the distribution of newer pan-Islamic literature throughout Arabia. Responding to the Battle of Dubai, Zwemer offered a 'probable explanation of the Arabs' resistance to HMS *Hyacinth*'s forces.

> He declares that recently pro-British sentiment on the Pirate Coast has grown weaker. Misled by the Egyptian Pan-Islamic Press, the Arabs believe that a departure from our policy of non-interference is contemplated and they fear a partition of Persia, followed by annexation in Arabia. These apprehensions have induced an increasing anti-foreign feeling, which has been intensified by the belief that our measures against the arms traffic are intended to lead to the disarmament of the Arabs who cling to their rifles as their most cherished possessions. Nothing else would have produced this unprecedented opposition to the British forces.[19]

The alternative secular explanation was that 'the Arabs of the Pirate Coast are rather poor and extremely avaricious, and they probably resented the sudden interference with a trade which promised them much wealth'.[20]

Despite the loss of life, the outbreak of open fighting could be regarded as a positive development. 'One result of the engagement', the *Times* opined, 'will probably be the prompt termination of any further attempt to develop the arms traffic at the ports of the Trucial chiefs. The incident ... shows that Great Britain can greatly restrict, and possibly suppress, the traffic but it also reveals the remarkable persistence of the gun-runners.'[21] Slade himself thought that the battle and its aftermath had been decisive.

> The arms traffic is almost entirely at a standstill, and large consignments have not been landed in Persian territory for some considerable time ... arms and ammunition landed there do not find their way into the hands of the Afghan traders, but are all absorbed in Persia itself, where there is a large demand at the present moment. Small consignments are smuggled across from time to time, but even these have ceased to a

18 Zwemer, 'Oman and Eastern Arabia', pp. 597–606.

19 'Gun-Running in the Persian Gulf: The Arabs and British Action from our Correspondent, Bombay, Dec. 29', *Times*, 30 December 1910.

20 'Leader – Gun-Running in the Persian Gulf', *Times*, 28 December 1910, p. 7.

21 Ibid.

> great extent since I authorised the burning of such dhows as were notoriously engaged in smuggling.[22]

He was determined to finish the job, and was emboldened to carry out further and more ambitious operations on the other side of the Gulf. In the spring of 1911 Slade landed the Indian Army's Mekran Field Force in Persia. Its operations culminated with the Battle of the Pashak Pass against the Baluchi Tahirzai tribe on 28 April 1911. Victory made it 'virtually impossible for the *nakhudas* [dhow captains] to run a cargo of rifles' from Oman.[23]

The Battle of the Pashak Pass, although without British casualties, was more controversial than the Battle of Dubai. The Liberal Secretary of State for India, John Morley, complained that 'I am at a loss to understand how proposals regarding arms traffic ... are reconcilable with military operations which [the] admiral is now undertaking.' The government of India confirmed that 'the admiral was exceeding his instructions'.[24] On the other hand, and despite Morley's best efforts, the Gulf of Oman blockade had aroused enough interest to ensure that Slade's Persian campaign was covered by an 'embedded' *Times* special correspondent, the Hon. Arnold Keppel.[25] Keppel's laudatory reports were later turned into a book, published in late 1911 by John Murray.[26]

Slade's naval and military success eventually forced the Sultan of Muscat to the negotiating table. When Slade hauled down his flag in March 1912, his Gulf operations were hailed as an exemplary application of British power. 'His regime', the *Times* commented, 'will be memorable for the vigorous and effective measures he has taken to suppress the arms traffic between Muscat and Mekran. These were so thorough that the traffic is now practically dead. The British preventive measures have now been appropriately crowned by the decision of the Sultan of Muscat to control the traffic himself.'[27] In June 1912, in return for a 100,000 rupee per annum subsidy, the sultan finally agreed to the creation of a

---

22 Slade (Bombay) to Admiralty, 18 February 1911, NAUK, FO428/8.

23 'With the Mekran Field Force: Review of the Expedition from our Special Correspondent, Sirik, May 4', *Times*, 27 June 1911, p. 5.

24 Morley to Government of India, 28 April 1911; Government of India to Morley, 1 May 1911, NAUK, FO428/8.

25 Morley to Government of India, 11 April 1911, NAUK, FO428/8.

26 Keppel, *Gun-Running*. Keppel sailed out to India to begin his investigation in October 1910. In December 1910, at the time of the battle of Dubai, he was on a tour of the North-West Frontier. He left Karachi for Muscat on 9 February 1911, spending a week in the town. He then made his way up the Gulf, but was unable to land at Dubai. News of the Mekran Field Force reached him at Basra and he returned to India to cover its activities. Keppel left Bombay with the Force on 6 April 1911. He finished the manuscript of his book on 7 October 1911.

27 'The Suppression of Gun Running', *Times*, 23 March 1912, p. 5.

bonded warehouse under strict government supervision. The warehouse began operation in September 1912.[28]

That the Mekran Field Force had been an unalloyed success became accepted wisdom. It was often cited by those who favoured the proper application of military force in the face of pusillanimous politicians. Notably, however, British officers in the Gulf, Slade himself included, were more nuanced. They had only solved the Mekran/Gulf of Oman problem as it pertained to the North-West Frontier trade. They had not solved the Persian Gulf problem: trade for Arabia and Persia merely moved to more northerly ports. Indeed one of the chief sponsors of the expedition, Percy Cox, the Political Resident in the Gulf, had warned at the start that this was likely to be the outcome of their efforts. 'I cannot go so far as to believe', he had written in 1906, 'that the closure of Muscat as the Gulf emporium is likely to prove a permanent check on the import of arms ... because there is now such a well-established demand for arms throughout the Arabian Peninsula, in Mesopotamia, parts of Luristan and Kurdistan, Arabistan, and Mohammerah territory, that I cannot suppose that, were Muscat suppressed, other bases for the supply of the market would ere long be devised.'[29]

Although military action had solved the strategic problem for India, it had created, or maintained, problems on the Arabian frontier. In August 1911 officials reported that

> the Katr Peninsula is the weak point in our chain of defences against the arms traffic in the Persian Gulf, and has recently been attracting the imports to a greater extent than formerly ... these imports are not re-exported to Persia, but are absorbed in Central Arabia, and are consequently a danger to Turkey rather than to ourselves, [but] it cannot be denied that the arming of the Arab tribes in the hinterland may have consequences to the tribes on the coast under our protection in which we may ourselves become involved.[30]

It was at this stage that the Battle of Dubai took on its full significance. In the aftermath of Pashak Pass, the Sultan of Oman 'unmistakeably identified himself with the arms dealing fraternity'.[31] The direct trade to the Mekran coast had come to a standstill. Instead large quantities of arms were transferred to the Omani port of Sohar, often in the sultan's personal steam yacht, to be taken by overland caravan to Dubai and other ports on the Arabian Gulf. Large inland gun depots were created to facilitate the trade. This plan was undermined by the Sheikh of Dubai. Chastened by the British attack, he intervened to seize

---

28 Burrell, 'Arms and Afghans in the Makrān', pp. 8–24.

29 Cox (Bushire) to Government of India, 2 December 1906, NAUK, FO428/1.

30 India Office to Foreign Office, 3 August 1911, NAUK, FO428/9.

31 Cox to Government of India, 11 June 1911, NAUK, FO428/9.

weapons and turned them over to the British himself. 'It has transpired', Percy Cox reported, 'that the arms and ammunition which were seized by the Sheikh of Debai were specifically for the Katar family.'[32] The British representative in the northern Gulf reported that the ruler of Dubai resented the possibility of Katar becoming the principal arms distribution centre for Arabia and Persia, fearing that traders would leave his port in order to set up in Doha.[33] British naval observers reported that 'the Trucial Coast trade has sustained a severe blow by the action of the Sheikh of Debai'.[34]

That was by no means the end of the matter, however. In January 1912 Slade reported that 'the unruliness of the tribes in Fars and Luristan is to be greatly accounted for by the facility with which arms and ammunition can be obtained from the Trucial Coast and Al Bida and Koweit. It is known that large quantities are going to these places from Muscat, partly by land and partly by sea'.[35] The Sheikh of Dubai visited Muscat for talks with the sultan in March 1912. He complained that he had not received sufficient reward from the British for the rifles he had seized. He intimated that he was now ready once more to receive land caravans of arms from Sohar.[36] The 'inconvenient' truth was 'that the bulk of the traffic is still going from Sohar by land to the Trucial Coast, and thence direct, or via Katar, to the Persian littoral ... Debai itself is being used, and the sheikh's disaffected relatives are directly interested'.[37]

The tipping point came with Cox's month-long negotiations in Muscat. In the wake of the promulgation of the warehouse agreement on 4 June 1912, he commented that 'the Sultan ... seems to feel that he has crossed the Rubicon'. Cox himself believed that the agreement was a major step forward, but that there were many loose ends to tie up in the Gulf before the matter was truly resolved.[38] In March 1913 he finally made it through to Qatar for talks with Sheikh Jasim, and his son Abdullah of Doha. He found that 'the question for them is merely one of price'. The subsidies offered to the key players on the Trucial Coast led to the reduction of the trade to levels to which the British could accede.[39]

32 Cox to Government of India, 27 November 1911, NAUK, FO428/9.
33 Captain Shakespear to Cox, 20 December 1910, NAUK, FO428/8.
34 Commander Beaty-Pownall to C-in-C East Indies, 20 October 1911, NAUK, FO428/9.
35 C-in-C East Indies to Admiralty, 20 January 1912, NAUK, FO428/10.
36 Knox (Muscat) to Cox, March 1912, NAUK, FO428/10.
37 Cox to Government of India, 13 May 1912, NAUK, FO428/10.
38 Cox (Muscat) to Government of India, 5 June 1912, NAUK, FO428/11.
39 Cox to Government of India, 2 March 1913, NAUK, FO428/12.

International Politics

Between 1906 and 1914 arms on the Arabian frontier remained a matter for constant great power diplomacy. The traffic touched on some of the most vexed issues in international politics. This era of the arms trade was defined by two significant great power arms accords: the December 1906 Anglo-Italian-French agreement on Abyssinia and the February 1914 Anglo-French agreement on Muscat. Sandwiched between these two concords was the failed international conference on arms traffic, held in Brussels in 1908 and 1909. The Brussels Conference was dominated by Anglo-French disagreement over the Persian Gulf.

Each of these negotiations, in turn, was shaped by differing interpretations of the 1890 Brussels Convention. The Brussels Convention had been an attempt to suppress the African slave trade. Embedded within the treaty, in articles eight to fourteen, was an agreement to control the small arms trade. Importation of firearms, especially 'rifles and improved weapons', was prohibited south of the '20th parallel of north latitude'.[40] The 1906 tripartite agreement sought to strengthen the Brussels provisions. In particular it focused on the role of the ports of Aden, Jibuti, and Berbera. The powers were to prevent transit of weapons to Ethiopia, unless there was a demonstrably legitimate end user, who had undertaken not to sell on the weapons. The three powers also promised to take measures to prevent dhows trading for arms from these ports 'to points outside the zone of protection defined by the Act of Brussels'.[41]

In late 1907 the British government broached a plan to extend the Brussels Convention itself to the Red Sea and the Persian Gulf.[42] The British negotiators, Arthur Hardinge and Walrond Clarke, were not, however, confident of success. There were, they argued, six important players: Britain, Italy, Turkey, Persia, France and Belgium: 'the first four', Clarke wrote, 'have only one object in view: the suppression of the trade as far as that might be possible. The last two have equally only one object: to minimize as far as lies in their power any repressive measures which the other four governments may try to get carried.' In the 'battle' Persia and Turkey would be 'quite useless', so it would boil down into a straight fight between an Anglo-Italian alliance and France. 'There can be little doubt', Clarke concluded, 'that the French will do everything they can to prevent the Conference having a successful result.' France favoured 'the present old and useless regime of the Brussels Act ... the only thing they have done has

---

40 *General Act of the Brussels Conference Relative to the African Slave Trade*, 2 July 1890. The signatories were Britain, Germany, Austro-Hungary, Belgium and the Congo, Denmark, Spain, the USA, France, Italy, the Netherlands, Luxemburg, Persia, Portugal, Russia, Sweden, Turkey and Zanzibar.

41 *Agreement between the United Kingdom, France and Italy, respecting the Importation of Arms and Ammunition into Abyssinia*, signed at London, 13 December 1906.

42 Memorandum communicated to Baron Gericke, 12 September 1907, NAUK, FO428/1.

been to make an insidious proposal to the effect that if the Conference could not agree even the old regime would come to an end, so that there would be no restraint at all. As to agree to this would have been simply to encourage them to wreck the Conference, we refused.' 'The real truth of the matter', he continued, 'is that Jibuti lives entirely upon its traffic in arms, and that if it was seriously interfered with not only would Jibuti itself suffer seriously but also the French manufacturers of arms and ammunition, who do a roaring trade and are, what is more to the point, extremely influential in the Chamber.'[43] As a subsequent report noted, '[W]e do not of course overlook the fact that Jibuti … fell within the zone defined by Brussels Act of 1890, and this fact has not prevented the French, by a quibbling interpretation of one of the clauses of the Act, from permitting the arms traffic to proceed unchecked.'[44]

The French were lukewarm even on the recently signed 1906 agreement. They opposed the Italian 'Colli scheme', designed to make the control of imports into Ethiopia work in practice by accepting orders only from the Emperor Menelik. The Belgians supported the French, because 'they too do a great trade in arms with Jibuti and their Liège people would be furious if it were interfered with'. The way around the French roadblock was a deal with the Germans, 'only too happy to give us their support if merely to *embêter* the French'. This was not palatable to supporters of the 1904 Anglo-French Entente. Hardinge, a hard-line Germanophobe, suspected a sinister Teutonic 'inclination to pose … as the champion of Moslem potentates against European neighbours on whom she wishes to exercise pressure in other quarters'. He had a feeling 'that the Germans, in their quest for an ostensibly commercial foothold in these waters, might at any moment attempt to conclude a Commercial Treaty with [the Sultan of Muscat] as a prelude to the acquisition of political interests in his dominions'. Hardinge admitted that 'so far there is no sign of any such intention on their part'.[45] The best he was willing to do was a bilateral secret deal with the French to give the conference the appearance of success. The alternative was 'the very barren and Pyrrhic satisfaction of having placed the French in a somewhat odious light, as preferring selfish objects of a peculiarly petty and sordid character to the wider interests of all civilized communities'.[46] The outcome of the Brussels Conference was much as predicted. Talks about Red Sea and the Persian Gulf dragged on through 1908 and 1909. All parties admitted failure in December 1909.[47] In Muscat, 'as soon as it was ascertained that no injury to the trade would result

---

43 Memorandum by [E.]A.W. Clarke, 28 January 1908, NAUK, FO428/2.

44 Committee of Imperial Defence: Report of Sub-committee on Arms Traffic, February 1917, NAUK, CAB16/44.

45 Hardinge to Grey, 9 April 1909, NAUK, FO428/4.

46 Hardinge to Grey, 3 May 1909, NAUK, FO428/4.

47 *Times*, Leader, 14 December 1909; British Delegates to Brussels Conference to Grey, 30 December 1909, NAUK, FO428/5.

from the Conference large supplies were ordered'.[48]It was French intransigence at Brussels that finally prompted the Indian government to support direct military action by Slade.[49]

The Battle of Dubai sent Britain and France back to the negotiating table. Lord Crewe, the Secretary of State for India, urged that the loss of life at Dubai, to say nothing of the cost of the blockade, made a limited deal with France the most sensible way forward.[50] Within days talks were underway about 'some arrangement of a local character, less extensive than the settlement previously proposed'.[51] Not that any rapid progress was made. In October 1911 Arnold Keppel wrote that 'it may as well be said at once that, short of a cession of colonial territory to France – an act against which sentiment revolts – or the possession of some powerful diplomatic lever which chance may put into our hand, we shall never secure the abrogation of the trading treaties'.[52] In the event, the main arms traffic issue that caused tension between Britain and France was the role of Jibuti, rather than Muscat.[53] The vigour of Slade's unilateral action at Dubai and Pashak reduced the Persian Gulf negotiations to a long drawn out haggle about price – France abandoned its earlier search for colonial advantage in favour of discussing the level of compensation for French businessmen subsequent to the closure of the Muscat arms market. In February 1914 a final settlement was marked by an exchange of diplomatic notes. France undertook to 'cease to oppose the application to their nation of the Muscat edict of the 4th June, 1912'.[54]

The International Arms Trade and the Arabian Frontier

The arms trade to the Arabian frontier had two main dimensions. The first was the 'rules of the game', the specific regime created by the 1906 trilateral agreement's reading of the Brussels Treaty, and the subsequent failure to expand or crystallise that regime by further agreement until 1914. The rules were significantly changed by unilateral British naval and military action on parts of the frontier from the end of 1909 onwards.

---

48 Holland (Muscat) to Political Resident (Bushire), 16 January 1910, NAUK, FO428/6.

49 Government of India (Minto and Kitchener) to Morley, 4 September 1909, NAUK, FO428/5.

50 India Office to Foreign Office, 29 December 1910, NAUK, FO428/7.

51 Grey to Bertie, 18 January 1911, NAUK, FO428/8.

52 Keppel, *Gun-Running*, p. 52.

53 Grey to Bertie, 28 August 1911, NAUK, FO428/9.

54 *Exchange of Notes between the Government of Great Britain and the Government of the French Republic Respecting the Trade in Arms and Ammunition at Muscat*, London, 4 February 1914.

The second dimension was the specific state of the commercial arms trade in the decade-and-a-half before the outbreak of the First World War. The firearms industry occupied a highly ambivalent position. On the one hand it was the very acme of modernity, and often used as a measure of national competitive advantage; on the other it was decried as little more than a deadly 'rag-and-bone' trade. Both identities were the flip sides of the same coin. It was the very rapid modernisation of the firearms industry, and its centrality to great power relations, that produced a vast second-hand trade in obsolescent, but still highly effective, weapons.

In the 1870s and 1880s Europe had undergone a 'breech-loading revolution'. That revolution accelerated in the 1890s. The first phase of the revolution was based on the perfection of the integrated cartridge, comprising primer, propellant and bullet. The integrated cartridge, and the weapons that fired it, *could* still be produced in the industrial workshops and arsenals created earlier in the century. They were more effectively produced, however, in factories equipped with advanced machine tools. The second, and arguably more profound, phase of the revolution turned upon the invention, and mass production, of 'smokeless' propellant to replace gunpowder, a technology that had been in universal use since the sixteenth century.

These technical developments had serious knock-on effects. First, the European firearms trade threw up new industrial companies capable of profiting from the recently developed advanced technology. Firms like the Österreichische *Waffenfabriks-Gesellschaft* at Steyr and *Gebrüder Mauser & Cie* at Oberndorf were the products of the first phase of the revolution. Major combines, notably *Fabrique Nationale* in Belgium and *Deutsche Waffen-und-Munitionsfabriken* in Germany, were the fruit of the second phase. In the 1870s all armies re-equipped their troops with single-shot cartridge breech-loaders such as the British Martini-Henry or the French Fusil Gras. Such systems had a high degree of functional commonality and were produced by, and for, an international market. The Martini was designed by a Swiss, and was just as likely to be produced in Liège as in Birmingham. Steyr in Austria was one of the primary manufacturers of the Fusil Gras. In the 1880s the single-shot rifles were replaced by types with multi-shot magazines. Within a decade these rifles had, in turn, to be replaced by new weapons capable of handling high-pressure smokeless powder. These weapons then had to be modified and upgraded with quick-loading systems. There was thus a new generation of weapons developed roughly every decade for nearly four decades. In 1909 Count von Schlieffen remarked that the development of small arms over the previous forty years had been so profound as to be 'barely conceivable'.[55]

[55] Antulio Echevarria, 'The "Cult of the Offensive" Revisited: Confronting Technological Change before the Great War', *Journal of Strategic Studies*, 25/1 (2002), pp. 199–214.

Small arms were seen as 'decisive weapons' in modern warfare. According to the controversial, but widely discussed, theorist, Jean de Bloch, the Boer War proved the superiority of 'smokeless powder, and long-range quick-firing rifles' over ineffective artillery.[56] In a twist in the argument particularly relevant to frontier warfare, Bloch also pointed out that 'the modern magazine rifle, owing to its precision and long range, was a weapon which afforded the best opportunities for the carrying on of guerrilla warfare'.[57] The reputation of the rifle as a weapon of war, whether regular or irregular, stood at its highest between the end of the Boer War and beginning of the First World War.

Britain's 'Arabian frontier' was defined by trade in both advanced weapons and 'junk'. The British made strenuous efforts to gather information about the trade. British customs monitored the weapons shipped from the port of London. British consuls in Antwerp, Hamburg and Marseilles sent regular reports on exports. On occasion the consuls were assisted by secret service agents. Intelligence work went on around Jibuti. There the main arms dealer was usually the governor himself, or one of his agents. The agent who registered most often on the British consciousness was Guignony, the French consul at Harrar.[58] In 1912 the British vice-consul in Harrar suborned Guignony's Indian clerk to get hold of his shipping manifests.[59] Guignony had brought 25,000 Gras and 50,000 Italian Vetterli rifles from Jibuti to Harrar during 1911.[60] In 1911 the French minister in Addis Ababa calculated that, 'within the past ten years, no less than 7,000,000 of rifles have passed through the Jibuti port'.[61] The business was not without its risks: the head of the *Comptoir de Djibouti* was shot by Somalis to whom he was selling rifles in 1907.[62] Under the Barthou government in Paris there appeared to be more willingness to rein in the Jibuti trade. But this attitude had little impact on the arms trade. Indeed the arms dealers in Jibuti created a more formal syndicate.[63] The Consul-General in Antwerp, who had been instructed to keep a careful watch on shipments, reported that 'the prohibition of the passage of arms through Jibuti appears to have made little difference to the quantities exported from Antwerp'.[64] It was 'perfectly evident that, whatever instructions may be sent from Paris, the Jibuti authorities ... are determined not

[56] Jean de Bloch, 'The Transvaal War: Its Lessons in Regard to Militarism and Army Re-organisation', Lecture to the Royal United Services Institution, London, Monday, 24 June 1901.

[57] Jean de Bloch, 'A Reply to Some Criticism', *National Review*, 37 (May 1901), pp. 370–382.

[58] H.F. Ward (Harrar) to G.R. Clerk, 30 June 1908, NAUK, FO428/3.

[59] Thesiger (Addis Ababa) to Sir Edward Grey, 12 September 1912, NAUK, FO428/10.

[60] Vice-Consul H.H. Dodds (Harrar) to Commissioner Byatt, 29 February 1912, NAUK, FO428/10.

[61] Doughty-Wylie (Abyssinia) to Sir Edward Grey, 12 December 1912, NAUK, FO428/10.

[62] Marquis di San Giuliano to Sir Edward Grey, 13 August 1907, NAUK, FO428/1.

[63] Thesiger (Addis Ababa) to Sir Edward Grey, 6 May 1913, NAUK, FO428/12.

[64] Sir Edward Grey to Thesiger (Addis Ababa), 17 May 1913, NAUK, FO428/12.

to stop the traffic'.[65] Significantly, HMS *Proserpine* was 'practically despatched for service on the Somali coast' from Slade's squadron in the Gulf of Oman, thus confirming the northern Horn as part of the 'Arabian frontier'.[66]

One element of the trade that the British only periodically acknowledged was that Britain itself remained one of the main proliferators of small arms to the Arabian frontier. In his 2007 global study Jonathan Grant could write that 'distinct national differences in attitudes toward the arms business emerged on the part of the Great Power governments. Britain offered the least support for arms sales diplomatically and financially.'[67] This was certainly not the view of the Ottoman government.[68] The first checks on the trade had been made in 1897–1898, when the Persian government, and subsequently the Sultan of Muscat, were persuaded to give the Royal Navy the authority to stop and search vessels in the Gulf.[69] At that time the Birmingham gun trade had complained that the measures were an 'illegal and unjustifiable' interference with legitimate trade. It claimed to have supplied 30,000 weapons per year to the Gulf region since 1880.[70] Twenty-five thousand rifles from Britain were still being landed at Muscat in 1910. By 1910 'the trade has long ceased to be, if it ever was, a French monopoly. In fact, the value of arms imported into Muscat ... from France was considerably smaller than that of the importations from ... Germany, Belgium, or the UK itself.'[71]

The British political resident in Aden argued that trade there was as great as in Muscat. Brigadier Bell observed that 'a regular gun-running business is carried on by the Arabs between Jibuti and certain ports in the Aden protectorate ... [W]e can do nothing to check this gun-running unless we have our own armed dhow service.'[72] An armed dhow patrol was instituted in March 1911. Even then most of the intelligence reports Bell received concerned gun-running across the Gulf of Aden: that 'illicit traffic in arms is rampant in our protectorate is notorious', he observed sadly.[73] The British complained to the Ottomans about their 'arbitrary seizure' of British rifles in the Red Sea.[74] Bell, however, was

65 Thesiger to Sir Edward Grey, 17 September 1913, NAUK, FO428/13.

66 'British Interests in the Persian Gulf: III – The Operations against Gun-Running', *Times*, 8 July 1911, p. 7.

67 Grant, *Rulers, Guns and Money*, p. 232.

68 Gökhan Çetinsaya, 'The Ottoman view of the British Presence in Iraq and the Gulf: The Era of Abdulhamid II', *Middle Eastern Studies*, 39/2 (2003), pp. 194–203.

69 Burrell, 'Arms and Afghans in the Makrān', pp. 8–24.

70 *The Times*, 18 July 1898.

71 'British Interests in the Persian Gulf: II – The Arms Traffic', *Times*, 6 July 1911, p. 5.

72 Brigadier-General Bell to Government of Bombay, 22 April 1911, NAUK, FO428/8.

73 Major-General J.A. Bell to Government of India, 18 December 1912, NAUK, FO428/12.

74 Vice Consul G.A. Richardson (Hodeidah) to Sir G. Lowther, 8 July 1911, NAUK, FO428/9.

embarrassed about 'how little we have done to carry out our obligations under the Brussels Convention'.[75]

The most spectacular act of small arms proliferation by Britain occurred in Somaliland in 1910. Despairing of suppressing the Islamic insurgency launched by the 'Mad Mullah' in 1898, a new commissioner with a new policy was installed. The British disbanded and disarmed the Somaliland battalion of the King's African Rifles, withdrew from the interior to a few coastal towns, and supplied a large quantity of modern firearms to 'friendly' tribes. Within months these tribes were massively supplementing their supply from the market at Jibuti/Harrar.[76]

When Sir Edward Grey tried to persuade the French prime minister, Georges Clemenceau, on the basis of intelligence gathered in Hamburg, that France was up to its ears in illegal arms smuggling, the Tiger replied that his people were blameless: the worst arms traffickers were British subjects.[77] *The Times* admitted at the end of 1910 that 'we cannot ... approach France without also taking heed of our own shortcomings in this grave matter. A Consular report shows that out of 85,820 rifles landed at Muscat in 1908–9, 43,280 were Martinis from Belgium and 25,600 came from Great Britain ... A case tried in London on March 15 last revealed fresh indications that British rifles were reaching Muscat ... It is useless to patrol the Mekran coast if rifles can be shipped in the Thames.'[78] 'It may be of interest to mention the fact', Arnold Keppel thought, 'that among the empty cartridge cases found in the Baluch trenches after the action of the 28th [1911] besides those of the principal Continental manufacturing states, were many of English make, specimens of which I retained as souvenirs.'[79]

The emerging problem, as seen by the trade and its observers, was not so much that Britain was ceasing to supply a large volume of weapons, but that these guns were at the junk end of the market, whereas others moved into the top end. The Birmingham trade was still organised in a series of relatively small plants. The Birmingham Small Arms Company of Small Heath was the largest rifle producer. 'During the period when the small arm was developing from a comparative crudely made rifle into a weapon of precision', it was reported in 1912, 'the capacity of the Small Heath shops gradually increased from 500 to 2,000 rifles a week.'[80] By comparison the Steyr works was, according to a British visitor, already manufacturing modern Mauser and Mannlicher-designed rifles

---

75 Major-General J. Bell to Government of India, 13 April 1912, NAUK, FO428/10.

76 *Cd. 5000 Correspondence relating to Affairs in Somaliland*, March 1910 and *Cd. 5132 Somaliland. Further correspondence relating to affairs in Somaliland*, April 1910; Robert Hess, 'The "Mad Mullah" and Northern Somalia', *Journal of African History*, 5/3 (1964), pp. 415–433.

77 Bertie to Grey, 8 August 1908, NAUK, FO428/3.

78 'Leader – Gun-Running in the Persian Gulf', *Times*, 28 December 1910, p. 7.

79 Keppel, *Gun-Running*, p. 182.

80 'Arms and Ammunition: The Birmingham Gun Trade', *Times*, 2 October 1912.

at the rate of 14,000 weapons per week by 1901.[81] In 1910 the Liège testing station stamped 1,175,723 weapons, or 22,600 weapons each week.[82] The largest concern in the city, FN, had been manufacturing modern Mausers since 1889.

British exports of old rifles were handled by disreputable dealers, notably the Birmingham companies of Clabrough & Johnstone and Laubenburg. Clabrough & Johnstone maintained a massive store of Martini cartridges at West India Dock for supply to the Gulf. In a three month period in mid-1907 it shipped 10,000 Martini rifles and seven million cartridges from London to Muscat. The company was prosecuted for sending ammunition to Bushire and was generally regarded as a 'danger to British interests'.[83] Percy Cox denounced the 'unscrupulous attitude of British arms manufacturers'.[84]

The official report on the aftermath of Slade's operations concluded that 'Germany has sent out costly weapons, and has beaten the UK in value and quality though not in quantity'.[85] Consul William Ward in Hamburg, who was working with the secret service, calculated that the *Hamburg-Amerika* line's service to the Red Sea and the Persian Gulf, started in 1906, carried 70 tons of arms and ammunition each month.[86] In 1909 it was estimated that German dealers had imported two thousand cases – over 100,000 rifles – into the frontier via Muscat.[87] By the time the products reached the Gulf 'it was impossible to say with accuracy from what specific source in Europe the captured rifles came because the Muscat merchants buy from one another, and the supplies from Europe are often ordered through agents ... who purchase from manufacturers'.[88]

Amongst the companies certainly involved was Gustav Genschow of Hamburg. Under the trade name of Geco, the Genschow concern was a very well-known large-scale supplier. Another favoured source was Meffert of Suhl, the traditional Thuringian gun-making centre. Genschow and Meffert made and converted rifles, as well as buying in weapons and branding them. The rifles going to the Arabian frontier were often of a modern 'Mannlicher'

81 W.A. Baillie-Grohman, 'An Unarmed People', *Fortnightly Review*, 69/411 (March 1901), pp. 527–542. The author, a well-known Anglo-Austrian big game hunter and imperial entrepreneur, owned an estate near the plant and was bilingual in English and German.

82 Verbatim Report of the Fourth Plenary Meeting of the Conference for the Control of the International Trade in Arms, Munitions and Implements of War at Geneva, 6 May 1925, NAUK, FO428/22.

83 Home Office to Foreign Office, 23 November 1907, NAUK, FO428/1.

84 Cox to Townley (Teheran), 8 December 1912, NAUK, FO428/12.

85 General Staff, India, 'Report on the Arms Traffic, 1st July 1911 to 30th June 1913', Simla, 1913, NAUK, WO106/6322.

86 Ward (Hamburg) to Sir Edward Grey, 26 January 1909, NAUK, FO428/4.

87 Commissioner H.E.S. Cordeaux (Camp Sheikh) to Lord Crewe, 21 June 1909, NAUK, FO428/5.

88 Holland (Muscat) to Political Resident (Bushire), 16 January 1910, NAUK, FO428/6

type, specifically the German 1888 Commission rifle.[89] The 1888 rifle was the first smokeless powder small-bore rifle of the German Army and used licensed Austrian Mannlicher technology. It lasted as a front-line weapon for a mere 10 years before being displaced by a superior Mauser design. Observers at Marseilles confirmed that regular shipments of 7.9 mm Mauser (used in the 1888 rifle) and 6.5 mm Mannlicher-Schoenauer cartridges were passing through the port on their way from Austria.[90]

The unravelling of affairs in Muscat provided considerable detail about the frontier's dealer network. The pioneering dealer in the Gulf had been Antoine Goguyer. He had arrived in Muscat in March 1899 with a new business model. Goguyer used native vessels flying the French flag, thus protected from any interference, to ply a trade in the Persian Gulf, the Gulf of Oman, the Horn of Africa and the Red Sea. When Goguyer returned to France, the business was run by his nephew, Ibrahim Elbaz, 'well dressed in Arab clothing, wears sandals, has light hair, long and heavy moustached, clipped beard ... about 5 ft. 3 in. height ... addicted to the use of scent'.[91] By the time Goguyer died, in 1909, Elbaz was losing out to the agents of another French dealer, Louis Dieu, and the German, Robert Wönckhaus.[92] Louis Dieu only started trading out of Muscat in 1909. In the each of the three years before the closure of Muscat the firm made £6,800 per annum profit from its Gulf trade.[93] Wönckhaus, who had arrived in the Gulf in 1897, from Hamburg via Zanzibar, had close links with the *Hamburg-Amerika* line.[94] Wönckhaus and Dieu used elaborate methods to avoid the British blockade. Their most notorious technique was the 'sugar method'. Sugar was loaded at Antwerp for shipment to Kuwait, via Bombay. 'The sugar was packed in layers and the rifles were cleverly concealed in the middle of the packages, the stocks and barrels being packed separately.' By 1911 'this method of smuggling arms to the Persian Gulf ... [was] now being employed in a systematic manner.' The method was pioneered by Wönckhaus, although it was Louis Dieu who was rumbled. He had, investigators discovered, ordered Mausers from the Loewe company in Berlin but 'the whole transaction has purposely been complicated,

89 Major Trevor (Bushire) to Government of India, 30 January 1910, NAUK, FO428/6.

90 Consul-General Gurney (Marseilles) to Sir Edward Grey, 15 July 1908, NAUK, FO428/3.

91 Knox (Kuwait) to Major Cox, 29 September 1907, NAUK, FO428/1.

92 Holland (Muscat) to Political Resident, Bushire, 16 January 1910, NAUK, FO428/6.

93 Colonel Tisdall to Bertie, 17 November 1913, NAUK, FO428/13.

94 Robert Machray, 'The Germans in Persia', *Fortnightly Review*, 590 (February 1916), pp. 342–353; Eugene Staley, 'Business and Politics in the Persian Gulf: The Story of the Wönckhaus Firm', *Political Science Quarterly*, 48 (September 1933), pp. 367–385; C.J. Edmonds, 'The Persian Gulf Prelude to the Zimmermann Telegram', *Journal of the Royal Central Asian Society*, 47/1 (1960), pp. 58–67; and Sean McMeekin, *The Berlin-Baghdad Express: The Ottoman Empire and Germany's Bid for World Power, 1898–1918* (London, 2010), pp. 89–91 .

and the Paris firm ... covered their tracks in such a way as to make legal conviction most difficult'.[95]

Consumption and Consequences

In 1907 the Bushire Residency was disturbed by a report of 'a Persian servant ... poring over the price list of the Army and Navy Stores, in which he quite able to distinguish by the illustration between a Browning and a Mauser pistol'.[96] The servant was thus one of many potential customers on the Arabian frontier demonstrating significant brand awareness. The Mauser and the Browning were examples of the most technologically advanced modern firearms. Both were easily available at Muscat. The German Mauser was the first commercially successful self-loading pistol, introduced in 1896. The Browning, manufactured by FN in Liège, established a standard design format for self-loading pistols when it was introduced as a competing product in 1900. Browning, like Hoover, became a brand that stood for an entire class of products, as in the phrase '*Wenn ich Kultur ... entsichere ich meinen Browning*'.[97] Each had high-profile users: Churchill carried a Mauser in the Boer War; Gavrilo Princip assassinated Franz Ferdinand with a Browning in 1914. At the moment that the Persian servant was weighing up the merits of the two designs, the armies of the world were doing the same. The Browning design, for instance, was adopted by the US Army in 1911 – although the British military stuck with the old-fashioned revolver.

Observers around the Arabian frontier noticed specific brand preferences. A February 1913 report commented that

> Afghans preferred Lee-Metfords to Mausers, whereas Persian opinion seen equally divided regarding the two makes, but preferred carbines to rifles. Persian tribesmen of Fars and south-west Persia buy the best modern rifles available, and prefer Mausers and Mannlichers. Arabs of Arabistan prefer Martini-Henrys, as do the Arabs of Arabia generally. Arabs of Kuwait, however, are chiefly armed with Martini-Metfords and Mausers; while those of the Pirate Coast are armed with Mausers and soft-nosed bullets.

On the Khyber, a Lee-Metford, a British bolt-action, small bore, black powder rifle, an *owah dazah*, was readily distinguishable from a Mauser, a German

95 Government of India to Government of Bombay, 28 September 1911, NAUK, FO428/9.

96 Extract from the Political Diary of the Persian Gulf Residency for the week ending 29 September 1907, NAUK, FO428/1.

97 'Whenever I hear the word culture I reach for my pistol.'

bolt-action, small-bore, smokeless powder rifle, a *pinza daza*.[98] Preparing for his seizure of territory on the Persian Gulf in May 1913 Ibn Saud demanded back rifles seized at Sharjah by brand.[99] In February 1911, the army officer and traveller G.E. Leachman observed the different armaments carried by tribes at the wells of Samit. He noticed that Madan Arabs were, 'as a rule, better armed than the Beduin ... most of them carrying a good type of Martini carbine; they invariably, however, knock the sights off, as having no use for them.'[100] 'The Le Gras pattern of rifle', the British Somaliland Protectorate reported in 1907, 'has now practically disappeared, having been ousted from the market by a well-finished looking Martini-Enfield weapon ... ammunition .303 black powder ball, nickel coated (with soft lead nose).'[101] In Persia, 'the tribes in Fars and Luristan ... are at present mostly armed with small bore modern rifles, the ammunition for which cannot be refilled. The older rifles, the cartridge cases of which could be filled with black powder many times, are mostly sold to tribes further north and east.'[102] Some of these selections could be ascribed to availability or to practical issues, most notably the availability of smokeless ammunition noted by Admiral Slade. There was, however, a genuine range of choice.

A study of Muscat in 1911 revealed specific local terms, and specific values, ascribed to weapons, depending on their brand.

Table 8.1 Firearms in Muscat, 1911

British designation	**Muscat designation**	**Description**	**Price (rupees)***
Lee-Enfield	Abu-Ashar	Small-bore smokeless powder rifle with 10-shot magazine	60
Mauser 7 mm	Filsi or Panj-tiri	Small-bore smokeless powder rifle with 5-shot charger-loaded magazine	80
Mauser 7.9 mm	Damudari	Small-bore smokeless powder rifle with 5-shot charger-loaded magazine	60

98 J.L. Maffey (Political Agent, Khyber), 'Note on Adam Khel Gun-runners', August 1909, NAUK, FO428/5.

99 Captain W.H.I. Shakespear to Percy Cox, 29 April 1913, FO428/13; Jacob Goldberg, 'The 1913 Saudi Occupation of Hasa Reconsidered', *Middle Eastern Studies*, 18/1 (1982), pp. 21–29; 'Captain Shakespear and Ibn Saud: A Balanced Reappraisal', *Middle Eastern Studies*, 22/1 (1986), pp. 74–88.

100 Captain G.E. Leachman (Royal Sussex Regiment), 'A Journey in North-Eastern Arabia', *Geographical Journal*, 37 (1911), pp. 265–274.

101 British Somaliland Protectorate: Intelligence Report No. 14, August 1907, NAUK, FO428/1.

102 C-in-C, East Indies to Admiralty, 20 January 1912, NAUK, FO428/10.

British designation	Muscat designation	Description	Price (rupees)*
Romanian Mannlicher 6.5 mm	Roumani	Small-bore smokeless powder carbine with 5-shot clip-loaded magazine	55
English Martini-Metford	Mauser (Martini) Londoni**	Small-bore breech-loading single-shot black powder rifle	35
Foreign Martini-Metford	Mauser (Martini) Belgiqui	Small-bore breech-loading single-shot black powder rifle	17
English Martini-Henry	Martini Xibir Londoni	Full-bore breech-loading single-shot black powder rifle	25
English Martini-Henry	Martini Sagir	Full-bore breech-loading single-shot black powder carbine	20
Foreign Martini-Henry	Martini Belgiqui	Full-bore breech-loading single-shot black powder rifle	25
Foreign Martini-Henry	Martini Sagir	Full-bore single-shot breech-loading black powder carbine	20
Snider	Snider	Large-bore single-shot converted breech-loading rifle	10
Gras	Fransawi	Large-bore single-shot converted breech-loading rifle	12
Gras	Fransawi Sagir	Large-bore single-shot converted breech-loading carbine	10
Werndl	Soljeri	Large-bore single-shot converted breech-loading rifle	3

* 1 rupee was worth 1s, 4d; ** Martini-action single shots were often referred to as 'Mausers'. A 'true' Mauser had a different local designation. General Staff, India, 'Report on the Arms Traffic, 1st July 1911 to 30th June 1913', Simla, 1913, NAUK, WO106/6322.

The British were not necessarily enamoured by the vibrant gun culture of the Arabian frontier. 'For', *The Times* remarked, 'apart from the danger which the arms traffic constitutes for the security of our Indian borderland, it has imported into the Persian Gulf a new element of demoralization which is imperilling the fruits of our labours during the last half-century in the cause of peaceful and orderly progress.'[103] Major Grey, the British representative at Muscat, in calling for the suppression of the Gulf trade, wrote that

> my reason ... can hardly be expected to carry weight except from a purely local point of view, but I state it, being compelled from my position to take an interest in the welfare of the people of Oman. As things are at present, it is the ambition of every bucolic Arab to become possessed of a rifle and cartridges, and money that might profitably be expended upon improving his social condition is hoarded up until the necessary sum has been collected, while the universal possession of fire-arms by people increases the inevitable bloodshed in each case of an inter-tribal quarrel. Hence, for the Omani's own sake, it is better that the wholesale importation of arms should cease.[104]

The new commissioner of British Somaliland, H. A. Byatt, had a particularly bleak view of the gun culture his government had created by the arming of the tribes. He reported from Berbera that

> raiding and fighting is a matter of everyday occurrence, and the total amount of bloodshed that that has taken place is little short of alarming. The roads to the interior are no longer safe ... the main causes of this state of anarchy are two: one being the unchecked influx of firearms into the protectorate, and the other the feeling, which is steadily growing among all natives, that the Government is ... unable to intervene in their affairs for the enforcement of order.

He warned that the supply of firearms to 'allied' tribes would bring little benefit to the fight against the Islamicists.[105] 'Since the issue of arms to the Ishaak Somalis in British Somaliland,' his representative at Harrar reported,

> these natives have lost no opportunity to increase their armament. The arms and ammunition issued by HMG in 1910 were found to be infinitely superior weapons than the spears and shields of former days with which to carry out intertribal warfare. The supply of ammunition, however, in time gave out. A further issue of .303 and Martini was unobtainable. The .303 and Martini rifles became useless. The native appetite for firearms had been whetted. It became necessary to find arms elsewhere for which ammunition was always procurable. Markets, which had already been suggested to them, were found in French Somaliland and Abyssinia. The traffic then begun has been carried out with so much success that the tribes

---

103 'British Interests in the Persian Gulf: III – The Operations against Gun-Running', *Times*, 8 July 1911, p. 7.

104 Major Grey (Muscat) to Percy Cox (Bushire), 26 November 1906, NAUK, FO428/1.

105 Acting Commissioner H.A. Byatt to Harcourt, 28 August 1911, NAUK, FO428/9.

are now, two years afterwards, almost independent of the Government issue of 1910 ... the source of origin is, however, the same, through Jibuti.[106]

The result was massive tribal raids on trade caravans. 'The first of these raiding parties has recently returned', Commissioner Byatt reported in April 1912.

> [I]t is said to have numbered 1,991 rifles, nearly all of which have of course been obtained through Abyssinian intermediaries from Jibuti ... After the experience of the past two years, there can, I think, be no further question as to the futility of all efforts to negotiate with the Habr Yunis tribe, and to control them by moral influence. Pressure brought upon them in the coast towns has proved to be entirely ineffective. They are rich, and thanks to the stream of arms flowing through many channels from Jibuti, they are altogether independent of Government control. I cannot regard them as otherwise than a danger to the general peace no less serious than that threatened by the Mullah.[107]

The main focus of debate, however, remained Muscat. It seemed to some that Britain's complicity in the arms trade, followed by robust action in pursuit of goals on the Indian frontier, had caused dangerous blowback in the Arabian frontier. In July 1913 the tribal confederations of the Omani interior launched a serious revolt against the Sultan of Muscat.[108] Their cause was overtly Islamist and anti-Western, the declared aim being to restore the Imamate.[109] The Sultan took the offensive with disastrous effect. He could hold Muscat with British aid. The East Indies Squadron – in the form of two cruisers and a sloop – originally deployed by Slade, remained in the port. Troops were sent from Bushire and, subsequently, Bombay. The sultan's power, however, relied on control of fertile agricultural territory inland.[110] He attempted to recapture this land but his small army of Baluchi mercenaries were defeated by 'well armed' tribesmen.

The revolt contained the 'seeds of an awkward difficulty for Great Britain'. The sultan's agreement to set up the bonded warehouse was 'used as the pretext to inflame his subjects against him'. In December 1912 the leader of the revolt, Sheikh Isa bin Saleh, had written to the sultan denouncing his decision to strike a deal with the British to curb the arms traffic. 'He said that they had waited long enough to see His Highness get rid of the European elements from his country,

---

106 Vice-Consul H.H. Dodds (Harrar) to Commisssioner Byatt, 29 February 1912, NAUK, FO428/10.

107 Commissioner Byatt (Berbera) to Harcourt, 18 April 1912, NAUK, FO428/10.

108 'The Revolt in Oman', *Times*, 30 August 1913.

109 Nabil M. Kaylani, 'Politics and Religion in 'Umān: An Historical Overview', *International Journal of Middle East Studies*, 10/4 (1979), pp. 567–579.

110 Mark Speece, 'Aspects of Economic Dualism in Oman, 1830–1930', *International Journal of Middle East Studies*, 21/4 (1989), pp. 495–515.

and since he has not done so, they must depend on themselves relying on God for help.'[111] At the same time, 'thanks to the arms traffic, modern rifles and ammunition have been strewn broadcast in Oman'.[112] The Indian Army report on the traffic commented that, 'through his previous encouragement of this nefarious trade, [the Sultan] had been gradually raising up trouble for himself ... these rebels have captured large depots of arms in the interior'.[113] The sultan was ineffective against such well-equipped foes: the British realised that they too would require a major land expeditionary force if they were to intervene – this they were unwilling to countenance. The result was stalemate. Sultan Feisal died unexpectedly in October 1913 and was succeeded by his son, Taymur. The revolt only reached its climax in January 1915 when the Islamist insurgents tried to seize Muscat by force: they were driven off by Indian Army troops with insurgent losses rising to 500 men.[114]

Conclusion

The British post-mortem on events in the Arabian frontier between the Battle of Dubai in December 1910 and the Battle of Muscat in January 1915 noted that unilateral operations, even when successful, incurred heavy costs, political as well as financial.[115] The report also concluded that the gun culture that had been created on the Arabian frontier before the outbreak of the First World War could do nothing but thrive further once War had broken out. 'Whether we control the arms traffic or no,' wrote the Middle East political fixer Mark Sykes in January 1917,

> it must be remembered that immense masses of modern arms have already been issued to natives in Persia, Arabia and Turkey, and these will spread in all directions ... there will be such a glut of rifles in all countries, neutral included, that it will be impossible to prevent their becoming very easy to get at for those who want them. If there are 10 or 12 million spare arms in the world, five or

111 Extract from Political Diary of the Persian Gulf Residency for the month of December 1912, NAUK, FO428/12.

112 'Leader: The Revolt in Oman', *Times*, 18 September 1913.

113 General Staff, India, 'Report on the Arms Traffic, 1st July 1911 to 30th June 1913', Simla, 1913, NAUK, WO106/6322.

114 'Arab Descent on Muscat', *Times*, 19 January 1915, p. 8.

115 Committee of Imperial Defence: Sub-Committee on the Arms Traffic, 1st meeting, India Office, 8 January 1917, NAUK, CAB16/44.

> hundred thousand will be trafficked, and such a number is enough to arm every black man who wants a rifle.[116]

The specific issues in Oman could be addressed. The revolt against Sultan Taymur eventually ended in an inter-tribal agreement at al-Sib.[117] In the 1930s the subsidy paid to the Sultan of Oman, in return for not engaging in the arms trade, was brought to an end.[118] Kuweit became a more significant arms centre.[119] By 1920 it was noted that 'Koweit is the chief place from which there is a steady stream of arms into Persia'. There were plenty of modern rifles and ammunition available, 'collected from the former battlefields of Mesopotamia, from friendly tribes of that country, to whom large issues of arms were made by the Government during the War'. Indeed there were so many weapons in the Gulf that there was no need to re-stock from Jibuti. The Persian Gulf trade was, however, 'trivial'.[120]

Much more powerful forces had, however, been unleashed on the frontier.[121] The major immediate problem was the unchecked influx of arms into the Ottoman Empire. These arms fuelled the Kemalist-Greek conflict and, ultimately, Britain's humiliation at Chanak in 1922. 'The free import of arms and munitions into Turkey and Greece by private firms', the Foreign Office regretted, 'continued unchecked until the late autumn of 1922 when, after the Greek defeat and the cessation of active hostilities between Greece and Turkey, it became apparent that any arms exported to Turkey could only be used against the Allies.'[122]

British post-war policy attempted to learn lessons from the Arabian frontier experience. These lessons were two-fold. First, if Britain itself was not to continue as a proliferator of small arms, then tight control of British exports

---

116 Committee of Imperial Defence: Sub-Committee on Arms Traffic, 'Some Considerations on the Traffic in Arms as a Post-War Problem', Memorandum by Lieutenant-Colonel Sir Mark Sykes, Bart, MP, 12 January 1917, NAUK, CAB16/44.

117 Kaylani, 'Politics and Religion in 'Umān'.

118 Political Resident, Bushire (T. C. Fowle) to HM Minister, Tehran, 24 January 1938, NAUK, FO371/21830.

119 Percy Sykes, 'South Persia and the Great War', *Geographical Journal*, 58 (1921), pp. 101–116; A.R. Lindt, 'Politics in the Persian Gulf', *Journal of the Royal Central Asian Society*, 26 (1939), pp. 619–633; Andrew Loewenstein, '"The Veiled Protectorate of Kowait": Liberalized Imperialism and British Efforts to Influence Kuwaiti Domestic Policy during the reign of Sheikh Ahmad al-Jaber, 1938–1950', *Middle Eastern Studies*, 36/2 (2000), pp. 103–123.

120 Senior Intelligence Officer, East Indies, Intelligence Report No. 10, 20 December 1920, NAUK, FO428/17.

121 John Fisher, '"The Safety of our Indian Empire": Lord Curzon and British Predominance in the Arabian Peninsula, 1919', *Middle Eastern Studies*, 33/3 (1997), pp. 494–520.

122 G.W. Rendel, Note on the Export of Arms and Munitions to Turkey, 1920 to 1924, 12 February 1924, NAUK, FO428/21.

was necessary.[123] Under the new export regime Laubenburgs ceased trading in 1919.[124] The inquiry into the arms trade on the Arabian frontier in turn fed through into domestic policy, crystallised in strict domestic firearms legislation introduced in 1920. More ambitiously, the British government committed itself to a multilateral small arms control regime. The attempt to construct this international regime foundered in 1925, largely because of the virulent opposition of the 'purchasers of death' most affected by Britain's pre-war efforts in the Straits of Hormuz.[125]

This conflict was encapsulated in the clash between Sir Percy Cox, representing the government of India, and the representatives of Reza Shah's Persian government at the 1925 Geneva Conference.[126] 'I speak', Cox declared, 'with lively personal recollections of a period of some eleven years' struggle ... to cope with a violent traffic in arms ... passing through small ports of the Gulf of Oman.' The trade had 'now happily been reduced to exiguous proportions'. The risk of recrudescence and the need for vigilance, however, was ever present.[127] The Persian delegation, by contrast, responded with a demand that the Persian Gulf and the Gulf of Oman should be immediately removed from any 'prohibited zone'. Cox replied that if that was done Britain and India had no interest in signing an arms convention. 'One of the main objects of the convention is to suppress or control the illicit traffic in arms, especially rifles and revolvers, in those parts of the world where their possession by local inhabitants constitute a menace to peace and order', Cox claimed. Persian recalcitrance was 'based mainly on sentiment ... even today, not only traffic in arms, but piracy and traffic in slaves, intermittently occur in the water which we are discussing. ... They flit across the Gulf to the Arab side, sell their human freight and bring back arms.'

123 Memorandum prepared by the Board of Trade in Collaboration with the Foreign Office, Admiralty, War Office, Air Ministry and Board of Customs and Excise: The Licensing System for the Control of the Export of Arms in Minutes of Evidence taken before the Royal Commission on the Private Manufacture of and Trading in Arms. Twelfth Day, Wednesday, 27 November 1935, NAUK, T181/112.

124 *Times*, 16 August 1919, p. 17.

125 David Stone, 'Imperialism and Sovereignty: The League of Nations' Drive to Control the Global Arms Trade', *Journal of Contemporary History*, 35/2 (2000), pp. 213–230; Andrew Webster, 'Making Disarmament Work: The Implementation of the International Disarmament Provisions in the League of Nations Covenant, 1919–1925', *Diplomacy and Statecraft*, 16/3 (2005), pp. 551–569, and, by the same author, 'From Versailles to Geneva: The Many Forms of Interwar Disarmament', *Journal of Strategic Studies*, 29/2 (2006), pp. 225–246.

126 John Townsend, 'Some Reflections on the Career of Sir Percy Cox', *Asian Affairs*, 24/3 (1993), pp. 259–272; Michael Zirinsky, 'Imperial Power and Dictatorship: Britain and the Rise of Reza Shah, 1921–1926', *International Journal of Middle East Studies*, 24/4 (1992), pp. 639–663.

127 Verbatim Report of the Seventh Meeting of the General Committee, held on Thursday, 14 May 1925, NAUK, FO428/22.

Faced with Cox's démarche, the Persians walked out of the Conference.[128] Sir Edmond Slade died in 1928. His operations were remembered more vividly on the Arabian frontier than in Britain.

[128] Verbatim Report of the 26th meeting of the General Committee, held on Monday, 15 June 1925, NAUK, FO428/22.

Chapter 9

'Give Him a Gun, NOW': Soldiers but Not Quite Soldiers in South Africa's Second World War, 1939–1945

Bill Nasson

Postal Interrogation

The notion of arming black servicemen in South Africa for combat duties in its Union Defence Forces (UDF) during the Second World War was stamped on by its political leadership, even when it came to stamps. For, in wartime, politics and the postal service intersected in some striking ways. A small story about a small Union of South Africa postage stamp provides a novel opening window through which to glimpse the meaning of firearms in a distinctive national war effort, that of an Allied country which mobilised its volunteers either as soldiering citizens or as soldiering subjects.

At the end of 1940, Captain Neville Lewis – an English UDF officer who had served in the British Army during the First World War, had trained in London at the Slade School of Art, and was a recognised portrait painter who had done both General Jan Smuts and General Bernard Montgomery – was appointed South Africa's first official war artist. One of his more leisurely early assignments was a commission by the Minister of Posts and Telegraphs to produce a patriotic stamp series in honour of the servicemen and servicewomen of the Union Defence Forces.[1] Already steeped in representations of steely, square-jawed masculinity and doughty femininity for his war art programme both on the home front and in the fighting theatres of East Africa and North Africa, Lewis was not one to keep his head down, nor his canvas narrow.

Favouring his customary broad brush, Neville Lewis produced five sinewy portraits that embodied the essence of wartime Springbok sturdiness – the fighter pilot, the sailor, the field hospital nurse, and the soldier. Or, to be more exact, the *soldiers*. Along with a white tank corps gunner, he submitted a portrait of an African soldier of the South African Native Military Corps. His model,

1 Allan Sinclair, 'The Use of the Neville Lewis Portraits for the Second World War Stamp Series', *Military History Journal*, 11/1 (1998), pp. 1–2.

a Sotho Lance-Corporal named Silas Molapo, was an agricultural tenant on an Orange Free State farm and a boxer of local reputation in his birthplace of Thaba Nchu.[2] On enlistment in the Non-European Army Services (NEAS) of the UDF at the end of 1939, Molapo had reportedly turned up with 'his native family treasure, something rather disturbing to see', according to a recruiting official.[3] That proud heirloom was a British Army Lee-Metford rifle, claimed by its bearer to have been captured by a relative who had seen war service as a mounted servant or *agterryer* with a Boer commando in the South African War of 1899–1902. It was swiftly confiscated.

Restricted to serving in Motor Transport as a lorry driver in the Native Military Corps (NMC), the personal role in armed conflict after which Molapo was hankering was never to materialise. A veteran of South Africa's 1940–1941 East African campaign against Italy at the time that Lewis painted him, his gun familiarity and dead-eye accuracy in downing antelope and bush-pig when handed a rifle impressed accompanying white officers.[4] One extreme martinet even mused that should 'the more exasperating incidents' of ill-discipline among NEAS troops ever warrant the formation of a firing-squad, Lance-Corporal Molapo would be 'ideal', the 'natural type for doing a clean job', in the event of such a necessity arising.[5]

Captain Lewis, who in pre-war years had been partial to romantic African subjects in Southern and Central Africa, considered his head-and-shoulders portrait of Silas Molapo to be the best of his postal services bunch, 'massively soldierly', and a 'dignified tribute' to the 'military spirit and sacrifice' of 'the men of our Native Military Corps'.[6] The Postmaster-General accepted the whole set as completely successful, complimenting the artist for so representative a range of men and women. His minister, though, was a little nervous. He knew that he would have to contend with Jan Smuts, Prime Minister, Minister of External Affairs and Commander-in-Chief of the armed forces all rolled into one. It was not for nothing that Smuts's official biographer, Sir Keith Hancock, would write some 25 years later that nowhere else in the wartime British Empire and Commonwealth lay so immense a concentration of military and political power in the hands of a single man.[7] Bossy, fussy, and forever poking his nose into everything, even the size of postage stamps required his approval. How would Smuts react to the intrusive representation of an NMC soldier?

---

[2] Tlali Makhele, author interview, Cape Town, South Africa, August 2011.

[3] R. Stevens to Col. A.J. Gordon, 19 December 1939, Gordon war correspondence (privately held).

[4] Capt. D. Cochrane to Gordon, 4 January 1941, Gordon war correspondence.

[5] Lt P. Westhuizen to Gordon, 26 January 1941, Gordon war correspondence.

[6] Col. O.E.F. Baker, author interview, Gordon's Bay, Western Cape, South Africa, June 1998.

[7] W. Keith Hancock, *Smuts, Vol. 1: The Sanguine Years, 1870–1919* (Cambridge, 1962), p. 371; Bill Nasson, *South Africa at War 1939–1945* (Johannesburg, 2012), p. 9.

Anticipating what that would be, Lewis was asked by Posts and Telegraphs officials to amend – or, to be more precise, to soften – his portrait of Molapo to give it less of the look of the combatant or the warrior. Could it be turned into something less martial, so as not to convey an image of armed implacability, with the implied stiffening of a rifle at the side? A peeved Neville Lewis consented to one emblematic compromise alteration. A helmet was replaced by a softer slouch-hat. Among other effects, it turned Basutoland into Australia. More to the point, the artist grumbled, was the effect of having to turn his official war portrait into a damp squib. For it diminished a bellicose masculine image, converting Silas Molapo from a thrusting soldier into a 'common native batman', or 'just an orderly'.[8]

The point was, of course, that of ensuring that volunteers of the NEAS appeared as what they were meant to be – soldiering auxiliaries of various sorts, including drivers, blacksmiths, cooks, clerks and stretcher-bearers. If on that subdued basis, Posts and Telegraphs hoped, Captain Lewis's full UDF stamp set might pass muster with the notoriously picky Smuts. In the event, it did not. Unsurprisingly, the prime minister scrutinised the postal issue personally for his approval. No less unsurprisingly, he vetoed the use of the NMC portrait.[9] The UDF stamp series that appeared thus commemorated a white volunteer war effort. It went on to become so successful that Lewis's portraits were still putting in periodic appearances on envelopes until 1972.

This little philatelic trope opera, as it were, serves as something of a metaphor for the burning question of guns and black hands in South Africa's preferred European or white war effort of the 1940s. There are, perhaps, one or two other perfect illuminations of what that political convention entailed, again involving the vigilant gaze of Jan Smuts, and the realisation by his officers on campaign of what had to be kept under Pretoria's radar. Recalling the UDF's Abyssinian offensive, Colonel Oswald 'Ossie' Baker, remembered as 'a far-sighted and caring commanding officer', acknowledged that Carcano rifles and even some Beretta submachine guns jettisoned by retreating Italian infantry were at times handed out to NMC and Coloured Cape Corps servicemen of forward motor transport and pioneer companies. Scratch training in their use was provided, as 'after all, it stood to reason, all they'd had a chance to have a go at out there was footballs and maybe dart boards'. It was, as Baker observed, 'entirely against the rules, no question, but to be frank, man, we were a hell of a long way from Pretoria'.[10]

For such brokers of a more liberal military culture, what mattered clearly was that 'those in the Corps units were *our men*, the Coloured chaps were even reading what we were reading. They were facing whatever we were facing, the

8 Mrs L. Freint, personal communication to author, London, February 2011.

9 Sinclair, 'Neville Lewis Portraits', p. 3.

10 Col. O.E.F. Baker, author interview, Gordon's Bay, Western Cape, South Africa, June 1998.

Ities (Italians) were not all finished. We had to defend our side, you never knew what could happen next. They needed decent weapons to defend themselves and we saw to it, out there, in that situation.'[11] At the same time, the worry, there, too, was of their Commander-in-Chief and his politically observant generals, and of the severe consequences of their ever catching wind of such European South African slackness downwind of Addis Ababa. 'Well, let me tell you, if that issuing of the Italian stuff we grabbed had ever got out to him, the *Oubaas* ['old boss', a common nickname for Jan Smuts] would've had our balls, sure as night and day, no question. Probably court-martialled, maybe even shot, knowing him and his stickiness, rules, regulations, endless.'[12]

Remembering Guns for Some

In making some sense of all this, it is worth recalling briefly the background story of what constituted soldiering in modern segregationist South Africa following its independent formation as a Union in 1910, joining the other British dominions. In creating the Union Defence Forces two years later, the 1912 South Africa Defence Act confined military duty to eligible white men, identified as those being fully 'of European descent'. Blacks were excluded from army service without special emergency parliamentary authorisation in exceptional national circumstances.[13] A first 'anxiety was that having Africans bearing arms and sharing duties with whites would erode the colour-bar by condoning equality in relations between white and black'. A second 'turned on the inflammatory political consequences of enlisting as soldiers even a fraction of the rightless African population'.[14] As a historically minded white correspondent to *The Diamond Fields Advertiser* in December 1912 reminded that paper's readers, the 1879 Peace Preservation Act and the ensuing 1880–1881 Gun War or 'Basuto War' fought by Cape colonial forces to disarm upstart Africans had left a crucial legacy. It remained 'essential' to ensure total and 'permanent' African disarmament.[15]

In local mobilisation for the First World War, Pretoria was adamant that while Africans would be recruited for war work, they would not be doing any

[11] Capt. B. Stern, author interview, Cape Town, South Africa, September 2008.

[12] Baker, author interview, 1998.

[13] Bill Nasson, 'A Great Divide: South African Responses to the Great War, 1914–1918', *War & Society*, 12/1 (1994), p. 51.

[14] Bill Nasson, *Springboks on the Somme: South Africa in the Great War, 1914–1918* (Johannesburg, 2007), p. 3.

[15] *Diamond Fields Advertiser*, 15 December 1912. For a definitive account of the Cape-Basutoland Gun War, see Peter Sanders, *'Throwing Down White Man': Cape Rule and Misrule in Colonial Lesotho, 1871–1884* (Pontypool, 2011), esp. part 2.

shooting in any of the theatres to which the UDF would be despatched. Then, Walter Rubusana of the South African Native National Congress (forerunner of the African National Congress or ANC) had offered personally to assist the Louis Botha government by recruiting 5,000 African volunteers for training in the methods of modern warfare. Rebuffing him, in November 1914 the Secretary of Defence advised Rubusana that the authorities viewed 'the present war' as 'one which has its origins among the white people of Europe', and was accordingly 'anxious to avoid the employment of its native citizens in warfare against whites'.[16]

That said, from another quarter a minor hue and cry over the recognition of patriotism, citizenship and an old common right to bear arms had produced a partial dilution of the 1912 Defence establishment gun colour-bar. Although all African and most Coloured recruits were enrolled as unarmed members of labour contingents and labour battalions, 'the eighteen-thousand strong Cape Corps' mustered combat battalions that 'served meritoriously in East Africa and Palestine',[17] while in Egypt, following specialist instruction at weapons centres in Alexandria and Cairo, trained Coloured gunners ended up crewing batteries as members of General Sir Edmund Allenby's expeditionary force. Segregated and second-class South African soldiers they certainly were, but for all that they were still officially and recognisably soldiers under arms, at home in rifle exercises in Egyptian field training. Participating in Allenby's offensive against the Turks in Palestine, Cape Corps infantry also endured their own honourable 'blood' sacrifice in the 1918 battle of Square Hill, which left a pile of dead and medals for bravery under fire to be clung to by the living.[18] Although, as Ian van der Waag has pointed out, the end of hostilities saw the Coloured contingent 'disbanded in the face of political pressure',[19] its veteran combatants retained a cocky symbolic affinity with guns through the inter-war decades. After 1918, their main public site of commemorative pilgrimage in the contingent's home town of Kimberley was a mount holding a German field gun, captured from the Turks by the Cape Corps and shipped back to South Africa with the aid of subscription funds.

---

16 Bill Nasson, 'War Opinion in South Africa, 1914', *Journal of Imperial and Commonwealth History*, 23/2 (1995), pp. 255–256.

17 Ian van der Waag, 'The Union Defence Force Between the Two World Wars, 1919–1940', *Scientia Militaria*, 30/2 (2000), p. 190.

18 Nasson, *Springboks*, p. 159.

19 Van der Waag, 'Union Defence Force', p. 190.

From Cartridges to Cooking to Spears

The revival of the Cape Corps in 1939 gingered up its veterans' association, which promptly summoned up its 1914–1918 fighting reputation and stressed the civic status of enfranchised Coloured men of the Cape, proud holders of an imaginative imperial citizenship as 'civilised' 'English' or 'British' loyalists, surely second to none.[20] Just as in 1917 and 1918 Cape Corps infantrymen had not been regarded as 'barbaric Chinese' or 'common Egyptian' carriers and bearers, were they not again even more civilised than the westernised New Zealand Maori of Auckland? Once more, Coloured volunteers pressed to be entrusted with firearms, to be formed into proper combat battalions and to serve under arms 'fully, as soldiers, in common with the sons of European people'; as Abe Desmore, Chairman of the Cape Corps Association, put it, 'it is no more than a right[;] we share alike in a great duty.'[21]

But this was altogether another time, and quite another war that was far more politically combustible than the Union's already tricky involvement in the first global war of the British Empire. Now, dismissing any of the latitude in UDF service terms for that earlier involvement in war, the government was emphatic that the armed ranks of its Defence Forces would not be open to hybridisation. Not only, as was already entrenched custom, would there be no relaxation of racialised regulations that specifically prohibited Africans from the possession of firearms. This prohibition would now apply also to Coloured, Indian and Cape Malay (Muslim) servicemen being recruited for units of the NEAS. Their non-combatant status was fixed, and would remain fixed absolutely. Thus, in July 1940, an officer of the Cape Peninsula garrison was nonplussed to hear of rumoured instructions from the Minister of Defence that Malay Pioneer Battalion members, Malay Corps Motor Transport men and Cape Corps drivers were not to be entrusted unaccompanied with the shipment of infantry firearms. Oversight should not be left to 'Non-European NCOs'. Rather, NCOs from 'European units' were to be posted to keep a watchful eye, lest the army's transport auxiliaries be tempted to develop 'an unwelcome familiarity with musketry'. For the Cape Town officer, the story 'beggared belief, incredible ... it was not even a matter of whether it was justified or not justified'.[22] Yet, whether fanciful or otherwise, the key point of the Defence Department missive was that it revealed a great deal about the state of the official mind at the exact time in which the first South African forces to serve in the war were leaving

---

20 Bill Nasson, 'Why They Fought: Black Cape Colonists and Imperial Wars, 1899–1918', *International Journal of African Historical Studies*, 37/1 (2004), p. 50.

21 Quoted in ibid., p. 67.

22 Lt J.L. Uys to D. Cochrane, 23 July 1940, Gordon war correspondence.

the country – preoccupations, priorities, obsessions, anxieties and a need for controlling limits on the experience of war.

Several weeks previously, during some sharp parliamentary exchanges with anti-war Afrikaner nationalists over the formation of the NMC, Smuts had made no bones of the government position on firearms. 'To', he stressed, 'forestall any misrepresentation', and to 'prevent any possible misunderstanding', it was 'to be clearly understood that natives will not, under any circumstances, be equipped with arms of precision'.[23] Inevitably, 'arms of precision' attracted immediate and probing political attention. In defining what that meant, the prime minister took a judicious line on what was appropriate and what was adept for a country of Europeans and Non-Europeans at war. Smuts's arms of precision meant modern European weaponry such as machine guns, mortars, grenades, and other small arms such as Lee-Enfield rifles and even Webley revolvers. Equally, it was not quite the end of what was always more than an industrial story. In a convoluted series of explanations and justifications, the Department of Defence addressed the issue of whether unarmed NMC personnel constituted volunteer *soldiers*, or would be entitled to the standard service provisions of the Union Citizen Force. By the early 1940s, the answer had been repeated many times: UDF soldiers comprised those who bore the charmed arms of precision. Under the colour-bar recruitment policy, all regulatory matters to do with firearms – issue, maintenance, operational use and the like – applied solely to enlisted European men.

So, if NMC servicemen were *not* soldiers, were they, as several nit-picking pro-war publications asked, employed simply as labourers, and if so as semi-skilled or skilled workers? Emphatically not, blustered the NEAS Directorate. In the previous war, Africans had served in the national war effort in defined labour contingents or labour battalions, of undeniable menial status. Now, however, the NMC was akin to the Cape Corps and therefore not a labour corps formation. In its powers of command Proclamation 15 of 1942, white personnel of the NEAS were reminded that the attainment of 'a maximum effort' by the Union of South Africa hinged upon 'the successful employment of Non-Europeans'. The 'best results' would be secured only:

> [i]f their sacrifice and wish to serve are recognised and if they are treated as soldiers. This does not condone pandering and intermingling socially. It does, however, demand justice under all conditions, a sharing of whatever alleviation of hardship and abstention from manhandling, swearing at and addressing the Non-Europeans in a non-military manner, unbefitting that of soldiers.[24]

---

23 *Union House of Assembly Debates*, 12 June 1940, col. 79.

24 Noëlle Cowling (ed.), 'Historical Survey of the Non-European Army Services Outside of the Union of South Africa (Part II)', *Scientia Militaria*, 24/2 (1994), p. 33.

In other words, as an essential instrument of the Union's modern industrial war effort, its uniformed African volunteers were soldiers and those in authority were expected to recognise them as soldiers. The repeated message from figures such as Captain J.C. Knoetze, Deputy Assistant Adjutant General of the NEAS, was that just as the UDF had to raise front-line combatants, so those needed the efficient reinforcement of men from other units in supply and other support roles, some of them absolutely critical to health and welfare, such as stretcher-bearers, nursing orderlies and sanitary workers.[25] The overall effect of this gloss was to make the identity of African soldiers highly ambiguous, or to invite confusing intimations. Officially, they were not army workers. Implicitly, they were uniformed soldiers, graded by rank (however narrow the range) from Private to Warrant Officer. Recognisably, they were not functioning soldiers in the sense of being ordinary combat infantry

If guns – or, rather, their conspicuous absence – stayed with constricted Corps servicemen as a buried wound or a missing limb, their relationship to firearms also went further than that. The direction it took assumed some complicated turns, either sardonic or faintly ludicrous. In January 1941, a British officer with an Indian Infantry Division fighting in Operation Compass in the Western Desert dubbed the non-combatant status of the NMC 'bone-headed ... simply ridiculous' and a handicap upon the South African campaign contribution in Egypt.[26] The 'nonsensical' situation that Lieutenant Charles Daniel was observing was, in its way, a text-book illustration of the wartime racial dilemma of the Union of South Africa and the limits of its military resolution. There, inflexible segregation was the instinctual contemporary orthodoxy, but in overseas expeditionary conditions its answer was clumsy, if not tortuous. Stuck with a limited capacity to conduct warfare, the Union was poorly prepared for hostilities at the end of the 1930s. Its white Permanent and Active Citizen Forces were skinny. There were, to be sure, tens of thousands of men in the reservist Commandos who considered themselves to be born sharp-shooters, but the great majority were rusty as well as being inadequately equipped. Even worse, 'being rurally based and overwhelmingly Afrikaans, many of these men did not support the war effort'.[27] More generally, a large portion of the country's dominant white minority was vigorously anti-war in sentiment, with some plumping for peace and for neutrality, while for others, mainly waspish Afrikaner nationalists, being anti-war meant 'being anti-British and thus, to one

25 Ibid, p.18.

26 Cited in J.L. Uys to D. Cochrane, 29 January 1941, Gordon war correspondence.

27 Van der Waag, 'Union Defence Force', p. 213. But for those loyalists who did, see Albert Grundlingh, '"The King's Afrikaners"? Enlistment and Ethnic Identity in the Union of South Africa's Defence Force during the Second World War, 1939–45', *Journal of African History*, 40/3 (1999), pp. 351–365.

or other degree, either implicitly or explicitly pro-German'.[28] Given so brittle a domestic atmosphere, just as in the First World War, white conscription was simply untenable politically.

The consequence of being compelled to rely on volunteer recruitment of troops was that South Africa started the conflict with a shortage of white combatants, a front-line fighting deficit that would remain chronic throughout the war years. With that predicament – and the impetus it provided – concentrating the mind of the Department of Defence and the UDF, maximising the effectiveness and endurance capability of the country's fewest of the few became an overriding objective. Thus, launching at Italy in 1940 was, as James Ambrose Brown characterised it in his overripe account of the Union's opening East African campaign, *The War of a Hundred Days*, a 'frontier war', demanding men who 'knew Africa' and who could be relied upon to 'lie up all day in the searing heat', taking it on the chin in acceptance that 'losses were unimportant'.[29] For white Springboks to be Spartans of that order, they would have to be adequately bolstered in the field. The solution to that lay in the provision of a second-string subsidy of black non-combatants. African and Coloured auxiliary Corps troops would service all essential operational needs, thereby freeing combat troops fully for front-line engagement. Indeed, they often came in for special praise in cases of straightforward substitution, when the availability of their domestic skills enabled white non-combatants to be plucked from mess kitchens and trained as real soldiers with guns. The culinary proficiency of NMC and Cape Corps cooks had been, the Union War Histories Committee noted in August 1945,'legendary', and in both the air force as well as the army they had acquired 'an enviable reputation, second to none'. Indeed, as Captain Knoetze recorded, 'most African cooks' ended up occupying 'the posts of master cooks which were actually intended for Whites only'.[30] When it came to the pots and pans of war, South Africa's 1940s colour-bar was certainly not without some novel flexibility.

At the same time, baffled observers from other British colonial forces, like Lieutenant Daniel, were witnessing something else entirely. Later, encountering a brigade of the 1st South African Division coming under heavy fire from its German enemy in November 1941, he witnessed 'a needless and wasteful' action in which white infantrymen were diverted to protect vulnerable NMC and Cape Corps soldiers on motor convoys and at fuel depots that were natural targets for General Erwin Rommel's attacking *Afrika Corps*. 'What a state of affairs it made', he reflected; everyone 'out here' was 'in it to fight Hitler', so why were 'these troops not being handed the equipment necessary for them to

---

28 Nasson, *South Africa at War*, p. 15.

29 James Ambrose Brown, *The War of a Hundred Days: Springboks in Somalia and Abyssinia, 1940–41* (Johannesburg, 1990), p. 62.

30 Cowling, 'Non-European Army Services', p. 16.

defend their own positions'?[31] This was a not uncommon refrain from officers of British colonial forces. Even though there, too, 'race was a determining element in the command and order' structure,[32] in general, it was applied comparatively less punctiliously in a number of areas, so that, for example, colonial African and white British soldiers could end up being treated together in common hospital wards. Equally, neither the Sierre Leone Battalion nor the 11th East African Division faced the Japanese in Burma as labourers in uniform.

Still, some of the Union's African labourers in uniform found themselves able to rise a little, if still not to be issued with guns. Saddled with a constant shortage of white recruits, for one security requirement the UDF had to cast about more widely for reliable, able-bodied men. Swallowing hard, the authorities drafted around 25,000 NEAS soldiers into 30 special guard battalions, commanded by a small band of white officers and NCOs. Organised into 70 sections, these battalions were stationed at vital strategic points such as airfields, harbours, chemicals, explosives and ammunition factories, placed on coastal watch patrols and deployed to secure prisoner-of-war camps in which over 70,000 Italian captives were interned. If those duties gave them more of a stance as soldiers, it was nonetheless a somewhat coy version of soldiering. Guards drawn from the NMC did their guarding in a manner pleasingly endorsed by the Minister of Defence as culturally fitting or, as reported by *The Natal Witness* in July 1941, as 'in a fashion drawing upon the admirably strong traditions and customs of the natives'.[33]

What that meant was that battalions would not be keeping Italian POWs in or invaders out with the menace of a Lee-Enfield 303, let alone the cold steel of a rifle bayonet. Instead, they were armed with traditional *knobkieries*, wooden clubs, and *assegaais*, light spears. At airfields, hangars, and bomb storage depots, Bristol Beaufighters of the South African Air Force were kept secure by NMC *assegaais*. In the greatest twentieth-century war of the machine age, it made for some vintage administrative record-keeping. Thus, the weapons inventory for a typical 1943 guard battalion of some 1,300 men became simplicity itself: instead of columns for rifles, ammunition, gun parts or other equipment, there was a typical list of, say, 1,687 spears. Or another table might provide a count of, say, 722 fighting sticks.[34] What it amounted to, in a way, was perhaps another idiosyncratic South African lesson – how to combine clashing cultures of warfare or, rather more to the point, how to see to it that they would not clash.

---

[31] Lt C.G. Daniel, encl. in Uys to Cochrane, 21 November 1941, Gordon war correspondence.

[32] David Killingray, *Fighting for Britain: African Soldiers in the Second World War* (Woodbridge, 2010), p. 84.

[33] *Natal Witness*, 26 July 1941.

[34] Ian Gleeson, *The Unknown Force: Black, Coloured and Indian Soldiers through Two World Wars* (Johannesburg, 1994), p. 150.

African Warriordom, White Womanhood and Rattling the Cage

This was all rather short of the patriotic expectations of African loyalists, most of all those singled out predictably by the Native Affairs Department in close liaison with the military authorities for especially energetic recruitment efforts. Pre-eminent in local 'martial race' discourse were, of course, those archetypal warriors of the British imperial imagination, the Zulu. Although lauded as 'a cut above' all other African men as ultra-masculine warriors, the soldierly response to NMC service enticements was arthritically slow.[35] While derisory levels of enlistment had many causes, two of the main deterrents were racially discriminatory service terms and the denial of guns to Africans. Stung by the cheek of official statements that the slackness of Zulu volunteering was due to cowardice, Chief Mshiyeni ka Dinuzulu, the Acting Paramount, nipped the hand that was feeding him for his support of the war effort. In a fabulous show of affectation, he instructed Pretoria, 'I desire to have established a Zulu Military Regiment trained in the manipulation of Big Guns, war tanks, armoured vehicles, motor cycles – all for combating the armed forces of the enemy. All necessary weapons and vehicles will be made available.'[36]

Other disenchantment was voiced in a less tongue-in-cheek manner. Dr A.B. Xuma, the Africanist-inclined president of the ANC, was scathing about press propaganda photographs and war documentary films that depicted guard battalions training for gas warfare while bearing spears, or guarding Ventura bombers with clubs. The firearms ban was not just denigrating; it was reducing and levelling the position of all African soldiers, whatever their status. 'They are', Xuma scoffed, 'expected to fight aeroplanes, tanks and artillery with *knobkieries* and *assegaais*. What mockery! It is demeaning. How degrading for a soldier to be reduced in standing to that of some common tribesman.'[37]

That standing, though, was elevated in the propaganda efforts of the Secretary of Native Affairs, his rural magistrates, and by some figures in UDF command, not least those white officers in Natal who grasped the Zulu language and assumed a more than working familiarity with its speakers. Their assertive tone was, if nothing else, respectfully martial. The *assegaai*, after all, was the ingrained tool of warfare, the ancient and preferred weapon of Africans, symbolically as

---

35 Henry J. Martin and Neil D. Orpen, *South Africa at War: Military and Industrial Organisation and Operations in Connection with the Conduct of the War, 1939–1945* (Cape Town, 1979), pp. 150–151; Killingray, *Fighting for Britain*, p. 71.

36 Cited in Louis Grundlingh, 'The Recruitment of South African Blacks for Participation in the Second World War', in D. Killingray and R. Rathbone (eds), *Africa and the Second World War* (Basingstoke, 1986), p. 191.

37 Cited in Hermann Giliomee and Bernard Mbenga (eds), *New History of South Africa* (Cape Town, 2007), p. 295.

Figure 9.1 'Native Military Corps soldiers in gas warfare training exercise, assegaais at the ready'
Courtesy of the South African Defence Force Archives, Pretoria.

traditional as the flint-lock to Europeans. Indeed, the brave Zulu had made do with even less, as in the titanic era of Shaka:

> In those days, the *impi* that formed his front-line troops and his personal bodyguard, hand-picked by Chaka himself from strapping young volunteers, proved their valour and strength in a death struggle with lions trapped in the Zululand mountains; the young warriors employed no other weapons than their bare hands and *riempies* to overcome and truss up the savage, rending beasts. That same fearlessness is being turned today to an even finer purpose on the battlefields.[38]

With those battlefields of today being secure sites such as the Swartkops air force base outside Pretoria, the Grand Magazine in the capital itself, or the Caprivi Zipfel or Caprivi Strip of northern South West Africa (Namibia), the foundations of an ancient fighting past were unlikely to be milked again. That notwithstanding, whether bare-fisted or behind metal and wood, the notion of a natural African fighting culture had gained a new illumination in the 1940s, and

[38] 'Zulu Hero: Old Fighting Spirit Lives On', *Libertas*, 4/8 (1944), p. 58. *Riempies* (Afrikaans) – leather thongs or straps.

in circumstances perhaps more quixotic than exotic. No longer expansive Zulu conquest, now it was well-drilled home defence. And, as illustrated magazines such as *Libertas* emphasised, holding the line with a harvest of *knobkieries* was peerlessly traditional, summoning up a proud heritage of Xhosa and Zulu stick-fighting agility. Assuredly, Military Corps guards were not being discriminated against nor being disadvantaged by not being equipped with modern firearms.

For, to the contrary, these soldiers were being *advantaged*. Freed from the burden of having to undergo instruction on unfamiliar European firearms, a patriotic honour different from that bestowed by the army gun was accruing to African guardsmen. It came from a crash course in recovering a lost warrior tradition, the natural or customary close-combat fighter re-born, shorn of European mechanical affectations. In this respect, it need hardly be said that this differed markedly from the standard British imperial 'martial race' ideology of South Asia in the later nineteenth and earlier twentieth centuries. What the Indian Army required, it was asserted, were northern recruits or 'men who were racially or environmentally closer to Europeans', such as Sikhs 'with their light skins' and '"European" features', the only troops who, their proponents argued, 'could hope to defeat Europeans in war'.[39] The NEAS guard battalions represented the antithesis of such beliefs. It was by not being akin to Europeans, by not being equipped as European soldiers, and by being true to their own implements of combat that African guards would form a most naturally strong barrier against the enemy.

In that sense, if one part of the Union war effort was the intensification of its industrial capacity for arms and other fighting equipment, another was turning the clock back not only to pre-capitalism but also to pre-colonialism. Indeed, it was the latter which was celebrated by press and other commentators for whom something ancient and African was being restored, including pride. For, in turn, the gun-free experience of NMC service became a kind of wartime decolonisation. As UDF soldiers capable of spearing chests and cracking skulls, Africans were being granted the opportunity of returning to a more elemental and more authentic world of pre-colonial warrior habits and instincts.[40]

Barely able to restrain himself, one retired old soldier in Kimberley, as effusive as he was disingenuous, declared battalion guards to be 'truly First Class soldierly men. In this present war, amid treachery at home and with the enemy now in our waters, they are surely proof, if ever that were still needed, that on their dawn watch the aboriginal arm and the aboriginal eye is more effective than that of the European.' Their whole purpose, he continued, would be 'gravely jeopardised'

---

39 Heather Streets, *Martial Races: The Military, Race and Masculinity in British Imperial Culture, 1857–1914* (Manchester, 2004), p. 95.

40 Nasson, *South Africa at War*, p. 121.

if they were to be 'loaded with heavy rifles' or 'other unnatural weapons'.[41] In similar vein, the 'Modern Warriors' captured in the picture strips of *The Springbok Record* were 'Bantu soldiers' whose '*assegaais* instead of rifles' ensured that with 'pomp and ceremony' they would 'march with pride in the footsteps of their warrior forebears'.[42] Such bizarre – or plainly surreal – rationalisations were certainly not shared by more reformist or modern-minded lights within the UDF white officer corps. In a view which captured well the tone of their disapproval, an officer who was also a member of the Communist Party of South Africa (CPSA) dismissed the carrying of *assegaais* as 'an humiliation, the vanished brand of some savage warrior ... [F]or this class of disciplined Bantu men, drilled on parade grounds, it is lamentable.'[43]

The issue of assigned inferiority in non-combatant status also raised its head on another front, that of gender rights. Throughout the conflict, white female volunteers of the South African Women's Auxiliary engaged in mild sniping over their terms of service as artisans, mechanics, lorry drivers and motorcycle riders. Some grievances were, naturally, the same as those of servicewomen in other Allied countries, such as unequal pay for wartime work. The tone of others, though, was specifically different and distinctively South African. As *The Women's Auxiliary* observed in May 1943, the war was not an adventurous finishing school for skirted patriotic volunteers. For 'the women of South Africa' had been coming under fire repeatedly while on transport duties in East Africa and in North Africa, were now making up for a chronic shortage of regular gunners at coastal batteries, were running artillery range-finding drills, and were also arming fighter and bomber aircraft.[44] This was to say nothing of women taking on factory munitions assembly and other vital armaments jobs that were usually the prerogative of skilled white male workers.

Granted, being female they could not be UDF combatants. Nonetheless, members of units like the Women's Auxiliary Defence Corps had their feet planted in the world of firearms, with responsibilities and duties which exceeded those carried by soldiers of the NMC. Moreover, with 'the men of South Africa', they shared European citizenship of the Union and its franchise rights. Did that not raise the status of white women volunteers, making them more aspirationally *soldierly*? Or, as readers of the left-wing *Guardian* were reminded, 'the thousands' of Defence Corps volunteers were 'pulling together' with UDF combatants, 'right up in the line', so that the SAAF Hurricane pilot flew because of the grease-smeared airwoman in drab overalls.[45] If theirs was still a non-

41 *Diamond Fields Advertiser*, 28 September 1942.

42 Harry Klein, *Springbok Record* (Johannesburg, 1946), pp. 32–33.

43 *Mentor*, August 1942, p.11.

44 *The Women's Auxiliary*, January 1941, pp. 16–17; Martin and Orpen, *South Africa*, pp. 285–288.

45 *The Guardian*, 25 October 1942.

combatant world, it was a big step up from the circumscribed *assegaai* universe of the NMC. For some of the writers of *The Women's Auxiliary,* the focus was, therefore, reassurance. In depictions of female effort and in the shaping of service imagery, there could be no sliding or diminishing of the validation of women's part in the common struggle. There was a certain raw sensitivity over making sure that the lesser status of the UDF's white women did not become conflated with the low status of the UDF's black men.

That nervousness was caught in 1943 by the vigilant South African League of Women Voters. In that year, the authorities announced the introduction of several perks, including free rail travel warrants and postal vouchers, for the women and men of the Women's Auxiliary Services and the Non-European Army Services, respectively. Suddenly, workmanlike white femininity in overalls and blunted black masculinity in tunics were inserted into the same pool of auxiliary war service – those undertaking lorry driving, refuelling, parachute-packing, stretcher-bearing, airfield guarding, despatch-riding and so on were being treated as all of a piece. Lumping them in together, or creating 'an unpalatable juxtaposition of two classes', was, according to Dorothy Kirby, Secretary of the League of Women Voters, 'very ill-advised'. Such disregarding of customary racial and gender hierarchies was 'an abomination', one 'deeply offensive to the women of South Africa'.[46]

The vehemence of that objection, with its axiomatic vision of ethnicity and hierarchy, was a prickly illustration of how wartime relationships made themselves felt. Predictably, the country's war propaganda made much of everyone pulling together, and of its female Auxiliary volunteers being an equal 'shareholder' in the UDF in its defence of 'her country's security'.[47] Yet binding together for the greater good did not mean, as the Union's League of Women Voters warned, any acceptance of commonality or reciprocation. There could be no symbiosis of service entitlement between the UDF's range of auxiliaries. To be equally non-combatant was emphatically not to be equal.

The festering issue of the right to be armed and the discriminatory imposition of being kept disarmed remained throughout the war, at times complicating a sense of preoccupations and priorities between a government with tender corns and its more calloused UDF high command, as well as between the Union and Allied Army command, particularly during the last years of hostilities. In preparations for the Italian campaign, for instance, British leadership was either baffled or frustrated by UDF responses and explanations. Try as it might, South Africa would be unable to meet requested quotas of combat infantry as it was still seriously short of front-line white soldiers. Equally, the UDF would be unable

---

46 See Suryakanthie Chetty, 'Gender under Fire: Interrogating War in South Africa, 1939–1945', unpublished MA dissertation, University of Natal, 2001, p. 134.

47 *The Women's Auxiliary*, June 1942, p. 5.

to arm Coloured soldiers of the Cape Corps as an alternative. Granted, in 1944, there had been a solitary fleeting shot in the dark. Under mounting pressure from British 8th Army command, Smuts agreed grudgingly to the contingency of organising experienced Coloured soldiers into a Cape Corps infantry brigade to increase the South African complement of fighting troops for the invasion of Italy. Caught on the hop while in Europe, this concession was something that the prime minister had always striven to avoid. To Smuts's probable relief, back at home, his Cabinet rejected the proposal of an armed Cape Corps.[48]

A couple of years earlier, the special political target of guns in the war had produced a much larger heart-stopping national moment. In 1941, increasingly edgy over anti-war nationalist Afrikaner subversion and rebellion, the authorities embarked on a massive trawl of civilian weapons mainly in rural areas, netting privately owned rifles. In a largely calm and compliant exercise, ordinary firearms were confiscated. In effect, potentially troublesome white citizens had disarmed themselves. But that domestic dampening was not quite the end of the story. In 1942, the country was throbbing with rumours of a pending Japanese strike or even invasion. What stretched white nerves even further was domestic uncertainty. Apprehensive intelligence monitors were in little doubt that many disgruntled rural Africans would be likely to welcome the Japanese as invading liberators. In Cape Town, the Coloured Teachers' League of South Africa declared that should Japan, 'a Coloured nation', invade the Union, it should be welcomed as control of the country would then pass naturally into the hands of its Coloured people.[49]

Scrambling to try to forge a more united country and to strike a stronger sense of common purpose in the face of an Oriental menace from the Indian Ocean, Smuts swallowed hard. In an effort to bolster the home front morale of Africans, the government suspended influx control (the detested pass laws that denied most Africans the right to live permanently in urban areas) for the duration of the Japanese scare.[50] Then, in Parliament, invasion worries were sufficiently disconcerting for Smuts to raise the prospect of a greater official taboo having to be abandoned.

In the event of the country finding itself confronting the national emergency of an invasion, all of the UDF's 'loyal Non-European soldiers' might have to be armed.[51] Some government supporters went further, raising the option of

---

48 Martin and Orpen, *South Africa*, p. 298.

49 Dick van der Ross, *A Blow to the Hoop: The Story of My Life and Times* (Cape Town, 2010), p. 79.

50 Ian Phimister, 'Union of South Africa', in M.R.D. Foot and I.C.B. Dear (eds), *The Oxford Companion to World War II* (Oxford, 2005), p. 798.

51 Ken Vernon, *Penpricks: The Drawing of South Africa's Political Battlelines* (Cape Town, 2000), p. 84.

an emergency 'national patriotic campaign'[52] to have firearms distributed to *all* loyal volunteers, an issuing of rifles that would turn not-quite soldiers fully into soldiers. Naturally, the response from radical anti-war Afrikaner nationalists was apoplectic. Correspondents to papers such as *Die Vaderland* and *Die Volkstem* denounced a situation in which peaceable Afrikaner farmers had been obliged to relinquish their cherished rifles, intrinsic to their fighting Boer character, many of them family heirlooms like commando Mausers from the 1899–1902 South African War.[53]

Being left defenceless against Tokyo was a bad enough nightmare. Even worse was the equivocation of Jan Smuts. What loomed was the spectacle of open season for armed and unpredictable black savagery – it would be, as editorials, cartoons and letters declared, armed mayhem and domestic breakdown.[54] For any devious African, Coloured or Indian person would then be able to exploit the circumstances of war – a gun would become their right. A good measure of how this was seen was an excitable cartoon in *Die Vaderland* in April 1942. Discarding *assegaais* and *knobkieries* for lethal modern firearms, cavorting barefooted warriors swarmed forward to be waved on by the prime minister to parliamentary assurance that, in the event of any Japanese incursion, 'any Coloured or African person who asked for a weapon would receive a weapon'.[55] The 1942 crisis was no more than a passing moment. But Pretoria's fleeting equivocation over the arming of black patriots was music – however vicarious its sound – to the ears of those who since South Africa's declaration of war on Germany had been pressing for blacks to be accorded the right of war service on equal terms with whites. For anti-segregationist bodies such as the ANC, given the perils of a dominant white minority that was so split politically over war participation, trusty African loyalists were the real backbone of any domestic survival under arms.[56] Taking a similar cue, in its March 1942 demand for 'Arms for All Non-Europeans', the left-wing Non-European United Front imagined the formation of a new armed citizenry, a mass home guard of '8,000,000 Non-Europeans, Fully Mobilised and Armed', who would draw up as the most reliable shield available to 'Lessen the Danger for All'.[57]

---

52 *The Star*, 19 June, 1942; *Natal Witness*, 8 July 1942.

53 *Die Vaderland*, 22 July 1942; *Die Volkstem*, 25 July 1942.

54 For the general issue of the containment of black troops, see Louis Grundlingh, '"Non-Europeans Should Be Kept away from the Temptations of Towns": Controlling Black South African Soldiers during the Second World War', *International Journal of African Historical Studies*, 25/3 (1992), pp. 539–560.

55 J.D. Pretorius, 'Ideology and Identities: Printed Graphic Propaganda of the Communist Party of South Africa, 1921–1950', unpublished DLit. et Phil thesis, University of Johannesburg, 2011, p. 213.

56 *The Guardian*, 13 August 1942.

57 Nasson, *South Africa at War*, p. 115.

Figure 9.2 'Communist Party of South Africa Wartime Campaign Poster, 1942'
Poster Collection. Courtesy of the National Library of South Africa, Cape Town.

For the Communist Party and its noisy 'Give Him a Gun, NOW' campaign, it was also as close as it would ever get in its propaganda drive to have African soldiers issued with firearms. After June 1941, the now pro-war party dropped its earlier derogatory line on volunteer African soldiers as gullible dupes of European imperialists, as boy-servants of white masters, or as emasculated adherents of a foreign cause. In posters and in publications such as *Inkululeko*, Communists promoted not merely the romanticised imagery of the Red Army Soldier, but also the heroism of the local, spear-carrying African Soldier. Pamphlets such as *They Served Their Country* toasted NMC members for their steadfast stand against the desecration of their national soil. Depictions of African soldiers in khaki and holding spears highlighted a simple message. For the final attainment of 'Defence and Victory', the means for all patriots had to be the gun.

All the while, armed masculinity remained white masculinity, an insistent convention that was reinforced even by its ironies. In recreation periods on the Somme in 1916, Natal soldiers of the 1st South African Infantry Brigade had been partial to indulging in mock Zulu war cries and war dances, 'sometimes brandishing patches of stretched cattle-hide as fake shields'. As *le Zulu blanc*, this martial pantomime spread a hot-blooded message – the regimental combat will of the Union Brigade 'could match that of a nineteenth-century Shakan impi or war party'.[58] And so it would be again in the Second World War, when, swopping the bullet for the bayonet, South Africa's white infantry would demonstrate their combat mettle by dipping into their colonial country's African past. As *Libertas* assured its readers, in the 'push up north', the 'deeds of our fighters, who charge with fixed bayonets, singing and shouting native war cries as they go, have astonished their friends and terrified their foes'.[59] It turned out that for those who were fully constructed soldiers, at times the blade could even eclipse the gun.

If it is hard to miss the bizarre quality of contradictions like these, there is also a last nuance that we ought not to lose sight of, one that may, perhaps, remind us of the ways in which the historical imagination of colonial armed service were continuing to exercise a potent and passionate grip upon subjects across many generations. Charles Adams of the Cape Corps felt intense pride in his regiment's long history as a 'proper' infantry force, acknowledged belligerents 'since 1793' and resented his prescribed non-combatant status, recalling in 1998 that 'they would not allow us to fight', purely 'to keep white people happy'. As he told the author, Christopher Somerville, it was difficult to stomach being stuck 'in the supply corps' when what Coloured soldiers wanted 'was fighting'. Some pro-war parliamentarians, too, were not averse in principle to unleashing the armed ferocity of black soldiers. A disgruntled

---

58 Nasson, *Springboks*, p. 128.

59 'Steady', *Libertas*, 2/1 (1941), p. 18.

gathering in November 1942 asked how it was 'possible to work up the spirit of the offensive among natives if the only training given them was that of watchdogs, patrolling a fence with assegaais', whereas 'if given arms and trained as soldiers', it would be 'a different story'.[60]

At the same time, from non-combatant soldiers themselves there was also a parallel aspiration to be immersed in an unconditionally common war service. However martial some of their protestations, it was never possession of the gun for itself that was cherished and elevated, as in, say, the iconic allure of the Winchester rifle in the cinematically mythic American West. The crucial factor for those who yearned to be under firearms was that they wanted to be run-of-the-mill UDF soldiers, with the acknowledgement of their trusty patriotism being the gun. In a sense, their frustration was not merely over some military denial of *esprit de corps*, but over a political denial of *esprit de citoyenneté*.

[60] Christopher Somerville, *Our War: How The British Commonwealth Fought the Second World War* (London, 1998), pp. 68–69.

Chapter 10

'Better Die Fighting against Injustice than to Die Like a Dog': African-Americans and Guns, 1866–1941

Kevin Yuill

Introduction

A small dispute involving fist-fights broke out between blacks and whites in Memphis, Tennessee, in 1892. After the case was dismissed from court with nominal fines, the whites involved were evidently not satisfied. A rumour spread that these white men were coming to destroy a local black-owned shop, The People's Grocery Store. The owners of the company consulted a lawyer, who advised them that, as they were outside of city limits and thus could not avail themselves of police protection, the proprietors would be justified in defending themselves if attacked. On Saturday, 6 March 1892, the shop posted several armed men at the rear of the store, where any attack was like to begin. Saturday night, at the time, was the time when men tended to congregate in groceries.

As Thomas Moss and Calvin McDowell, two of the owners of the grocery, updated the accounts books, they heard shots from the backroom. Three white men who were attempting to enter the premises through the rear door were wounded and the other white men fled and sounded the alarm. The next morning's Sunday newspapers referred to the wounded men as 'officers of the law' who were wounded in the discharge of their duties, and contained lurid details of 'a dive in which drinking and gambling were carried on: a resort of thieves and thugs'. According to contemporary accounts, over a hundred black men were dragged from their homes in the middle of the night in the search for those who had fired the shots. Several were held in jail.[1]

Though there had been no lynchings in Memphis since the Civil War, the situation appeared very tense, so the Tennessee Rifles, a black militia that survived the end of Reconstruction, posted guard at the jail where those

[1] Ida B. Wells, *Crusade for Justice: The Autobiography of Ida B. Wells* (Chicago, 1970), pp. 48–52. See also Linda O. McMurry, *To Keep the Waters Troubled: The Life of Ida B. Wells* (Oxford, 1998), pp. 131–134, and Paula J. Giddings, *Ida: A Sword Among Lions. Ida B. Wells and the Campaign against Lynching* (New York, 2008), pp. 176–180.

suspected of firing the shots were held. On the Tuesday morning, however, the papers announced that all of those who had been wounded were out of danger. The Tennessee Rifles assumed that the crisis had passed and dismissed the guard on the third night. While the Tennessee Rifles slept, a body of men was admitted to the jail, according to press reports, by telling the turnkey they had brought in a prisoner. Once in, the mob seized Thomas Moss, Calvin McDowell and Henry Stewart, the three who ran the People's Grocery. Loaded onto a switch engine and carried a mile and a half out of town, the three men were then coldly executed; McDowell's body was found with his eyes gouged out.

When African-Americans, shocked and stunned at the news, gathered at the grocery, an order was issued by Judge DuBose of the criminal court. Sheriff McLinden was to 'take a hundred men, go out to the Curve [where the grocery store was located] at once and shoot down on sight any Negro who appears to be making trouble'.[2] On 10 March, Judge DuBose ordered the sheriff to 'take charge of the arms of the Tennessee Rifles, a Negro guard, whose armoury is near Hernando and Union Streets'. Refusing to drill with wooden sticks, the Tennessee Rifles disbanded, never to reform. Frank Schumann, a Jewish gunsmith, was jailed for selling guns to African-Americans.[3]

This event inspired Ida B. Wells, who knew the lynched men and was then the editor of the *Free Speech*, a local paper popular amongst African-Americans at the time, to take up a lifelong campaign against the evils of lynching. Part of her complaint, published in the *Free Speech*, was the fact that black men could not defend themselves adequately against the white mob:

> The city of Memphis has demonstrated that neither character nor standing avails the Negro if he dares to protect himself against the white man or become his rival [the grocery store had rivalled the trade of a local white merchant who had the monopoly on trade before the People's Grocery opened]. There is nothing we can do about the lynching now, as we are outnumbered and without arms. The white mob could help itself to ammunition without pay, but the order was rigidly enforced against the selling of guns to Negroes.[4]

On the night of 27 May 1892, 'leading citizens' entered the office of the *Free Speech*, ran the business manager J.L. Fleming out of town, destroyed the type and left a note saying that anyone attempting to publish the paper again would be punished by death. Wells, the proprietor, remembered her response to these events:

---

2 'Mob of Negroes Gathering: Triple Lynching at Memphis May Be Followed By a Riot', *The Evening World* (New York), 9 March 1892, p. 1.

3 See Wells, *Crusade for Justice*, pp. 48–51.

4 Ibid., p. 52.

> I had bought a pistol the first thing after Tim Moss was lynched, because I expected some cowardly retaliation from the lynchers. I felt that one had better die fighting against injustice than to die like a dog or a rat in a trap. I had already determined to sell my life as dearly as possible if attacked. I felt if I could take one lyncher with me, this would even up the score a little bit.[5]

The fact that Wells advocated armed self-defence might surprise those familiar with the civil rights struggles of the 1950s and 1960s. Whereas the postwar era conjures up images of black school children facing hoses and dogs in Birmingham, Alabama, and college students patiently putting up with physical attacks as they sat waiting to be served in the sit-in movement in the early 1960s, before World War II, guns meant something different for African-Americans. Disarmament meant helplessness. Whereas few wished for a fight, African-Americans objected as strongly as possible to the campaign of disarmament waged against them through the years of Reconstruction and Jim Crow. Moreover, in an era characterised by lynching, possessing a weapon could make the difference between life and death.

The United States' relationship with guns stands in contrast with most European countries. Born of a revolution accomplished, in part, because of a well-armed citizenry, the right to bear arms is written into the Constitution and remains a symbol, for many Americans, of the freedom and power enjoyed by a sovereign people. As a Democratic paper complained in the Reconstruction period, faced with threatened disarmament by Union troops: 'We pour scorn upon the craven, the pusillanimous notion that freemen may not vindicate their rights by arms. Courage to resist oppression is the ultimate security for good government ... Deny the right [to forcible resistance] and you give full license to any unscrupulous minority ... to render its authority perpetual.'[6]

The bitter irony of such a statement by Southern Democrats, who were only too ready to seize weapons from black hands, would have been felt by African-Americans at the time. Between 1866 and 1941, African-Americans were deprived of their weapons, robbed of their dignity and property. Disarming blacks was usually the first step in stripping them of further rights, such as the right to free speech and the right to vote. By disarming them, Southern whites were able to ensure that blacks remained an inferior race.

Contrasting with the nonviolent campaigns of the more recent civil rights movement, radicals like Ida Wells understood that gun controls were simply a legal method to accomplish what the Ku Klux Klan had gone some way to accomplishing in the first place – disarming blacks. Whereas ostensibly race-

---

5 Ibid., p. 62.

6 Editorial, *The New York World,* reprinted in *The Evening Telegraph* (Philadelphia), 16 July 1868, p. 2.

neutral gun controls existed from 1870 onwards in the South, they were seldom enforced against whites. Though few African-Americans wanted armed conflict between blacks and whites, guns were occasionally needed as an immediate defence against white mobs. Wells also understood the principle later elucidated by postwar armed self-defence advocate Robert F. Williams – that whites valued their lives more than those of African-Americans and would seldom trade shots with blacks.[7]

Whereas accommodationist African-American leaders like Booker T. Washington took on face value Southern whites' advocacy of prohibition on weapons and supported the campaign for gun controls, radical black voices resisted curbs on their Second Amendment rights, sardonically noting their real purpose. Northern liberals also resisted calls for gun controls in Southern states, only too aware of the context of white violence against African-Americans. Only in the early years of the twentieth century, after the assassination of President McKinley in 1901, armed confrontations between company goons and strikers and fears of immigrant violence, did their resistance falter.

In order to understand the interrelationship between guns and African-Americans, we might divide the era 1866–1941 into three periods. In the Reconstruction era of the immediate post-Civil War years, the right to bear arms became an important symbol of black equality, and the infamous Black Codes initiated by Southern states were designed to keep African-Americans in bondage. A key feature of these codes was the ban on possessing weapons. After the passage of the Fourteenth Amendment banning race-discriminatory legislation was adopted, the Ku Klux Klan (KKK) and other 'night riders' seized weapons from African-Americans, so that they might be more easily prevented from voting. State militias, which became effectively black after they were formed by Republican governments in the years after the Civil War, mounted stern defences of black rights but were outgunned and out-soldiered by better equipped and more experienced local white unofficial militias.

After 1877, when the last troops were moved out of Southern capitols, Democrats maintained power and increasingly institutionalised white supremacy. Circumscribed by the Fourteenth Amendment and the Civil Rights Acts, they implemented gun controls that, while ostensibly neutral, were aimed at disarming African-Americans; few whites were ever charged. Rather than the mass seizing of weapons and terrorism inflicted by the Klan, which had by then been banned, Southern whites periodically reminded African-Americans by lynching that, whatever the law said, white supremacy remained intact.

During the Progressive period, the restrictions on gun possession in the Southern states spread throughout the nation. Northern liberals, who had in earlier periods resisted gun laws on the basis that they were aimed at blacks,

---

[7] Robert F. Williams, *Negroes with Guns* (Detroit, 1993; 1st edn, 1962).

stopped the opposition that they had mounted to Southern gun laws during Reconstruction. In the early years of the twentieth century, more weapons bans accompanied the petty apartheid Jim Crow laws. Strengthened by the Supreme Court's infamous 1896 decision in *Plessy v. Ferguson*, where the judges ruled that 'separate but equal' was defensible, Southerners now argued openly against black suffrage and against the rights of blacks to possess weapons on the basis that, as a race, they were incapable. National and international race consciousness also emboldened them. Whereas gun laws of earlier periods disguised their racial intent, those calling for gun control legislation at this time justified their support simply: 'disarm the negro'. Though the Fourteenth Amendment continued to ensure that no legal statutes could be race specific, justifications for race-neutral legislation were increasingly made on blatantly racial bases. Moreover, as this chapter indicates, the ostensibly race-neutral legislation was enforced almost exclusively against, as one newspaper put it, the 'ignorant and inferior race'.[8]

Black Gun Ownership before the Civil War

Before the Civil War, the gun played a unique historical role in America, as settlers battled against Native Americans and the wilderness. During the colonial period, African-Americans, both slaves and freemen, faced only sporadic legal restrictions to possession of guns. Restrictions based on skin colour were normally enacted in specific colonies at times when the fear of insurrection outweighed the fear of attack from outside. In 1639, Virginia issued the first prohibitions against blacks being issued guns: 'All persons except negroes' were provided with arms and ammunition. In 1648, an Act Preventing Negroes from Bearing Arms, directed at slaves, was passed in the same legislature. By 1680, all blacks, slave and free, were prevented from bearing arms in Virginia. But even then, Virginia still allowed free blacks to keep a gun in their home, and, on frontier plantations, both free blacks and slaves were permitted weapons. South Carolina allowed free blacks to both serve in the militia and own weapons, only reversing the policy in the eighteenth century.[9]

In the Northern colonies, free blacks faced similar restrictions to Catholics in England; they were permitted to own weapons but were not to serve in the militia in Massachusetts, Connecticut and New Jersey. Massachusetts, indicating the changing needs of the colonists, enacted a statute in 1652 that specified that militia enlistments were to include 'all Scotsmen, Negers and Indians inhabiting

---

8 'Knife and Pistol', *Nashville Union and American*, 25 September 1874, p. 2.

9 Benjamin Quarles, 'The Colonial Militia and Negro Manpower', *Mississippi Valley Historical Review*, 45/4 (1959), p. 644; Joyce Lee Malcolm, *To Keep and Bear Arms: The Origins of an Anglo-American Right* (Cambridge, MA, 1994), p. 141.

with or servants to the English'. Four years later, however, the legislature prohibited the mustering in of African-Americans or Native Americans, explaining the step as necessary in the interests of 'the better ordering and settling of severall cases in the military companies'. However, it is worth noting that in nearly every state in the Northern colonies (and in 1756 the formerly Catholic colony of Maryland) Catholics faced almost identical restrictions to blacks.[10]

Nor were restrictions set for all time. Whereas, in 1680, the Virginia assembly prohibited 'negroes' from carrying clubs, swords, guns, or other weapons of defence or offense, in 1705, a restatement of the law substituted 'slave' for 'negroe or other slave'. Some 20 years later, though, an enactment insisted that arms were to be carried by 'no negro, mulatto, or Indian whatsoever' under penalty of a whipping not to exceed 29 lashes. When African-Americans faced restrictions on bearing arms, it reflected the fear that, as one group of North Carolina merchants noted, 'our slaves when armed might become our masters'.[11]

After the founding of the Republic, the Second Amendment guaranteed the right to bear arms to the citizenry, but the concept of who constituted a citizen was left vague.[12] Robert J. Cottrol and Raymond T. Diamond note: 'Despite the prejudices of the day, lawmakers in late eighteenth-century America were significantly less willing to write racial restrictions into constitutions and other laws guaranteeing fundamental rights than were their counterparts a generation or so later in the nineteenth century.' Tennessee, for instance, in its original Constitution of 1796, stated 'That the freemen of this State have a right to keep and to bear arms for their common defence'. In 1834, however, amidst fear of slave uprisings, it replaced 'freemen' with 'free *white* men'. Free hitherto enfranchised black males were also disbarred from voting. Similar amendments were made by Arkansas in 1836 and Florida in 1838.[13]

However, the clearest demarcation of the relationship between African-Americans, possession of arms and other rights came in the notorious Supreme Court judgment, *Dred Scott v. Sanford*, delivered by Roger B. Taney in 1857. Dred Scott was an African-American slave whose owner, U.S. Army Surgeon Dr John Emerson, took him to Illinois, which prohibited slavery, and then to Wisconsin territory, where it was also prohibited. He and his wife, whom he had legally married in Illinois with the permission of his owner, sued for their

---

[10] Cited in Quarles, 'The Colonial Militia and Negro Manpower', pp. 644–645.

[11] Cited in ibid., pp. 647, 644.

[12] As Joan Gundersen notes, because of the property qualification for voting, free blacks and women who met the property qualification were occasionally allowed the vote during the eighteenth century. See Joan R. Gundersen, 'Independence, Citizenship, and the American Revolution', *Signs*, 13/1 (1987), pp. 59–77.

[13] Robert J. Cottrol and Raymond T. Diamond, 'Never Intended to Be Applied to the White Population: Firearms Regulation and Racial Disparity – The Redeemed South's Legacy to a National Jurisprudence', *Chicago-Kent Law Review*, 70 (1994 1995), p. 1317.

freedom after Emerson died. The infamous opinion handed down by Taney explicitly indicated the connection between gun rights and citizenship:

> For if they were so received, and entitled to the privileges and immunities of citizens, it would exempt them from the operation of the special laws and from the police regulations which they considered to be necessary for their own safety. It would give to persons of the negro race, who were recognized as citizens in any one State of the Union, the right to enter every other State whenever they pleased, singly or in companies, without pass or passport, and without obstruction, to sojourn there as long as they pleased, to go where they pleased at every hour of the day or night without molestation, unless they committed some violation of law for which a white man would be punished; and it would give them the full liberty of speech in public and in private upon all subjects upon which its own citizens might speak; to hold public meetings upon political affairs, and to keep and carry arms wherever they went. And all of this would be done in the face of the subject race of the same color, both free and slaves, and inevitably producing discontent and insubordination among them, and endangering the peace and safety of the State.

The connection between freedom of speech, freedom of movement, freedom to hold public meetings and 'to keep and carry arms' stands out; it implies, as various authors have noted, that a right exists for American citizens to keep and carry arms wherever they went and that these various freedoms are invariably linked.[14]

Militias and the Ku Klux Klan

Black militias formed after the Civil War and some, such as the Tennessee Rifles, survived long after Reconstruction had become a dead letter. They were formed when the existing state militias, which made no secret of their anti-Republican views and terrorised freedmen across the region, were replaced by those with Congressional backing in 1867. The reconstituted militias, controlled by Republican governors, recruited heavily within black communities. 'To arms! To arms! To arms! Colored men to the front', exclaimed one handbill.[15]

The militias were hampered by the fact that they were largely segregated, poorly led and poorly provisioned, and progressively debilitated by lack of

---

[14] See ibid.; Clayton E. Cramer 'The Racist Roots of Gun Control', *Kansas Journal of Law and Public Policy*, 17 (1994–1995), pp. 17–33; and Stefan Tahmassebbi, 'Gun Control and Racism', *George Mason University Civil Rights Law Journal*, 2/1 (1991), p. 72.

[15] Otis A. Singletary, 'The Negro Militia during Radical Reconstruction', *Military Affairs*, 19/4 (1955), pp. 177–186.

federal support on the part of the Republican administration. Local whites, affronted by the site of blacks in uniforms, boycotted the militias and ostracised all whites who cooperated with or, worse, joined the militias. Thefts of weapons by organised whites also hampered the militias. Thousands of rifles, hundreds of thousands of cartridges, and over a million percussion caps were stolen when they were transported between Memphis and Little Rock. Black militia captains were singled out for violence. But, in the end, as historian Otis Singletary noted, these militias were doomed because they became black militias. Starting out as mixed militias, they distilled into effectively black units. Though some militias survived into the late nineteenth century, they were progressively disarmed as Democrats gained back power from the radical Republicans.[16]

Many of these militias were physically defeated by heavily armed whites. The story of Hamburg, South Carolina, which lies across the Savannah River from Augusta, is indicative. Hamburg was occupied by black freedmen after the war and, by 1876, 80 men drilled in the streets with the best Winchester rifles. On 4 July 1876, a buggy attempted to pass when Doc Adams, commander of the militia, was drilling his men. A hostile exchange occurred, and Matthew C. Butler, a lawyer and former Confederate major-general, issued a demand that the militia surrender their arms and apologise to the men in the buggy. An Attorney General's Report concluded that: 'The primary object of the whites in pressing the issue, apparently, was the disarming of the Hamburg militia.' On 7 July, 200–300 heavily armed whites surrounded the town and forced the militia to take shelter within their armoury. The whites drew up a cannon and fired four shots. Finally, the black militia men fled; 20–25 were captured and most of the arms confiscated. Seven African-Americans and one white man were killed; the majority of African-American casualties occurred after the surrender.[17]

The Colfax massacre, inspiring books and articles, is perhaps the most notorious of the attacks on armed African-Americans. On Easter Sunday, 1873, a black State militia of about 150 men was faced by a white militia of about the same number, though the latter was armed with cannon as well as small arms. Both were contesting recent elections for the governorship and local offices, and the blacks had barricaded themselves in the courthouse. Whites surrounded the building and set it on fire, firing after the flag of truce had appeared and shooting African-Americans escaping from the flames. Around 40 blacks were killed – again, mostly after they were captured. In all, between 61 and 80 African-Americans and three whites died (though many reports note over 150

---

16 Ibid. and Alwyn Barr, 'The Black Militia of the New South: Texas as a Case Study', *Journal of Negro History*, 63/3 (1978), pp. 209–219.

17 'Official Report of Attorney General Stone', *Anderson Intelligencer* (South Carolina), 20 July 1876. Cf. Joel Williamson, *After Slavery: The Negro in South Carolina During Reconstruction*, 1861–1877 (Chapel Hill, 1965), pp. 268–270.

casualties). This act signalled an end-point for Reconstruction in Louisiana, as LeeAnna Keith notes.[18]

In all, it is estimated that the Ku Klux Klan killed more than 3,000 persons in the brief years it terrorised the South. The above incidents were merely some of the most significant battles in a protracted war against black rights. Of course, such a campaign was supplemented by more legal means of disarmament and, once whites controlled the Southern states again, such extra-legal methods became unnecessary and even threatening to white control.

A fear of black insurrection – never far from the surface – infected Southern whites and gave them frequent reasons to confiscate any arms held by African-Americans. Many of the 'riots' discussed hysterically in the press of the time turned out to be either inconsequential incidents blown up to immense proportions or based on nothing at all. Some incidents involved blacks resisting lynching attempts. It did not take actual incidents to precipitate seizures of black-owned weapons; mere rumours could spark confiscation.

Typifying the easy connection between fear of disquiet that haunted the South and consequent disarmament was an 1888 dispatch from New Orleans announcing that 'Rumors of Uprising of Negroes Seem Unfounded'. For three or four weeks 'wild rumors were spreading ... that the negroes were arming and that a conflict of races was imminent'. Whites organised for a possible defence but the first precaution they took was to disarm all negroes in the area. They met little resistance except for Albert Harris and his son, who fired upon the white men when they arrived at the Harris's cabin demanding their arms. Harris and his son were run out of town. The dispatch noted that the weapons found 'were mostly old shotguns'.[19]

Legal Disarmament

The infamous Black Codes, enacted by several Southern states before the Fourteenth Amendment outlawed explicit racial discrimination, pointedly removed weapons from black hands. These legal statutes and constitutional amendments sought to restrict the liberties of newly freed slaves and maintain a white-dominated hierarchy. Virtually all the Black Codes prevented blacks from possessing or carrying guns. The Florida code threatened '39 stripes' – whipping – for African-Americans who transgressed a new order that 'it shall not be lawful for any negro, mulatto, or other person of color, to own, use or keep in

18 See Charles Lane, *The Day Freedom Died: The Colfax Massacre, the Supreme Court, and the Betrayal of Reconstruction* (New York, 2008), and LeeAnna Keith, *The Colfax Massacre: The Untold Story of Black Power, White Power and the Death of Reconstruction* (New York, 2008).

19 'All Quiet at St. Martinsville', *The Atlanta Constitution*, 27 August 1888, p. 1.

his possession or under his control, any Bowie-knife, dirk, sword, firearms or ammunition of any kind, unless he first obtain a license to do so from the Judge of Probate of the county in which he may be a resident for the time being'.[20]

Contrasting with liberal sentiment today, anti-slavery activists were indignant that African-Americans had their rights to bear arms curtailed. At an Anti-Slavery Conference held in Paris on 26 and 27 August 1867, veteran anti-slavery activist William Lloyd Garrison complained that white state governments were 'factious as ever in spirit' and that, with the power to legislate in its hands, the white South enacted cruel and tyrannical laws against the defenceless freedmen, whereby the most grievous outrages could be inflicted upon them with impunity. Their testimony against their white assailants was not allowed in the courts. They were not permitted to own or bear firearms in self-defence. Vagrant laws were passed, under which they could be arrested, fined, sold for a limited time to pay the fine, hired out to service at the will of the court ...[21]

Political representatives objected loudly to the seizing of weapons. Debating the Civil Rights Act in 1866, Representative Henry J. Raymond, a Republican from New York, explained that the rights of citizenship entitled the freedman to all the rights of United States citizens: 'He has a defined status: he has a country and a home; a right to defend himself and his wife and children; a right to bear arms.' Democratic Senator William Salisbury of Delaware added that '[i]n most of the southern States, there has existed a law of the State based upon and founded in its police power, which declares that free Negroes shall not have the possession of firearms or ammunition'.[22]

As Cottrol and Diamond note, once the Fourteenth Amendment and Civil Rights Acts were passed, the new state constitutions made more explicit references to individuals – rather than militias – having the right to bear arms. South Carolina established in 1868 the 'right to keep and bear arms for the common defense'. Mississippi and North Carolina changed the language from 'citizens' having a right to bear arms to 'all persons' and including 'self-defense' in the Constitution of 1868. North Carolina replaced the 1776 constitutional provision '[t]hat the people have a right to bear arms for the defense of the State' with the language of the Second Amendment. Arguably, the right to bear arms became more explicit and less tailored to the needs of the state after the

---

20 Alexander DeConde, *Gun Violence in America: The Struggle for Control* (Boston, 2001), p. 72. See the Codes at http://home.gwu.edu/~jjhawkin/BlackCodes/rptBlackCodes.pdf (accessed 24 April 2012). See also Joel M. Richardson, 'Florida Black Codes', *Florida Historical Quarterly*, 47/4 (1969), pp. 365–379.

21 Cited in the *Special Report of the Anti-Slavery Conference held in Paris, 26 and 27 August 1867* (London, the Committee of the British and Foreign Anti-Slavery Society, 1867), p. 82.

22 Cited in Tahmassebbi, 'Gun Control and Racism', p. 72.

abolition of slavery by the Thirteenth and the guaranteeing of racial equality in the Fourteenth Amendments to the Constitution.[23]

Consequently, the South became expert at using race-neutral legislation to deny weapons to a particular race. A code enacted in 1870 in Tennessee, for instance, punished those in possession of a pistol or other proscribed weapon with a fine of between $10 and $50 or with imprisonment of 30 to 60 days, at the discretion of the court.[24] But the enforcement was directed at blacks, despite the ostensibly neutral language of the law. Reviewing the Georgia Statute against concealed weapons in 1878, the *Atlanta Constitution* noted that 'Most of these cases have been against colored persons'.[25] The *Hartwell Sun*, advocating hanging for those who carried pistols, left no one in doubt about whom they spoke of: 'After the war the ku kluxers were a necessity and did much good in keeping down lawless, murderous villains.'[26] Indicating, at least, how the law was enforced in South Carolina since a statute enacted in 1881, one paper complained: 'The hardware dealers say that they have sold more pistols this fall than ever before in the same length of time. The majority of the buyers are negroes, who are becoming more and more addicted to carrying pistols every year.' The law, this paper averred, was a 'dead letter': 'A prosecution under the concealed weapon law is never heard of except when some unfortunate negro falls into the hands of the law officers and a pistol is found on his person.'[27]

Southern states pioneered laws that, while ostensibly neutral, effectively kept guns out of black hands but in the hands of the 'better sort'. In the aforementioned 1870 Tennessee code, all pistols were banned except expensive 'Army and Navy model' handguns, likely to be in the possession solely of whites and beyond the means of most African-Americans. In 1881, Arkansas enacted an almost identical ban on the sale of cheap revolvers. In 1902, South Carolina banned the sale of handguns to all but sheriffs and their special deputies. In 1893 Alabama pioneered attempts to tax guns out of existence, a method that would be used in the 1934 National Firearms Act.

What about the constitutionally protected 'right to bear arms'? One of the rare discussions of the Second Amendment occurred in relation to the Colfax massacre. Through the efforts of J.R. Beckwith, the US attorney in New Orleans, who became obsessed with bringing the perpetrators to justice, some of them appeared before the court. In *US v. Cruikshank* (1876), the United States Supreme Court ruled that William Cruikshank – who was charged with 32 counts of depriving blacks of their constitutional rights, including two claiming

23 Cottrol and Diamond, 'Never Intended to Be Applied to the White Population', p. 1320.

24 'Knife and Pistol', *Nashville Union and American*, 25 September 1874, p. 4.

25 'Carrying Concealed Weapons', *The (Atlanta) Daily Constitution*, 13 February 1878, p. 1.

26 Cited in 'Concealed Weapons', *The Daily Constitution*, 27 March 1879, p. 2.

27 Untitled, *The Watchman and Southron*, 28 October 1896, p. 1.

that he had deprived them of firearms possession – was not in violation of the Force Act of 1870.[28] The Court noted that 'bearing arms for a lawful purpose' was not a right guaranteed by the Constitution; using tortuous logic, it declared that the Second Amendment dealt only with federal government oppression, not oppression by private citizens nor by state authorities.[29]

In Florida, a candid admission by a Supreme Court judge in 1941 indicated the roots of a gun control law enacted in 1893 and revised in 1901 and 1906 that prohibited the carrying of handguns and repeating rifles, openly or concealed, with exceptions for peace officers and persons licensed by a county commissioner. Justice Rivers Buford explained why the carry ban was enacted and how it had actually been enforced:

> I know something of the history of this legislation. The original Act of 1893 was passed when there was a great influx of Negro laborers in this state drawn here for the purpose of working in turpentine and lumber camps. The same condition existed when the act was amended in 1901 and the act was passed for the purpose of disarming the negro laborers and to thereby reduce the unlawful homicides that were prevalent in turpentine and saw-mill camps and to give the white citizens in sparsely settled areas a better feeling of security. The statute was never intended to be applied to the white population and in practice has never been so applied. We have no statistics available, but it is a safe guess that more than 80 percent of the white men living in rural sections of Florida have violated this statute. It is also a safe guess to say that not more than five percent of the men in Florida who own pistols and repeating rifles have ever applied to the Board of County Commissioners for a permit to have the same in their possession and there has never been, within my knowledge, any effort to enforce the provisions of this statute as to white people, because it has been generally conceded to be in contravention of the Constitution and non-enforceable if contested.[30]

Justice Buford indicates clearly the purpose of these gun controls and, as we have seen, Florida was not alone in wishing to prevent African-Americans from possessing weapons. The lack of constitutional protection for African-Americans *only* stands out. These statutes never used the word 'negro', and the cases upholding those statutes scrupulously avoided any racial language. Yet the purpose and application of those laws was well known.

---

[28] The Force Acts refer to three Acts passed in 1870–1871 with the express purpose of limiting the activities of the Ku Klux Klan.

[29] Leslie F. Goldman, 'The Second Amendment, the Slaughter-House Cases (1873), and United States v. Cruikshank (1876)', *Albany Government Law Review*, 1/2 (2008), pp. 365–418, and Lane, *Day Freedom Died*.

[30] Cottrol and Diamond, 'Never Intended to Be Applied to the White Population', p. 1307.

'Disarm the Negroes'

The racial consciousness that characterised the United States at the turn of the century affected the South, sending it into paroxysms of fear and paranoia that too often resulted in violence, felt by African-Americans, and calls for pistol bans, also only for blacks. Often coupled with alcohol prohibition, the cause of pistol prohibition came to be heard more often, particularly after riots and tensions between blacks and whites, but also in relation to strikers and immigrants.

Meanwhile rumours of race war circulated. Hysterical stories about possible insurrection in Oklahoma occupied headlines of such Oklahoma dailies as the *Daily Ardmoreite*, as well as Southern newspapers.[31] However, it was with the Atlanta riot in 1906 that any ostensibly neutral language was dropped. Whereas Northern papers complained of the inherent violence of Italians – particularly Sicilians – and campaigned for more strenuous gun controls for aliens, Southerners took aim directly at African-Americans.

The *Atlanta Constitution* and other newspapers kept up a steady volley of fire against concealed weapons, condemning the 'pistol toter' and the ruffian. 'Punish the Pistol Toters' and 'Suppress the Pistol Toters' were the regular headlines of its numerous anti-gun campaigns. Of course, there was a hint of Progressivist spirit in the editorial stance, but there was no question that it was fear – and fear of armed African-Americans, in particular – that inspired these editorials. In 1901 the *Constitution* declared: 'In Charleston [South Carolina] pistols can be so easily obtained by negroes and for such trifling amounts that almost every black man, with fighting implications, goes about fully armed, and the killings follow as a matter of course. Every pawn shop in the city – and there are many of these places – offer [*sic*] pistols and dirks at remarkably low prices.'[32]

The Atlanta Race Riot of 1906 brought out the *Constitution*'s true colours, however. On Saturday, 22 September 1906, a riot began in Atlanta, lasting for four days. The tension had been building as the *Atlanta Journal* and the *Atlanta Constitution* vied for superiority in sales by publishing a lurid series of articles on the defiling of white women by black men. That day, both Atlanta newspapers reported four alleged black male assaults on local white women and, later than night, an estimated five to ten thousand white men and boys gathered at Five Points in downtown Atlanta. They rampaged, killing numerous African-Americans.

On Sunday, the black community armed themselves. Although there was then no law against blacks purchasing arms, the pawnshops and hardware stores refused to sell to them. Armed African-Americans patrolled their neighbourhoods. On Monday, 24 September, in a confrontation in Brownsville

---

31 See, for example, 'Armed Negroes Defy Marshalls', *Daily Ardmoreite*, 27 July 1897, p. 1.

32 'Weapons are Numerous', *The Atlanta Constitution*, 4 March 1901, p. 3.

with armed African-Americans, a policeman died and several whites were wounded. In response, a military unit raided Brownsville for weapons; 257 blacks were arrested in the raids enforced by soldiers manning a Gatling gun. The total casualties suffered were 3 whites and at least 25 blacks.[33]

The *Atlanta Journal* published an article entitled 'Disarm the Negroes' on 25 September that was widely republished. Proving all vestiges of Reconstruction to be utterly dead, the paper noted:

> The course adopted by the military Monday night in searching every negro on the streets and every negro house for deadly weapons is a step in the right direction. It is a solar plexus blow to the menace of retaliation that has thrown the city in a quiver of anxiety.
>
> Disarm the negroes. This is the keynote to the crisis. A good negro is contaminated by the possession of a weapon in a time like this; a bad negro is made very much worse the moment he places a pistol in his pocket.[34]

The emphasis on retaliation gives an indication that the most important motivation behind gun controls was fear of a race war. The fear was unrelated to any actual threat, but it was undoubtedly heightened by more frequent discussions of 'race war' both in the United States and internationally; the recent war between Japan and Russia, in which a 'coloured' nation resoundingly defeated a white nation, and the pronouncements of President Roosevelt on the importance of the issue of 'race suicide' ramped up fears. Whereas much of the scholarship directly blames the media for whipping up concern about black crime and assault on white women, a backdrop of race paranoia was necessary for the eruption of such blatantly racial attacks, of which disarmament was only one part.

The *Journal* continued: 'Should a collision between the races occur it would be too late to deplore the fact that the negroes had been allowed to arm themselves.' Self-deceptively, the journal assured readers: 'With the negroes without firearms there is little to be feared, for the white people are calm and quiet and there will be no more violence unless the rioting is started by the blacks.'[35]

Atlanta proved important in influencing the course of what would soon be referred to as 'race relations' over the next few years. Certainly, race conflicts occurring over the next period held the attention of the media and the elite throughout the South. The *Daily Ardmoreite*, a paper published in Oklahoma,

---

33 David F. Godshalk, *Visions: The 1906 Atlanta Race Riot and the Reshaping of Race Relations* (Greensboro, 2005).

34 'Disarm the Negroes', *The Atlanta Journal*, 25 September 1906, p. 6.

35 Ibid.

no doubt inspired panic when its headline announced in 1907: 'Situation Desperate ... A Race War is Feared'. Following the lynching of James Gardin, blacks congregated defensively in a black area. 'More than thirty armed blacks have gone from Weleetka to Clearview, one of the thickest negro settlements in the coal fields. About 50 stands of small arms were purchased by negroes in Weleetka before the hardware stores quit selling to blacks.' But, despite its hysterical prognosis of the situation, the paper also announced that 'The negroes, who came into town from the country, were disarmed without trouble'.[36]

At this time, fears of insurrection merged with a more general fear of crime. The inevitable response of Southern whites was to both arm themselves and disarm African-Americans. The need to arm themselves made it necessary to drop the pretence that gun controls were part of a Progressive attempt at rationalising Southern life in the twentieth century. In the North, generalised fear of immigrants and concern with criminal activities made racial justifications for a general disarmament more acceptable, which also emboldened Southern commentators. Removing any doubt that gun control measures were part of Southern determination to neutralise and neuter African-Americans, the Louisiana *True Democrat* carried a simple suggestion in a syndicated column entitled 'The Negro and the Pistol':

> It is estimated by a man who should know what he is talking about that of the young bucks between the ages of twenty and thirty, seven out of ten carry revolvers. As long as they use them to shoot up each other there will be no general complaint, perhaps but what a menace to law and order it is to have so many negroes loose in the community bearing arms. If it is true that seven out of ten young negroes carry guns, is the average white man doing himself and his family justice if he fails to break the law and go unarmed?
>
> What is the alternative?
>
> Disarm the negro, of course.
>
> But how are we going to do it? Here is a problem that presses for solution. No man, white or black, unless he is an officer of the law in the discharge of his duty, should be permitted to carry a deadly weapon concealed on his person.
>
> Certainly the irresponsible young negro is the last person that ever should be given that privilege.[37]

---

[36] 'Situation Desperate ... A Race War is Feared', *The Daily Ardmoreite*, 27 December 1907, p. 1.

[37] 'The Negro and the Pistol', *The True Democrat* (St Francisville, West Feliciana Parish), 24 October 1908, p. 1.

Black Response

How did African-Americans respond to the frequent attacks made on them, lynchings and attempts to disarm them? There is no evidence that any insurrections were ever planned and, as many of the reports note, African-Americans, while many retained weapons of various sorts, were simply outgunned by Southern whites. Even in fights involving similar numbers, old shotguns were no match for repeating rifles in terms of deadliness or accuracy. Moreover, African-Americans who did not cooperate with whites risked financial ruin.

Many who were outspoken, like Ida Wells, armed themselves for practical reasons, determined to sell their lives as dearly as possible, if attacked. On a personal level, the South remained a heavily armed part of the country, and there is evidence that blacks, unwilling or unable to avail themselves of the law, possessed weapons for personal security. As Swedish sociologist Gunnar Myrdal observed in his magisterial study of African-Americans in the early 1940s:

> In this region, the custom of going armed continually or having weapons within easy reach at home was retained from *ante-bellum* days. This custom was taken over also by the Negroes during Reconstruction days. The writer has been astonished to see how firearms and slashing knives are part of the equipment of many lower class whites and Negroes in the South, the laws against carrying 'concealed weapons' are not efficient, as they do not – and for constitutional reasons cannot – forbid the owning, buying, and selling of arms. ... In the Negro community, where personal security is most lacking, the dangerous pattern of having knives and guns around is most widespread. It undoubtedly contributes to the high record of violent actions, most of the time directed against other Negroes.[38]

It is difficult to determine exactly the prevalence of the ownership of weapons amongst African-Americans, but it is likely that those who lived in the country and had to eke out livings as sharecroppers would augment their meat supplies by hunting. Many, understandably, probably kept a weapon, perhaps hidden somewhere, given that blacks often lived in the poorest and most dangerous areas.[39]

---

38 Gunnar Myrdal, *An American Dilemma: The Negro Problem and American Democracy* (London, 1944), p. 560.

39 'Contrary to general belief, Negroes seldom use razors as a means of committing murder. For the years 1924–1926 Negro homicide victims were slain with firearms in 72.7 per cent of the cases, while for the same period whites were slain with fire-arms in only 68.3 per cent of the cases. For the entire Registration Area in continental United States the percentage was 70.2 use of firearms in slayings.' H.C. Brearley 'The Negro and Homicide', *Social Forces*, 9/2 (1930), p. 252.

The response of civil rights leaders to issues of black gun ownership varied. Generally, the more radical the leader, the more insistent he or she was on black equality, the more hostile to gun controls. Bishop Henry McNeil Turner, of the African-Methodist Episcopal Church, called, in response to two lynchings in 1897, for blacks to get guns:

> Let every negro in the country who has a spark of manhood in him supply his house with one, two or three guns or with a seven or sixteen shooter, and I advise him to keep them loaded and ready for immediate use and when his domicile is invaded by bloodthirsty lynchers or any mob, day or night, Sabbath or week day, turn loose your missiles of death and blow the fiendish invaders into a thousand giblets.[40]

Conversely, conservative black leader Booker T. Washington urged blacks to give up their guns. As the *Washington Post* observed, though the problem was not confined to 'negroes', 'the colored educator frankly admits that his own people are the greatest enemies to society in this respect'. Washington added that he had no need to carry a gun:

> During all my years that I lived in Alabama I have never carried a concealed weapon. I have never kept one in my house and I have never felt the need of one. I have travelled all through the south by night and by day and have never felt that I have been in the least danger and if I had I am quite sure that I would not have been protected by reason of carrying a pistol.[41]

During this period, however, Washington's popularity as a black leader was falling and those who were creating the National Association for the Advancement of Colored People (NAACP) felt differently. Mary White Ovington protested after the Atlanta riot that Atlanta had given guns to a rampaging white mob, and had taken guns away from respectable black residents defending their homes and families.[42]

W.E.B. Du Bois, another founder of the NAACP, also praised black self-defence during the Atlanta riot. As the author of a book on the Civil War 'martyr' John Brown and as editor of the NAACP's paper, Du Bois consistently praised blacks who defended themselves against the depredations of whites. In relation to the Chicago and Washington riots of 1919, Du Bois wrote: 'We are no longer

---

40 'Get Guns, Negroes!', *Milwaukee Journal*, 17 March 1897, p. 4.

41 'A Word to Gun-Toters', *The Washington Post*, 15 January 1912, p. 6; Booker T. Washington, 'B. Washington Asks Negroes to Suppress the Gun-Toter', *The Atlanta Constitution*, 10 January 1912, p. 6.

42 Carolyn Wedon, *Inheritors of the Spirit: Mary White Ovington and the Founding of the NAACP* (New York, 1999), p. 89.

depending upon our friends; we are depending upon ourselves. If mobs attack us we are prepared to defend ourselves and we are going to defend ourselves as Washington and Chicago know.' In 1921, he approvingly wrote of the resistance of blacks in Tulsa, Oklahoma: 'Black Tulsa fights! It fights mobs with firearms and it fights economic oppression with cooperation. It has appealed to the colored people of the United States for funds. The National Association for the Advancement of Colored People will be glad to act as its agent ... Strength to their arms!'[43]

Perhaps, though, the attitude amongst ordinary blacks was not so hopeful or crusading as Du Bois' spirited writings. As the Baltimore *Afro-American* sardonically noted about impending gun legislation in Louisiana: 'Hereafter if you are a colored man you will be jailed if you carry a gun here. A white man is not molested. A relentless drive to arrest and disarm all Negroes found in possession of firearms has been ordered by Superintendent of Police Theodore A. Ray.'[44]

In Jefferson County, Alabama, in 1936, James Cooper, an African-American, became the first to be prosecuted under a new law that required licences for pistols to be carried. Found with a pistol, he was arrested and jailed with a $100 bail bond set. This was despite the fact that no licences had yet been issued when he was arrested.[45]

Conclusion

The relationship of African-Americans with guns during the period under discussion is complex and best understood as part of a process of disarmament through both legal and extra-legal means beginning at Reconstruction and complete by the 1930s. Personal security – the reason most people own pistols today – was less important than the political significance of disarmament. Guns, to African-Americans, were symbols of freedom as well as ways to secure it. When guns were removed from black hands, it was often only as a prelude to the removal of other important privileges of citizenship such as the right to vote and the right to free speech. As many of the stories told above illustrate, when African-Americans dared to resist these attempts at disarmament whites inflicted an extraordinary amount of violence upon them.

Blacks were not only not trusted with arms; their lives, evidently, were regarded as so cheap that, as Myrdal depressively observed nearly 80 years after

---

43 Dominic J. Capeci Jr and Jack C. Knight, 'Reckoning with Violence: W.E.B. Du Bois and the 1906 Atlanta Race Riot', *Journal of Southern History*, 62/4 (1996), pp. 727–766.

44 'Disarm Negroes, but Leave Whites Alone', *Afro-American*, 11 January 1930, p. 13.

45 'Colored Man Is First Victim of New Law', *Atlanta Daily World*, 27 October 1936, p. 2.

the passage of the Fourteenth Amendment: 'Any white man can strike or beat a Negro, steal or destroy his property, cheat him in a transaction or even take his life, without much fear of legal reprisal.'[46] What this chapter demonstrates is that the disarming of blacks took place because of an irrational fear of black insurrection and a less irrational fear that blacks might more stridently insist on equality if armed. Moreover, the anti-pistol movement, strongly supported by Southern whites, belied a concern about firearms only when the finger on the trigger was black.

46 Myrdal, *American Dilemma*, p. 559.

PART IV
Celebrating Guns: Firearms in Popular and Military Cultures

Chapter 11
Retrospective Icon: The Martini-Henry

Ian F.W. Beckett

I

The scene is familiar to many. On the morning of 23 January 1879, Colour Sergeant Bourne reports to Lieutenant Chard that the Zulu have disappeared from Rorke's Drift: 'It's a miracle', Chard replies. 'If it's a miracle, Colour Sergeant, it's a short-chamber Boxer-Henry point four-five calibre miracle.' The rejoinder, 'And a bayonet, Sir, with some guts behind it.' The exchange comes, of course, from the screenplay for the 1964 film, *Zulu*, starring Stanley Baker as Chard and Michael Caine as Lieutenant Bromhead.[1] *Zulu* has consistently ranked high in polls of Britain's favourite films. It was placed eighteenth – the second highest ranked British film – by the 60,000 subscribers of Sky's 'Millenium Movies' poll in 2000, and eighth by website users for Channel Four's '100 Greatest War Movies' in 2005. It was also the most popular film of all among Conservative MPs polled in 2004.[2] *Zulu* also inspired a popular publishing boom that has continued unabated for almost 50 years. A casual survey suggests that, since 2000 alone, Rorke's Drift and Isandlwana that immediately preceded it have resulted in the publication of at least a dozen titles, including one by this author.[3]

Part of the mystique surrounding Isandlwana and Rorke's Drift is undoubtedly the role of the Mk II .450 calibre Martini-Henry rifle that fired the Boxer-Henry cartridges. The single-shot breech-loading Martini-Henry was the British infantryman's principal weapon from 1874 until 1888. It was certainly expected that a Zulu polity that largely eschewed firearms for cultural reasons would be easy prey.[4] The Xhosa had been simply swept away by the firepower of the Martini during the Ninth Cape Frontier War of 1877–1878, not least at Centane on 7 February 1878. Lieutenant General Lord Chelmsford predicted

---

1 Originally, Bourne's lines were given to another character deleted from the final version. Sheldon Hall, *Zulu: With Some Guts Behind It. The Making of the Epic Movie* (Sheffield, 2005), p. 106.

2 Ibid., pp. 369–370.

3 Ian F.W. Beckett, *Isandlwana* (London, 2003).

4 John Laband, 'Firearms and the Zulu Kingdom: The Cultural Ambiguities of Transferring Weapons Technology', paper to the 'Tutü te Puhu: New Zealand's Wars in the Nineteenth Century Conference', Wellington, 13 February 2011. For a more detailed consideration of this issue, see Hogan, this volume.

on 23 November 1878: 'I am inclined to think that the first experience of the power of the Martini Henrys will be such a surprise to the Zulus that they will not be formidable after the first effort.'[5] As *Zulu* suggested, the impact of the Martini-Henry contributed materially to the defence of Rorke's Drift by 139 defenders confronted with between 3,000 and 4,000 Zulu over the course of about 10 hours. How then could the Zulu – admittedly some 20,000 in total – have inflicted such a crushing defeat on an infinitely larger force of 1,174 officers and men, with over 400,000 rounds of ammunition available to them, at Isandlwana earlier on 22 January in not much more than two hours? It was the greatest single day's loss of British troops between 1815 and 1914: 854 Europeans died, including 706 regulars. Only 78 Europeans of those in the camp are known to have survived, of whom just five were regular officers.[6]

It has become popularly supposed that a failure to be able to open the ammunition boxes of Boxer-Henry cartridges was a major contribution to the disaster. The myth of the ammunition boxes at Isandlwana, however, is itself of relatively recent origin. It was perpetuated in the film, *Zulu Dawn*, in 1979, and it is arguably through film that the Martini-Henry has become one of the most instantly recognisable of all firearms. Not only did it appear prominently in *Zulu* – though some extras were equipped with poorly disguised bolt-action rifles – and *Zulu Dawn*, but it has also been seen in *The Man Who Would Be King* (1975) and *The Four Feathers* (2002).

Martinis have become highly desirable, one major antique arms dealer in Brighton currently offering two Mark I Martinis at £1,350 and £1,275 respectively.[7] The same dealer attracted considerable media interest in 2008 when offering, for £1,100 each, two Martinis captured from Taliban insurgents in Afghanistan. They were believed to have been taken originally from the bodies of British troops killed at Maiwand in July 1880. As the dealer remarked at the time, 'The Martini-Henry is a very, very collectable gun – almost entirely due to Michael Caine and the film *Zulu*.'[8] Equally, spent Martini-Henry cartridge cases and bullets are eagerly sought relics. Bullets that may or may not be from Zulu War sites currently sell on eBay for between £64 and £78, and cartridge cases for between £90 and £131, one website noting: 'It is nothing short of astonishing what collectors – mostly British – are willing to pay for the remains of even a single bullet from a South African battle.'[9]

---

5 Ian Knight and Ian Castle, *Zulu War* (Oxford, 2004), p. 26; Philip Gon, *The Road to Isandhlwana* (Johannesburg, 1979), pp. 139–140; KwaZulu-Natal Archives Depot, Pietermaritzburg, Wood Mss, II/2/2, Chelmsford to Wood, 23 November 1878.

6 Beckett, *Isandlwana*, pp. 7–8, 88–89.

7 www.thelanesarmoury.co.uk/shop/shop.php [accessed 28 April 2011].

8 *The Independent*, 8 June 2008.

9 http://goldiproductions.com/angloboerwarmuseum/Boer92s_pastdis_relics_paar.html [accessed 29 April 2011].

The resonance of the Martini-Henry in popular culture can be illustrated in other ways. There was, for example, a 1978 novel by Barrie Hughes, *The Martini Henry*. Somewhat curiously, this was not a novel of colonial warfare as might be imagined, but a 'powerful story of a man's love for a boy, set on a deserted island off the coast of Australia'.[10] Briefly, between 2001 and 2006, there was also a Cardiff-based group, 'The Martini Henry Rifles', characterised both as an 'alternative new wave punk' and as a 'noise-punk' band.

In one respect at least, the Martini-Henry enjoyed something of a similar status for the Victorians, Kipling twice mentioning the weapon in his poetry.[11] Thus, from 'The Young British Soldier', set in Afghanistan,

> When 'arf your bullets fly wide in the ditch,
> Don't call your Martini a cross-eyed old bitch;
> She's human as you are – you treat her as sich,
> An' she'll fight for the young British soldier.

and from 'Fuzzy Wuzzy' set in the eastern Sudan,

> We sloshed you with Martinis, an' it wasn't 'ardly fair;
> But for all the odds agin' you, Fuzzy-Wuz, you broke the square.

The Martini-Henry features in Kipling's original short story, *The Man Who Would Be King* (1888), and also in 'The Black Jack' from *Soldiers Three and Other Stories* (1888). There are similar references to the Martini-Henry in Bram Stoker's *The Watter's Mou* (1895), O Henry's short story, 'The Admiral' from *Cabbages and Kings* (1896), and Joseph Conrad's *Heart of Darkness* (1902). The Australian policeman, William Hughes Willshire, who wrote extensively on Aboriginal society, is known for one particularly purple passage from *The Land of the Dawning* (1896): 'the Martini-Henry carbines at the critical moment were talking English in the silent majesty of these eternal rocks'.[12] In a different cultural association, a celebrated New Zealand-bred mare, Martini-Henry, won both the Victoria Derby and the Melbourne Cup in 1883.

For the most part, however, the status of the Martini-Henry as cultural icon is an entirely modern phenomenon for, in the late Victorian period, it was a highly controversial and problematic weapon. First, it will be necessary to outline the role of the Martini-Henry in the Zulu War, before, secondly, discussing its general development as the army's principal weapon. Thirdly, discussion of the final replacement of the Martini-Henry will lead to a general conclusion.

---

10 Barrie Hughes, *The Martini Henry: A Novel* (London, 1978).

11 *Rudyard Kipling's Verse: Inclusive Edition, 1885–1918* (London, n.d.), pp. 455, 474.

12 William H. Willshire, *The Land of the Dawning* (Adelaide, 1896).

II

To some extent, the problematic reputation of the Martini-Henry is reflected in the popular assumptions about ammunition supply at Isandlwana, but these are entirely unfounded. The official history prepared for the War Office in 1881 blamed the defeat on the collapse of the locally recruited Natal Native Contingent (NNC) at a crucial point in the battle.[13] It did not mention any failure in ammunition distribution. Nor did it mention the wider contemporary controversy surrounding the decision of Chelmsford to divide his force in the early hours of 22 January, and whether or not his instructions to defend the camp were fulfilled properly by Brevet Colonel Anthony Durnford. The latter arrived at Isandlwana later that morning and thus became senior officer. The first professional historian to study the battle, Sir Reginald Coupland, writing in 1948, then added the failure of ammunition supply to the collapse of the NNC as a primary cause of the disaster, alleging that no proper system of distribution was organised and that the 'tight-screwed lids' of the boxes had not been removed in advance. Coupland believed this a 'safe inference' from Chelmsford's pre-campaign instructions on the point, and from the 1925 memoir by one of the five regulars who escaped, Lieutenant (later General Sir) Horace Smith-Dorrien.[14] Smith-Dorrien stated that he was breaking open boxes and helping to send ammunition out to the firing line when he was seen by Quartermaster Bloomfield of the 2/24th, who exclaimed: 'For heaven's sake, don't take that, man, for it belongs to our Battalion.' Smith-Dorrien replied: 'Hang it all, you don't want a requisition now, do you?'[15]

The popular account of the war was begun before *Zulu* was conceived, but benefited immensely from its near simultaneous publication. Donald Morris's *The Washing of the Spears* (1965) considerably embroidered the story of hidebound quartermasters from Smith-Dorrien's account. The Mark V ammunition box, containing 600 rounds in packets of 10, was enclosed by two copper bands, each fastened by nine screws. In addition, the wedge-shaped sliding panel on the box that gave access to the contents was secured by another screw, after which the tin lining of the box could be torn back by its attached wire handle. According to Morris, six of the frequently rusted screws had to be removed to gain access, and the quartermasters of the 1/24th and 2/24th in the camp had only one screwdriver each. Thus, desperate soldiers 'hacked at the copper bands with axes or thrust bayonets under them and attempted to snap

13 John Sutton Rothwell, *Narrative of Field Operations Connected with the Zulu War of 1879* (London, 1881), p. 36.

14 Reginald Coupland, *Zulu Battle Piece: Isandhlwana* (London, 1948), pp. 89, 141.

15 Horace Smith-Dorrien, *Memories of Forty-Eight Years' Service* (London, 1925), p. 14.

them or prize them up over the screwheads'.[16] Other popular accounts in the 1960s and 1970s generally followed Morris's lead, though a detailed refutation of the myth by F.W.D. Jackson was published as early as 1965.[17]

In fact, Smith-Dorrien recounted his exchange with Bloomfield specifically as an example of the 'coolness and discipline of the regiment'.[18] He made no mention of any ammunition shortage in a contemporary letter to his father, or to the official enquiry into Isandlwana that convened on 27 January 1879. The only contemporary mention of any possible shortage was in the evidence of another of the survivors, Captain Edward Essex, who merely recounted that, when the two companies of the 1/24th first engaged were 'becoming short of ammunition', he organised their re-supply.[19] It is generally acknowledged that some of the colonial troops, and also the NNC, who carried only 15 rounds per man, and of whom only one in ten was armed anyway, did run short of ammunition. The experienced soldiers of the 1/24th, however, were regarded as 'old, steady shots'.[20]

All the evidence suggests that the officers controlled the rate of fire in the approved, careful manner, a significant factor in judging how far ammunition was expended. Each man would have routinely carried 70 rounds and possibly the additional 30 rounds usually reserved for emergencies and kept separately from the regimental supplies in the boxes. Ammunition expenditure in other hotly contested actions averaged only 33 rounds per man in four hours' action at Khambula in March 1879, and only 10 rounds in an hour at Ulundi in July 1879. An average of 10 rounds was also expended at Gingindlovu in April 1879. Similarly, average expenditure at Laing's Nek and Ingogo during the Anglo-Transvaal War in January and February 1881 was 17 and 19 rounds respectively. At Tamai in the Sudan in March 1884, where the dervishes broke into the British square, the average expenditure was still only 50 rounds. Morris claims the Rorke's Drift defenders fired about 20,000 rounds over 10 hours, but this still equates to only 15 rounds per man per hour.[21]

More to the point, far from needing to remove the copper bands and six screws from each ammunition box, only the removal of the one screw on the

---

16 Donald Morris, *The Washing of the Spears* (London, 1965), pp. 371–373.

17 F.W.D. Jackson, 'Isandhlwana, 1879: The Sources Re-examined', *Journal of the Society for Army Historical Research* (hereafter *JSAHR*), 43 (1965), pp. 30–43, 113–132, 169–183.

18 Smith-Dorrien, *Memories*, p. 14.

19 Beckett, *Isandlwana*, pp. 108–111.

20 Frank Emery, *The Red Soldier* (London, 1977), p. 49.

21 Beckett, *Isandlwana*, pp. 66–70; Ian Knight, *Zulu Rising* (London, 2010), pp. 377–381; Ian Knight, 'Old Steady Shots: The Martini-Henry Rifles, Rates of Fire and Effectiveness in the Anglo-Zulu War', *Journal of the Anglo Zulu War Historical Society*, 11 (2002), pp. 1–5; Adrian Greaves (ed.), *Redcoats and Zulus* (Barnsley, 2004), pp. 182–191; Charles Callwell, *Small Wars: Their Principles and Practice* (London, 1906; 3rd edn), p. 396; Morris, *Washing of Spears*, p. 416.

sliding lid was required. In the absence of a screwdriver, a sharp kick or a blow from a rifle butt would more than suffice to break it open. Archaeological exploration at Isandlwana in 2000 revealed wire handles from boxes on the firing line and screws clearly bent by a blow. As Jackson concluded in 1965, there is 'no reliable contemporary evidence that the regulars ran out of ammunition while in the line'. Nor, as Ian Knight has remarked, is there any evidence that ammunition shortage informed any tactical decisions made by British officers.[22] This is not to deny that there were problems with the Martini-Henry, which can now be examined.

III

As its name implied, the Martini-Henry was a compromise weapon, incorporating the breech mechanism of a Swiss-Hungarian, Friedrich Martini, and the barrel of an Edinburgh-based Scot, Alexander Henry. With rapid developments taking place in weapons technology, as demonstrated by the perceived success of the Prussian Army's breech-loading Dreyse Needle-gun in the current campaign against Denmark, the War Office appointed an Ordnance Select Committee in June 1864 to begin the search for a new breech-loader to replace the existing muzzle-loading .577 calibre Enfield percussion rifle musket. A temporary solution was the adoption, in September 1866, of the breech-loading mechanism of an American inventor based in Philadelphia, Jacob Snider, which could convert the Enfield. When the supply of Mk I and II Snider-Enfield conversions was used up in 1869, a new Snider was introduced, though officially designated the Mk III Snider-Enfield. Firing a rolled-brass cartridge patented in 1866 by Colonel Edward Boxer, Superintendent of the Royal Laboratory at Woolwich, the Snider-Enfield was capable of about 10 rounds a minute up to a range of about 1,000 yards.[23]

Trials were already underway to find a better purpose-built breech-loader, the Secretary of State for War advertising a design prize competition on 22 October 1866. Specifications had been agreed previously by the Ordnance Select Committee, including a maximum weight of 9 lb 5 oz., and an overall maximum length of 51 inches. A Special Sub-Committee of the Ordnance Select

22 Jackson, 'Isandhlwana', p. 123; F.W.D. Jackson, *Hill of the Sphinx: The Battle of Isandlwana* (London, 2002), pp. 38, 40, 75; Ian Knight, *Zulu: Isandlwana and Rorke's Drift* (London, 1992), pp. 81–83. Of those who have written on Isandlwana recently, only Edmund Yorke and Lock and Quantrill give credence to ammunition supply problems. See Edmund Yorke, *Rorke's Drift, 1879* (Stroud, 2001), pp. 69–75, and, by the same author, 'Isandlwana, 1879: Reflections on the Ammunition Controversy', *JSAHR*, 72 (1994), pp. 205–218; Ian Knight, 'Ammunition at Isandlwana: A Reply', *JSAHR*, 73 (1995), pp. 237–250; Ron Lock and Peter Quantrill, *Zulu Victory* (London, 2002), pp. 322–327.

23 Charles Purdon, *The Snider-Enfield* (Alexandria Bay, 1963).

Committee under the chairmanship of Lieutenant Colonel Henry Fletcher began work in March 1867. Nine rifles were chosen for trial in June from 37 considered to have met the specifications. Eighty-three others were rejected as not meeting the desired specification, but with 11 of them set aside for further consideration.[24] Those chosen for trial included separate submissions by Martini and by Henry, as well as a rifle designed by an American, Henry Peabody, who claimed later that Martini had copied aspects of his design.[25]

Initially reporting on 12 February 1868, Fletcher's committee believed that none of the nine rifles merited the £1,000 prize, but that Henry's design came closest to the requirements and should be awarded £600. The committee also awarded £400 to the cartridge submitted by George Daw, 49 different kinds of ammunition having been submitted. With consideration now given to all other rifles submitted, including repeating weapons, a reconstituted Special Committee on Breech-Loading Rifles continued trials on all submitted breech mechanisms. Those of barrels, however, were confined to just seven that had performed best in the previous trials. Westley Richards, from the well-known Birmingham arms company, withdrew his guns in November 1868 when it was decided that the Royal Small Arms Factory (RSAF) at Enfield would modify the submitted weapons as it saw fit so that only Boxer cartridges could be used. His decision reflected the residual suspicion of the RSAF in the commercial sector.

The final recommendation on 11 February 1869 was to combine Henry's barrel and Martini's mechanism as the Martini-Henry, and to use a Boxer cartridge with a Henry bullet to be known as the Boxer-Henry. The determination of the cartridge swung the committee away from its original commendation of Henry's mechanism. The Secretary of State, Edward Cardwell, confirmed the decision on 1 April 1869.[26]

All versions of the Martini-Henry weighed eight to nine pounds and had an overall length of around 49 inches, the barrel length being some 33 inches. The breech mechanism was a hammerless falling block lever action. The lever behind the trigger guard opened the breech for the insertion of a round, and the raising of the lever cocked the weapon. The opening of the breech after firing would then automatically eject the cartridge, extracted by two retractor hooks

---

[24] B.A. Temple and I.D. Skennerton, *A Treatise on the British Military Martini: The Martini-Henry, 1869–1900* (Burbank, 1983), pp. 9–37, 167–173.

[25] Ibid., pp. 47–49. Subsequently, 'Peabody-Martinis' were manufactured in the US for the Ottoman government.

[26] *House of Commons Parliamentary Papers* (hereafter *HCPP*) 1867–1868 [4020], Reports on Breech-loading Arms by a Special Sub-Committee of the Ordnance Select Committee; ibid., 1868–1869 [4119] Reports of a Special Committee on Breech-loading Rifles; National Archives of the UK, Kew, London (hereafter NAUK), WO 32/22, 'Précis on Martini-Henry Rifles'; *Pall Mall Gazette*, 24 August 1868, 'The Coming Military Breechloader'; ibid., 19 November 1868; ibid., 10 March 1869.

that would engage the bottom of the cartridge rim. The rifle was sighted for 1,400 yards, though it was thought to be most effective at 400 yards, which was precisely the range at which the Zulu attack at Isandlwana was stalled for some time.[27] The muzzle velocity of 1,362 feet per second was greater than that of the Snider, whose velocity was 1,252 feet per second. At 100 yards, therefore, the Martini-Henry could penetrate a sandbag that the Snider would not at 10 yards; at 200 yards the Martini-Henry went through a 0.25 inch iron plate that the Snider would only penetrate at 75 yards.[28]

Despite improvements, a continued problem was the overheating of the barrel after about 10 rounds, making it difficult to hold the barrel with the left hand, and a vicious recoil that bruised shoulders and made it extremely uncomfortable to fire after about 40 rounds. Overheated barrels could be countered by wrapping or sewing ox-hide or cowhide around them, but little could remedy the recoil. As *The Sporting Gazette* remarked of prize shooting meetings in July 1874, '[b]lack eyes, bloody noses, bruised shoulders, and half-fractured collar bones have been the ordinary consequences of competition for any prizes in which the Martini-Henry has been the weapon used'.[29]

In fact, a number of reports when the Martini-Henry underwent trials suggested the recoil was not much greater than that of the Snider. The Superintendent of the RSAF remarked of one report from the 2/4th Foot in 1873 that '[t]here must be something exceptional in this regiment, as there are no other complaints as to recoil'.[30] In any case, as *The Pall Mall Gazette* noted, such problems had 'no more connection with the breech action than they have with the colour of the rifleman's coat', rendering much of the criticism of the Martini mechanism invalid.[31]

The periodical was essentially correct, in that the real difficulty was with the Boxer cartridge. One of the ironies was that Boxer had been dismissed from the Royal Laboratory in November 1869 for earlier assigning his patents to Messrs Eley Brothers, who had then received the contract for supplying the cartridge cases for the Snider.[32] The Boxer-Henry was undoubtedly a highly effective 'man-stopper'. It flattened on impact and smashed through bone, causing horrific injuries, as most accounts of the aftermath of actions such as Rorke's Drift, Khambula and Ulundi indicate. Visiting Zululand in 1882, Bertram Mitford commented on how few wounded Zulu he encountered, most hit having

27 Knight, *Zulu*, pp. 79–80.

28 Jackson, *Hill of Sphinx*, pp. 74–75; 'The Future Weapon of the British Soldier', *The Living Age*, May 1869; 'The Martini-Henry Rifle', *Illustrated London News*, 22 July 1871, p. 66.

29 *The Sporting Gazette*, 18 July 1874, p. 661.

30 *HCPP* 1876 [C.1452], Martini-Henry Rifles, p. 2.

31 'The Martini-Henry Rifle and Its Critics', *Pall Mall Gazette*, 21 August 1874.

32 *HCPP* 1870 (60), 'Colonel Boxer'.

perished. It is estimated that at least 1,000 Zulu died at Isandlwana, 600 at Rorke's Drift, 2,000 at Khambula, 1,200 at Gingindlovu, and 1,500 at Ulundi.[33]

The Boxer-Henry cartridge – technically the Mk III Boxer – as it emerged in August 1873 was of relatively thin and soft rolled-brass sheeting, while the charge of 85 grains of black powder was an especially heavy one. Prolonged firing of any weapon would have led to a greasy deposit left in the breech chamber. But the density of the Boxer-Henry charge, coupled with the thin cartridge, meant that there was an increasing tendency as the weapon got hotter for it to stick to the chamber. The retractors might break or, more commonly, they would tear off the base of the cartridge and jam the breech, making it necessary to remove the debris with a cleaning rod or a knife, though kicking the lever might also free the cartridge. This was not necessarily a problem for a trained and experienced infantryman: Henry Hook, who won the VC at Rorke's Drift, reported that his Martini had jammed several times. But it could be so for the inexperienced. There is certainly an account of Durnford removing jams for his Natal Native Horse at Isandlwana. An account by a survivor from the Rocket Battery suggests that one of the NNC companies was having significant difficulties with jammed weapons.[34] The cartridges could also be too easily deformed after a relatively short time being carried in pouches or bullet bags.[35] The use of black powder would additionally result in a considerable amount of smoke, potentially impairing visibility on the battlefield.

For these reasons, therefore, the Martini was to prove a controversial choice. As *The Times* commented later, the 'ordnance world ... rose in insurrection' at the original decision. Westley Richards in particular complained at what he saw as the committee moving from a 'judicial' role to a 'constructive' one. He was still complaining about the process when he appeared before the Stephen Commission on Warlike Stores in 1887, claiming his own rifle, incorporating a rifled barrel by William Ellis Metford, as far superior to the Martini-Henry.[36] Sir Walter Barttelot, a Conservative MP and a prominent member of the rifle volunteer force, particularly championed Richards's cause. Arguing that other rifles had consistently outperformed the Martini-Henry at the annual Wimbledon meetings of the National Rifle Association, Barttelot tried unsuccessfully to get parliamentary enquiries into the Martini-Henry's selection, and its continued adoption, on both 28 April 1871 and 9 June 1876. The first attempt went down

33 Bertram Mitford, *Through the Zulu Country* (London, 1883), p. 177; John Laband, *Kingdom in Crisis: The Zulu Response to the British Invasion of 1879* (Manchester, 1992), pp. 90–91, 106–108, 164–166, 178, 230–231.

34 Emery, *Red Soldier*, p. 130; Beckett, *Isandlwana*, pp. 56, 69.

35 'Another Sixty Man', *The Times*, 30 July 1878, p. 11.

36 *The Times*, 13 June 1876, p. 9; *Morning Post*, 31 July 1879; *HCPP* 1887 [C.5062-I], pp. xxi; ibid., [C.5062-II], p. 226; W.P.P. Marshall, 'The Comparative Merits of the Martini Rifle and the Westley Richards Rifle', *Engineering*, 27 April 1871.

by 137 votes to 72. He then withdrew his motion to reduce the vote on supply on the second occasion after the Secretary of State, Gathorne-Hardy, indicated all efforts were being made to remedy defects. Replying to comments by the Duke of St Albans in the House of Lords earlier in February 1876, Cardwell had wearily cautioned against listening to 'disappointed inventors'.[37]

Generally, the press was supportive. *The Times* proclaimed it 'the most magnificent weapon ever placed in the hands of a soldier'.[38] In particular, *The Pall Mall Gazette* constantly supported the rifle's introduction, believing the process to have been impartial and exhaustive. Indeed, once the sub-committee had recommended the Martini-Henry, there were yet further trials to ensure that the combination of mechanism and barrel was workable.[39] The first prototype batch of 200 rifles manufactured at Enfield was sent to selected units in May 1869, a Special Committee on Martini-Henry Breech-loading Rifles, again chaired by Fletcher, reporting in July 1870 and February 1871. Some 50 Martinis were tested in Britain and Ireland, 30 in Canada, 100 in India, five in South Africa, and 15 by the Royal Navy. The committee concluded that the Martini was 'admirably adapted for a military arm', and unanimously recommended its adoption. It was apparent, though, that there were problems with defective cartridges and in extracting them from the breech; with misfires due to a weak spring or coil; and with the position of the back sight.[40] Changes were recommended for the first production models issued on 3 June 1871, Cardwell having formally approved the Martini-Henry's adoption on 3 May 1871.

Yet more changes were made in September 1872, and a new series of trials undertaken in 1873. Approved by a War Office conference in April 1874, the final version of the Mk I appeared on 17 July 1874. Significant changes had been made to the screw that locked the breechblock axis pin in place; the trigger assembly; the striker that had previously caused broken firing pins; the safety mechanism; the cleaning rod; and the sights. [41] The Martini-Henry finally began to be issued on 12 October 1874.

---

37 *Birmingham Daily Post*, 23 July 1869; *Daily News*, 23 July 1869; *John Bull*, 29 April 1871, p. 288; *The Times*, 29 April 1871, p. 9; *Punch*, 6 May 1871; *Western Mail*, 14 May 1874; *The Times*, 10 June 1876, p. 8; *Hansard*, HC Deb. v. 205, cc. 1872–1910, debate of 28 April 1871; ibid., v. 229, cc. 1636–59, debate of 9 June 1876; ibid., HL Deb. v. 227, c. 673, Cardwell, 22 February 1876.

38 'The Martini-Henry Rifle', *The Times*, 19 August 1874, p. 12.

39 Temple and Skennerton, *Treatise*, pp. 38–47, 53–84, 196–225.

40 *HCPP* 1870 [C.198], Abstract of Reports on the Experimental Martini-Henry Arms issued for Trial in 1869; 1871 [C.299], Reports of Special Committee on Martini-Henry Breech-loading Rifles, p. 6 (also NAUK, SUPP 5/892); *Leeds Mercury*, 13 August 1870; *Morning Post*, 13 August 1870, p. 2.

41 *HCPP* 1874 (387), Copies of Reports of the Two Battalions of Infantry who have had the Martini-Henry Rifle in Use, p. 17; Dennis Lewis, *Martini-Henry .450 Rifles and Carbines* (Tucson, 1996), pp. 9–17; Temple and Skennerton, *Treatise*, pp. 85–99, 226–230.

With the controversies failing to subside, yet further trials were carried out in November 1874. Following another conference in October 1875, strikers and triggers were modified on a further 1,000 rifles sent for trials. Changes were approved by a conference chaired by Major General J.W. Armstrong in January 1876 and formally approved by Gathorne-Hardy on 18 February 1876. The resulting Mk II with new trigger assembly, and modifications to rear sight and breechblock, was approved on 25 April 1877. It was produced under contract until 1890 but, so far as the British Army was concerned, a MK III was also approved on 22 August 1879. This incorporated a modified rear sight and a further improvement to the breechblock as a result of experience of the production of a shorter carbine for cavalry and artillery in 1877. The Mk III continued into production well into the 1890s, primarily for colonial and auxiliary forces.[42]

The Mk III had been approved before the repetition of the Mk II's usual reported defects in South Africa, but these inevitably prompted further controversy. The War Office's Siege Operations Committee began to consider alternative weapons in service with other armies in 1879. It reported in December 1880, on the basis of trials at Dungeness, that the American-designed Berdan single-shot bolt-action rifle in service with the Russian Army was superior. The War Office Machine Gun Committee also examined existing magazine rifles in the course of 1880. The Director of Artillery, Lieutenant General Sir Frederick Campbell, had already convened a War Office Committee on 22 October 1880 to investigate possible improvements to ammunition and sights, and whether or not the Martini-Henry could be converted to a magazine or other quick-loading system. The Martini was judged to have the heaviest bullet, the lowest muzzle velocity, and the highest trajectory of any weapon in military use around the world. While not wishing to forego its advantage at longer ranges, there was accordingly a concern to reduce the trajectory at lower ranges.

Extensive reports were received on ammunition defects, including bullets shelling out of cartridges, and cartridges becoming deformed in pouches and bags. Brigadier General Sir Evelyn Wood VC, who had commanded at Khambula, suggested ammunition he had picked up on the field of Isandlwana when revisiting the site in 1880 remained perfectly serviceable. But his second in command, Brevet Colonel Redvers Buller VC, maintained that the Boxer-Henry cartridge had proven altogether too delicate. Another general complaint, echoed by the DAAG for Musketry at the War Office, was that the Martini was under-sighted 'to such a degree as to amount to a very grave defect'. Thus, the

---

42 *HCPP* 1876 [C.1452], Martini-Henry Rifles; NAUK, SUPP 5/893, Martini-Henry Rifles: Reports and Minutes of Conferences, 1873–1876; Lewis, *Martini-Henry*, pp. 18–24; Temple and Skennerton, *Treatise*, pp. 100–144, 230–238.

75th Foot reported that the error increased by 35 yards between the ranges of 500 and 800 yards. There was also a reported loss of velocity over distance.

The committee was satisfied that a change in the specification of the black powder charge had now solved the problem. Experiments had also been carried out with a long-range sight in India in 1878 and 1879. Although it was felt there was no real need to sight beyond 1,400 yards, the committee resolved in January 1881 on new trials for improved back sights to be undertaken by the 2nd Grenadier Guards and 1st East Kent Regt; and on long-range sights by the 2nd Highland Light Infantry and 2nd Essex Regiment. HMS *Excellent* and HMS *Cambridge* would test a 'quick-loading' system akin to a magazine rifle. In addition, it was decided to examine whether or not a Martini could be 'bushed' to use the same cartridge as a Gatling Gun. The latter merely increased the heating of barrels and fouling of the breech, and the experiment was abandoned in July 1882. On 29 June 1881 and, again on 16 January 1882, the Secretary of State for War, Hugh Childers, firmly ruled out re-examining the breech action 'unless in case of urgent necessity'. He believed there would be parliamentary demands for a new competition, and resulting unnecessary delay and inconvenience.

The trials of sights continued and trials were also carried out into the Austro-Hungarian Mannlicher, the Norwegian Jarmann and the Lee Experimental magazine rifles. In addition, investigation was also authorised in July 1882 into the lack of a half-cock mechanism: a Special Resident Magistrate in Ireland had complained in July 1882 that rifles were carried unloaded for they were 'always going off' when troops were on 'protection duties'.[43]

The improved Mk III, however, did not meet the criticism, the campaigns in the Sudan in 1884–85 bringing sufficient complaints about the army's equipment in general to prompt the appointment of the Stephen Commission in 1887. It was repeatedly claimed that rifles had jammed after a few rounds at crucial moments, as at Tamai in March 1884, Abu Klea in January 1885, and Tofrek in March 1885. At Tamai and Abu Klea, the dervishes had penetrated the British square before being driven off. Bennet Burleigh, the war correspondent of the *Daily Telegraph*, had his own Martini jam at Abu Klea. He also reported that it was the smoke generated by them that had allowed the dervishes to get too close to the square at Tamai.[44] The War Office had already established a special

---

[43] Centre for Buckinghamshire Studies, Aylesbury, Fremantle Mss, D/FR 165/5, Papers of the War Office Committee on Martini Henry Rifles and Ammunition, 1880–83; NAUK, SUPP 5/902, Martini Henry Rifles: Questions as to Jamming of Cartridges, and of Solid and Rolled Cases; ibid., SUPP 5/896, 'Martini-Henry Rifles and Ammunition'; ibid., SUPP 5/898, Martini-Henry Rifles and Ammunition.

[44] Philip Haythornthwaite, *The Colonial Wars Source Book* (London, 1995), p. 332; Bennet Burleigh, *Desert Warfare* (London, 1884), pp. 155–156, 169; *Hansard*, HC Deb. v. 295, cc. 1878–79, Lawrence and Hartington, 12 March 1885.

committee on cartridge jamming under Colonel Philip Smith in October 1885, and the Stephen Commission took further evidence.

Smith took testimony not only from Lieutenant Colonel the Hon. R.A.J. Talbot, who had commanded the Heavy Camel Regiment at Abu Klea, but also from non-commissioned officers and ordinary soldiers. Depending on the unit, jamming had occurred in between 25 per cent and 50 per cent of the rifles used, men picking up those from casualties in order to continue firing. While concluding that heat, sand and grit had not helped when rifles were being fired rapidly, Smith and his colleagues concluded that problems had also arisen from weak extractors, too heavy a charge, and over excited soldiers. A solid cartridge and a better extractor, therefore, would solve some of the problems.[45]

Having suggested to Stephen in November 1885 that the breech action had stood the test of time, Buller, who had commanded the Desert Column following Abu Klea, resurrected his comments from 1878. He also claimed that there was now a mistrust of the Martini 'as so many were temporarily disabled by the cartridge sticking'. Talbot, as well as other witnesses of Abu Klea, such as Lieutenant Colonel the Hon. Edward Boscawen of the Guards Camel Regiment and Brevet Lieutenant Colonel Charles Barrow of the Mounted Infantry, all suggested that sand was primarily to blame. It was concluded by Stephen that heat and sand alone could not have been the cause, or else it would have applied to earlier campaigns in India and South Africa, which had not experienced the issue to the same degree. Nor did the commission feel there was any merit in resurrecting the case of Boxer's dismissal in 1869, as one witness had done, since it could have no real bearing on the cartridge's performance. A combination of factors, therefore, accounted for the problems.[46]

Others such as *The Times* correspondent had concluded that the thin cartridge rather than the sand was to blame.[47] As it happened, General Lord Wolseley, commanding the Gordon Relief Expedition, had demanded in January 1885 that solid brass cartridges be substituted for rolled brass. Childers had approved this in March 1885 to the horror of Boxer, who claimed climate and conditions – and not the cartridge – had led to jamming: it was not the time for a 'hurried' adoption of an untried solid cartridge. The new cartridge was introduced on 9 June 1885, two million rounds being ordered at once with a million for Wolseley in Egypt, and a million for Sir Gerald Graham's expedition at Suakin, at an additional cost over the rolled cartridges of £1.0s 11d per 1,000. A solid cartridge had been rejected in October 1870 as too heavy and likely to be too costly. Another investigation of the possibility in 1875 also concluded that

---

45 NAUK, SUPP 5/904, Special Committee on Small Arms: Report on Jamming of Cartridges in Martini-Henry Rifles in Egypt.

46 *HCPP* 1887 [C.5062], pp. xcviii, cxxi–cxxii; ibid., [C.5062-I], pp. 112, 217.

47 *The Times*, 23 April 1885, p. 5; E Gambier-Parry, *Suakin 1885* (London, 1885), p. 194.

replacement would be too costly. Reusing returned solid cartridge cases would save only £4.14s 0d per 10,000 rounds, as well as being impracticable in action. A Mk II solid cartridge version was adopted in September 1885.[48]

A Mk IV Martini-Henry, known as the Enfield-Martini, was approved on 17 April 1886, with further changes in a new pattern issued on 13 May 1887 as an interim measure before a new magazine rifle – the Lee-Metford – was ready. It reflected the lessons of the Sudan with an improved action; a new extended operating lever; and a .402 calibre cartridge. The latter recognised the need for a smaller calibre high velocity bullet though they were then converted back to a .45 cartridge rather than have three separate cartridges in use once the .303 Lee-Metford was introduced. Three further patterns of the Mark IV were approved in September 1887, but it was clear that the introduction of the Lee-Metford was imminent. The Mark IV was largely issued for use by Indian troops. Subsequently, Mark IVs were converted to the .303 cartridge, these being designated Mark V and Mark VI Martini-Metfords in 1891.[49]

The desire for a superior weapon had become more widespread in the 1880s, notwithstanding the introduction of the Enfield-Martini. As a result of the deliberations of the War Office Committee on Martini Henry Rifles and Ammunition, a new Small Arms Committee had been formed in February 1883 under the chairmanship of Philip Smith to continue the examination of a number of experimental magazine rifles. A pattern was approved in October 1885 and issued for trials in June 1886.[50] Ultimately, the .303 Lee-Metford, a combination of the bolt-action system of the Scottish-born American inventor James Paris Lee, and the rifling of English engineer, William Ellis Metford, went into production in December 1888.

It needs to be noted that the Lee-Metford aroused as much controversy as the Martini-Henry. It had 37 more components than the Martini-Henry, required 620 more processes to manufacture, and it took 559 more workmen some 30,186 hours longer to produce 1,000 rifles.[51] Therefore, it was more costly. At the same time, there were fears that it might encourage troops to fire indiscriminately beyond the control that could be exercised by officers, leading to unacceptable expenditure of ammunition. In November 1886 the army's Commander-in-Chief, the Duke of Cambridge, expressed particular

48 Boxer, *Morning Post*, 1 April 1885, p. 2; *The Times*, 10 April 1885, p. 4; Lewis, *Martini-Henry*, p. 57; NAUK, SUPP 5/902, Martini Henry Rifles: Question as to Jamming of Cartridges, and of Solid and Rolled Cases; ibid., Précis on Solid Cases; *Hansard*, HC Deb. v. 296, c. 841, Elcho, 27 March 1885.

49 NAUK, SUPP 5/905, Special Committee on Small Arms; Lewis, *Martini-Henry*, pp. 26–34; Temple and Skennerton, *Treatise*, pp. 145–157; 'The New Martini-Enfield Rife', *The Saturday Review*, 16 February 1884; *The Times*, 5 October 1886, p. 4; 12 October 1886, p. 3; 16 October 1886, p. 7; 19 October 1886, p. 11; 23 October 1886, p. 7.

50 *HCPP* 1887 [C.5062-I], pp. 512–515.

51 *HCPP* 1890–91 (63), Army Rifles.

opposition to independent firing. In fact, the Lee-Metford was initially issued with a so-called magazine cut-off that provided officers with the opportunity to order troops to load only one round at a time into the magazine. By contrast, others such as Lieutenant Colonel Charles Slade, who sat on the Small Arms Committee, believed fostering greater tactical flexibility was not incompatible with maintaining control. The Royal Navy also favoured a magazine rifle that would generate a higher rate of fire for ship-to-ship action, not least against a new generation of motor torpedo boats.[52] In the end, what mattered was that the Lee-Metford was far more effective than the Martini. At Omdurman in September 1898, Egyptian and Sudanese troops still armed with the Martini-Henry opened fire at 1,000 yards and stopped the dervishes opposite them at 500 yards. British troops' Lee-Metfords opened fire at 2,000 yards, and stopped their opponents at 800 yards. [53]

IV

The introduction of the Lee-Metford was not the end of the Martini as its use at Omdurman by the Anglo-Egyptian Army suggests. It had been axiomatic ever since the Indian Mutiny for Indian troops to be armed with weapons already discounted by British troops. In the Second Afghan War of 1878–1881, Indian troops still used the Snider. There was considerable disquiet among the Indian high command when the 8,000-strong Indian Expeditionary Force sent to Malta in May 1878, and then on to occupy Cyprus in July, was given the Martini-Henry on arrival on Malta in June: the weapon had not yet been fully distributed to all British troops in India. The issue of the Martini to Indian troops was briefly considered in 1886 but not then implemented. By 1891, only half the Indian troops in India had them. Ironically, the Afghans at Maiwand had captured over 400 Martinis.[54] The King's African Rifles still had Martinis in 1902.[55] Similarly, the rifle volunteer movement in Britain was not fully equipped with the Martini-Henry until 1885, and did not receive the Lee-Metford until

52 NAUK, SUPP 5/908, Trial of Magazine Rifles in England, 1879–87, p. 32; ibid., SUPP 5/903, Special Committee on Small Arms: Trials of Various Systems of Magazine Arms, Quick Loaders, etc., p. 9; ibid., ADM 116/349, The Magazine Rifle; *HCCP* 1881 (223), Report of the Result of Recent Experiments with Machine Guns at Shoeburyness, p. 2.

53 Haythornthwaite, *Colonial Source Book*, p. 37.

54 Ian F.W. Beckett, 'The Indian Expeditionary Force on Malta and Cyprus, 1878', *Soldiers of the Queen*, 76 (1994), pp. 6–11; *Hansard*, HC Deb. v. 256, c. 1288, Churchill, 4 September 1880; ibid., v. 303, cc. 107–108, Campbell Bannerman, 8 March 1886; ibid., v. 301, cc. 34–35, Churchill, 14 August 1885; ibid., v. 354, c. 143, Gorst, 11 June 1891.

55 *Hansard*, HC Deb. v. 113, c. 947, Cranborne, 28 October 1902.

1895. While English and Scottish militia received the Martini-Henry in 1882, Irish militia regiments only received the Martini in May 1887.[56]

In preparation for its possible armed defiance of the imposition of Home Rule, the Ulster Volunteer Force acquired 12,523 Enfield-Martinis among its other weaponry by 1914. The Royal Irish Constabulary still had Martini carbines in the 1920s, as did the IRA. The ineffectiveness of the Martin-armed Malappuram Special Force against Moplah insurgents in South India in 1921 led to the raising of a new 300-strong Malabar Special Police armed with Lee-Enfields.[57]

Martinis were handed out to white settlers on the Rand amid fears of an uprising by Chinese mine workers in 1905. Once out of service with the British Army, the Martini also began to appear in the hands of arms dealers and their clients.[58] There was growing concern from 1897 onwards that Martinis – fetching high prices and originating in Muscat – were reaching the North-West Frontier through Persia and Afghanistan.[59] Some at least had been disposed of by Australian forces after the South African War. [60]

Even earlier, the Ndebele chief, Lobengula, had been persuaded to grant the so-called Rudd mining concession in 1888 by the supply of 1,000 Martinis to be passed to him through Bechuanaland. The British government concluded that the concession itself was advantageous. It was not for its High Commissioner to express either approval or disapproval, or for it to take any view of the validity or otherwise of the deal. In the event, temporising, Lobengula did not take possession of all the rifles until June 1893. It still meant that the Ndebele used the Martinis during their uprising against the British South Africa Company in 1896, though, by this time, the company's forces had the latest Maxim machine guns. The Khama also received 400 Martinis a month until fully armed from the Bechuanaland Exploration Company, while there is a Kgatla praise poem celebrating their use of the Martini against the Boers during the South African

---

56 Ian F.W. Beckett, *Riflemen Form: A Study of the Rifle Volunteer Movement, 1859–1908* (Aldershot, 1982), pp. 134, 137; *Illustrated London News*, 22 January 1887, p. 108; *Hansard*, HC Deb. v. 273, c. 1036, Childers, 7 August 1882; ibid., v. 315, c. 905, Stanhope, 23 May 1887.

57 Timothy Bowman, *Carson's Army* (Manchester, 2007), p. 145; Charles Townshend, 'The IRA and the Development of Guerrilla Warfare, 1916–21', *English Historical Review*, 94/371 (1979), pp. 318–345, at pp. 324, 333; David Arnold, 'The Armed Police and Colonial Rule in South India, 1914–47', *Modern Asian Studies*, 11/1 (1977), pp. 101–125, at p. 111.

58 Gary Kynoch, 'Your Petitioners are in Mortal Terror: The Violent World of Chinese Mineworkers in South Africa, 1904–10', *Journal of Southern African Studies*, 31/3 (2005), pp. 531–546, at p. 544; R.E. Dunn, 'Bu Himara's European Connection: The Commercial Relations of a Moroccan Warlord', *Journal of African History* (hereafter *JAH*), 21/2 (1980), pp. 235–253, at p. 243.

59 See Ball, this volume.

60 Robert M. Burrell, 'Arms and Afghans in Makran: An Episode in Anglo-Persian Relations, 1905–12', *Bulletin of the School of Oriental and African Studies*, 49/1 (1986), pp. 8–24.

War.[61] The Martini-Henry survived, therefore, well after the introduction of the Lee-Metford. The solid case Boxer-Henry cartridge was even found useful as the basis for incendiary bullets against Zeppelins in 1915.

It may appear from this discussion that the Martini-Henry was always a problematic weapon. That downplays its general robustness, accuracy and acceptability to the troops that used it. As Captain Edward Hutton wrote of Gingindlovu, 'we all had the utmost confidence in our rifles, which were at that time the most perfect weapons on the world'. Similarly, as *The Times* correspondent still wrote after speaking to participants at Tofrek, 'It is considered a perfect weapon, whose deadly efficiency was once more proved by the terrible slaughter of that memorable Sunday.'[62] The Martini-Henry's defects resulted mostly from the Boxer-Henry cartridge rather than the weapon itself, and these did not materially lead to British military failures. Defeats such as those at Isandlwana, Maiwand and Majuba were attributable not to the Martini-Henry but to faulty leadership, and an underestimation of opponents' capabilities. Therefore, although its extraordinary iconic status derives largely from the cinema, and from entirely modern perceptions of the conduct of the Zulu War, it is not perhaps unreasonable to accord the Martini the prominence it now enjoys. Arguably, it won, if not 'The West', certainly the South and the East.

---

61 Arthur Keppel-Jones, *Rhodes and Rhodesia: The White Conquest of Zimbabwe, 1884–1902* (Kingston, 1983), pp. 76–82, 88–90, 92, 94, 252, 269; Anthony Atmore et al., 'Firearms in South Central Africa', *JAH*, 12/4 (1971), pp. 545–556, at pp. 552–553; Sue Miers, 'Notes on the Arms Trade and Government Policy in Southern Africa between 1870 and 1890', *JAH*, 12/4 (1971), pp. 571–577; *Hansard*, HC Deb. v. 333, cc. 254–56, de Worms, 25 February 1889; ibid., v. 350, cc. 1352–53, debate, 23 February 1891.

62 'Some Recollections of the Zulu War, Extracted from the Unpublished Reminiscences of the Late Lieut. General Sir Edward Hutton', *Army Quarterly*, 16 (1928), pp. 65–80; *The Times*, 23 April 1885, p. 5.

Chapter 12

'The Shooting of the Boers was Extraordinary': British Views of Boer Marksmanship in the Second Anglo-Boer War, 1899–1902

Spencer Jones

I

The British Army of the Victorian era possessed vast experience in colonial warfare. Over the course of Queen Victoria's reign the army fought an incredible 230 wars.[1] Many of these conflicts were quintessential 'small wars', fought by a handful of regular troops in a distant geographic location. Although these conflicts could feature hard fighting and occasional setbacks, the technological advantage possessed by the British Army was typically decisive. In colonial wars from 1857 to 1899, British forces only twice lost 100 men killed in a single action.[2] Yet, despite this considerable combat record, the British Army had surprisingly little experience in facing the fire of modern weapons. Few of the colonial foes encountered by the army possessed more than a handful of effective firearms. However, there was one major exception to this general rule, namely the Boers of South Africa. The Boers were almost universally armed with modern rifles and possessed a distinctive military culture that emphasised firepower.

The Boers were descendants of Dutch colonists who had founded Cape Town in the seventeenth century. The British annexed Cape Colony during the Napoleonic Wars,[3] but found the fiercely independent Boers to be difficult and antagonistic subjects. Divided by language, culture and heritage, the Boers chafed under British rule. This resentment provoked a large scale migration of

---

1 Andre Wessels, *Lord Roberts and the War in South Africa 1899–1902* (Stroud, 2000), p. xiii.

2 Ian F.W. Beckett, 'The South African War and the Late Victorian Army', in P. Dennis and J. Grey (eds), *The Boer War: Army, Nation and Empire* (Canberra, 2000), p. 33. The two battles were Isandlwana (1879) and Maiwand (1880).

3 Cape Town was captured by the British in 1806. The annexation was made formal at the end of the war in 1815.

Boer settlers in the 'Great Trek' of the 1830s. This process ultimately resulted in the founding of the independent Boer republics of Transvaal and Orange Free State in the 1850s. Relations between these fledging nations and the British were often tense. Small conflicts between Boers and British forces erupted on several occasions, and two full scale wars were fought between Britain and the Boer republics in 1880–1881 and 1899–1902.

The unusual heritage of the Boers was matched by a unique military culture that had no parallels with any of Britain's other colonial foes. In Colonel Charles Callwell's famous treatise on colonial warfare, *Small Wars*, it was noted that, as a military force, the Boers defied easy classification.[4] The Boers lacked a formally constituted army and instead relied upon a voluntary militia system, with volunteers being formed into units known as commandos.[5] Boer citizens responding to the rallying call were expected to bring their own firearm and horse, thus ensuring that the force was both well armed and highly mobile.[6] This combination of firepower and mobility was the defining feature of the Boer military system. The effectiveness of the Boers in combat had been demonstrated in regular conflicts with local Africans, where small numbers of burghers had often been able to defeat numerically superior opposition.[7]

The experience of the British Army in conflict with the Boers confirmed the reputation of the Afrikaners as formidable opponents. The British were particularly impressed with the quality of individual Boer shooting. As early as the Battle of Boomplaats in 1848, British commander Sir Harry Smith, a hardened veteran of the Napoleonic Wars, paid compliment to Boer musketry, stating '[a] more rapid, fierce and well-directed fire I have never seen maintained'.[8] By the time of the First Anglo-Boer War 1880–1881, the firepower of the Boer commandos had been further enhanced by the latest generation of weapons. The British Army suffered a humiliating defeat in this conflict, and observers were

4 Charles Callwell, *Small Wars: Their Principles and Practice* (London, 1906; 3rd edn,), p. 31. Callwell felt that the Boers had more in common with a European guerrilla movement than a typical colonial foe. F.H.E. Cunliffe, *The History of the Boer War* (2 vols, London, 1901), vol. 1, pp. 4–5, argued that the Boer fighting style was adopted from the autochthonous 'Hottentots', albeit with the benefit of modern weapons.

5 For a thorough study of the Boer commando system, see Fransjohan Pretorius, *Life on Commando during the Anglo-Boer War 1899–1902* (Cape Town, 1999).

6 Ibid., pp. 80–83. The Afrikaner governments provided rifles to those who did not possess their own.

7 Frederick Maurice, *History of the War in South Africa 1899–1902* (4 vols, London, 1906), vol. 1, pp. 68–71; Bill Nasson, *The South African War* (London, 1999), p. 64.

8 Quoted in George F.H. Berkley, 'Sir Harry Smith: A Reminiscence of the Boer War in 1848', *Fortnightly Review*, December 1899, p. 1034.

quick to identify superior Boer marksmanship as a key factor, as it allowed the commandos to inflict disproportionate casualties on attacking British units.[9]

The stark contrast between skilful individual Boer marksmanship and cumbersome British musketry had been cause for concern throughout the era. In the aftermath of an action against the Boers in 1842, one British officer wrote:

> From the encounter with the Boers, the following would seem more particularly deduced ... The necessity of practising the infantry soldier more frequently in the use of his musket with ball cartridge, more particularly in countries like the Cape, where he has so often to trust to his own individual correctness of aim and knowledge of his weapon.[10]

However, this warning appears to have had no meaningful influence prior to the First Anglo-Boer War. This latter conflict confirmed the image of Boers as talented marksmen and prompted a degree of introspection within the army. In the aftermath of defeat at Majuba Hill, the officers of the Northamptonshire Regiment were said to have taken an oath that they would not rest until their unit had improved its standard of shooting.[11] The defeat also prompted Ian Hamilton to write the book *The Fighting of the Future,* which identified the importance of effective marksmanship in modern warfare.[12] Hamilton would go on to put some of his early ideas into practice during his spell in charge of the School of Musketry in India.[13] However, the overall influence of the First Anglo-Boer War on the British Army was decidedly limited.

The greatest clash between the British and the Boers would come during the Second Anglo-Boer War 1899–1902. This conflict dwarfed earlier engagements in terms of scale, duration, cost and intensity. Boer firepower would prove to be an important battlefield factor, and the magnitude of the war ensured the experience left a deep and lasting impression upon the British Army. This chapter will study British impressions of Boer marksmanship during this major conflict. Although the popular press were quick to attribute success to natural Boer skills, thoughtful military commentators identified a variety of factors that contributed to the effectiveness of Boer rifle fire. This study will examine three key elements that contributed to Boer marksmanship, namely terrain, culture and equipment, demonstrating how they combined to produce unusually

9 For a concise study of the problems faced by the British attackers in the First Anglo-Boer War, see G. Tylden, 'A Study in Attack: Majuba 27 February 1881', *Journal of the Society for Army Historical Research,* 39/157 (1961), pp. 27–36.

10 Quoted in Ian Knight, *Go to Your God like a Soldier* (London, 1996), p. 173.

11 John Dunlop, *The Development of the British Army 1899–1914* (London, 1938), p. 37.

12 Ian Hamilton, *The Fighting of the Future* (London, 1885), p. 14

13 Matthew Ford, 'The British Army and the Politics of Rifle Development', unpublished PhD thesis, King's College London, 2008, pp. 81–82.

effective rifle fire. The chapter will also consider the British impression of Boer musketry in the aftermath of the war, showing how overall opinion was one of considerable admiration. This admiration would play an important role in the British Army's musketry reforms in the years 1902–1914, which, in turn, contributed to the famous rifle skills of the British Expeditionary Force in the opening battles of the First World War.

II

An immediate problem faced by British troops in the Second Anglo-Boer War was the nature of the terrain and climate. The sheer scale of the geography could be intimidating to inexperienced troops. Sweeping grass veldt in the east and scrub desert in the west stretched for miles, occasionally being broken by huge kopjes and wide rivers. Yet, despite the vastness of the country, effective cover on the veldt was spartan. Boulders, scrub vegetation and anthills offered some concealment for troops, but in many battles the attackers were forced to advance over disturbingly open terrain.[14]

The difficulties posed by the terrain were exacerbated by the incredibly clear atmosphere of the country. Troops who were unaccustomed to the conditions faced particular difficulty in estimating ranges correctly, but even veteran troops were known to make serious errors when judging distances.[15] This had dangerous implications when advancing to the attack, as it was easy to misinterpret the range to the enemy position. For example, confusion over the exact range to the Boer lines played a role in the destruction of Colonel Long's battery at the Battle of Colenso on 15 December 1899. On the other hand, the clear atmosphere could offer a great advantage for the defenders, especially if they occupied a kopje, as they could observe advancing foes at remarkable distances. Howard Hillegas, an American journalist attached to the Boer forces, expressed his amazement at the distance at which advancing British forces could be seen, noting that at long range they resembled 'huge ants more than human beings'.[16]

Afrikaner riflemen took full advantage of these conditions. Well adapted to the clear atmosphere, the quality of Boer eyesight was a source of much admiration amongst British troops. One officer commented that the average Boer had 'magnifying eyes', while General Sir Redvers Buller was said to have

---

[14] For an evocative discussion of the terrain in South Africa, see Count Adalbert Sternberg, *My Experiences of the Boer War* (London, 1901), pp. 204–206.

[15] 'Jack the Sniper' [Charles James O'Mahony], *A Peep over the Barleycorn: In the Firing Line with the P.W.O. 2nd West Yorkshire Regiment, Through the Relief of Ladysmith* (Dublin, 1911), pp. 135–136; G. Forbes, 'Experiences in South Africa with a New Range Finder', *Journal of the Royal United Services Institute*, 46/2 (1902), p. 1389.

[16] Howard Hillegas, *The Boers in War* (New York, 1900), p. 146.

stated that 'if a European and Boer were walking towards each other in an open country, the Boer would see the other two miles in advance'.[17] Making use of this natural advantage, the Boers often opened fire at ranges of well over a mile.[18] This long range rifle fire came as an unpleasant surprise to British troops, who were not trained to fire at ranges above 800 yards.[19] Furthermore, pre-war British tactics had assumed that it would be possible for infantry to advance to within approximately half a mile of the enemy's position before it became necessary to shake out into extended order, and did not anticipate receiving anything but desultory enemy fire beyond 1,500 yards' range.[20] This was not the case in South Africa, where British formations were often engaged at ranges of 2,000 yards or more.[21] Officers recorded their alarm at this tactical development, with one noting:

> War is not what it was when armies manoeuvred in sight of each other, and when 600 yards was the limit of artillery fire ... That was old-time fighting, and some sport about it too. Now Bill is killed at 2,400 yards, and Bill's pal hasn't an idea where the shot was fired. That is modern warfare[22]

Such long range fire could be especially problematic for cavalry, who were initially armed with carbines that had a maximum range of 1,200 yards. Lieutenant-General Sir Charles Warren complained that the 'Boers had only to keep at 2,000 yards from our cavalry in the hills and could shoot them down with impunity'.[23]

However, even in the clear atmosphere of South Africa, it took an exceptional marksman to hit the target reliably at long range. Observers noted that Boer long-distance fire tended to be erratic unless the range to the target had been established in some fashion. This could take the form of crack shots firing ranging shots and communicating the distance to their comrades. Artillery

---

17 Quoted in Jack, *Peep over the Barleycorn*, p. 192.

18 The Boers were also capable of holding their fire until close range. See Pretorius, *Life on Commando*, pp. 139–140.

19 H.R. Mead, 'Notes on Musketry Training of Troops', *Journal of the Royal United Services Institute*, 43/1 (1899), pp. 250–251.

20 War Office, *Infantry Drill Book 1896* (London, 1896), p. 131.

21 *The Official Records of the Guards' Brigade in South Africa* (London, 1904), p. 18; William Balck, 'Lessons of the Boer War and Battle Workings of the Three Arms', *Journal of the Royal United Services Institute*, 48/2 (1904), pp. 1273–1274.

22 'Not by a Staff Officer', 'Some Remarks on Recent Changes', *United Service Magazine*, October 1904, p. 47.

23 *Report of His Majesty's Commissioners Appointed to Inquire into the Military Preparations and Other Matters Connected with the War in South Africa* (London, 1903), Cmd no. 1789–1792, vol. 2, Q 15850, p. 233 (hereafter referred to as the *Elgin Commission*).

was also used to establish the range, and so were nearby geographical features.[24] Once the range had been established, the fire was considerably more effective. For example, at the Battle of Willow Grange on 22 November 1899, the West Yorkshire Regiment reported:

> for about one and a-half hours the Boers kept up an ineffective fire on our position, only one man being hit. The Boers then brought up a Vickers-Maxim at about 1,800 yards range, and very quickly found our range, and after that their musketry became very effective ... The position under this fire quickly became untenable.[25]

Long range Boer shooting was particularly dangerous to dense formations. When Lord Roberts took command of British forces in South Africa, his tactical 'Notes for Guidance' urged infantry to adopt extended formations between 1,500 and 1,800 yards from Boer positions, effectively doubling the distance set down in the pre-war regulations.[26] In practice a number of units chose to abandon close order at even greater distances. For example, Major-General Henry Colvile, commanding Guards' Brigade, favoured shaking into extended order at 2,500 yards.[27]

However, despite its capacity to cause losses at huge ranges, Boer long range fire was rarely decisive on its own. Casualties at such range were often more a matter of luck than judgement. For example, Lieutenant-General Sir Archibald Hunter commented that he believed the effectiveness of long range fire was 'mythical' and related that he had regularly patrolled the Ladysmith perimeter in full general's uniform, secure in the knowledge that none of the besieging Boers would be capable of hitting him![28]

The main battlefield function of Boer long range fire was to slow down the pace of the British advance by forcing them to adopt extended formations at great distances from the Boer position.[29] Once under fire, battlefield manoeuvre became considerably more difficult and any element of surprise was lost. A journalist attached to Lord Methuen's force described this kind of action, writing that the series of attacks during the attempt to relieve Kimberly in November 1899 consisted of 'no beastly strategy, or tactics, or outlandish tricks of any sort; nothing but an honest, straightforward British march up to a row of waiting

---

24 *Elgin Commission*, vol. 1, Q 6860, p. 294.

25 *Extract from the Digest of Service of the 2nd Battalion The Prince of Wales's Own (West Yorkshire Regt.) in South Africa* (York, 1903), p. 4. The Vickers-Maxim was an autocannon that fired small explosive shells. It was commonly referred to as a 'pom pom gun' due to its distinctive sound when firing.

26 National Archives of the UK (NAUK), Kew, London, WO 105/40, Lord Roberts Papers, 'Notes for Guidance in South African Warfare', 26 January 1900.

27 *Records of Guards' Brigade*, p. 19.

28 *Elgin Commission*, vol. 2, Q 14587, Q 14588, p. 138.

29 Ibid., vol. 2, Q 19200, p. 397.

rifles'.[30] This could be a trying experience for British troops, and it was worsened by the fact that the source of the fire was usually invisible. Part of the reason for this was the use of smokeless powder, which will be discussed in detail below, but it was also due in large part to the military culture of the Boers.

The commandos were essentially a force of individual riflemen, many of whom wielded their own personal weapons. Although officers were a key part of the commando, there was no drill or training to inculcate obedience to orders or the use of particular formations.[31] The Boers had neither the discipline nor the inclination to adopt formal European formations for either attack or defence. Instead, commandos tended to fight as a loose group of skirmishers, with individual burghers choosing their own cover and frequently picking their own targets.[32] The lack of formal organisation in the Boer fighting line allowed it to take advantage of available cover and thus blend into the countryside with remarkable skill. Ruminating on his combat experiences, Major-General Geoffrey Barton commented that the Boers were 'extraordinarily well trained by nature and habit to lie still'.[33]

The individualistic military culture of the Boers stood in stark contrast to the traditional British approach. Although attitudes differed from unit to unit, much of the British Army favoured close control, volley fire and strict discipline.[34] Although these ideas had proved useful in previous colonial wars, they required adaptation to make them effective in South Africa. Henry Colvile commented on his wartime experiences of conservative attitudes in Guards' Brigade:

> At first officers and men were very stupid about taking cover. I have seen men halted on a rise in full view of the enemy when a few paces forward or backward would have placed them in shelter, the reason being that to have taken this step would have broken the dressing of the line.[35]

A combination of British inexperience in taking cover and the relative invisibility of Boer positions magnified the effectiveness of Boer fire. The Boers were able to observe and engage the British forces without revealing themselves; for the British coming under fire from an unknown source was a disturbing experience and often necessitated a delay in the attack until its location could be

---

30 L.M. Phillips, *With Rimington*, (London, 1902), p. 10.

31 Maurice, *History of the War*, vol. 1, p. 86.

32 Balck, 'Lessons of the War', pp. 1272–1273.

33 *Elgin Commission*, vol. 2, Q 16215, p. 256.

34 For a discussion of this issue, see Edward Spiers, *The Late Victorian Army, 1868–1902* (Manchester, 1992), pp. 313–315. Reactionary pre-war attitudes were mercilessly lampooned in 'George D'Ordel' [Mark Sykes & Edmund Sandars], *Tactics and Military Training* (London, 1904).

35 *Elgin Commission*, vol. 2, Q 16974, p. 286.

pinpointed.[36] Furthermore, the British were troubled by the inability to gauge the effect of their own fire against relatively invisible opposition, especially as the evidence of the Boer's shooting was plain to see. Major-General Neville Lyttleton contrasted previous colonial experience with the new conditions, writing of the Battle of Colenso:

> Few people have seen two battles in succession in such startling contrast as Omdurman and Colenso. In the first 50,000 fanatics streamed across the open regardless of cover to a certain death, while at Colenso I never saw a Boer all day till the battle was over, and it was our men who were the victims.[37]

Colonel E.E. Carr echoed similar sentiments, noting that during most fire fights his troops were forced to shoot purely at geographic features to try and suppress enemy fire, whereas the Boer usually had a clear target:

> They do not fire unless they are pretty certain you are there; I do not say they always see you; although the difficulty is that we cannot see them and they can see us, they can see us for miles; but we seldom see them.[38]

A private soldier, Charles James O'Mahony, expressed his frustrations with such fighting after the defeat at the Battle of Willow Grange, writing:

> We were much handicapped for the Boers take cover in a manner never to be equalled ... we sprayed every nook, crevice, donga, spruit etc. on and surrounding the Boer position with lead as if from a watering can, rocks being splintered two miles in the kopjes rear.[39]

In stark contrast, Izak Meyer, a Boer veteran, described his experience of combat at the Battle of Modder River, 28 November 1899, in the following terms:

> Now I am deadly calm, and with deadly calm I pick my man, pick them one by one. I pick him, my Mauser drops, my left eye closes, I get him in my sights and my Mauser cracks. The Englishman totters, drops his rifle, grabs his chest ... I shoot them down, one after another, one after another.[40]

36 Jay Stone and Erwin Schmidl, *The Boer War and Military Reforms* (Lanham, 1988), p. 80.

37 Neville Lyttleton, *Eighty Years: Soldiering, Politics, Games* (London, 1927), p. 212.

38 *Elgin Commission*, vol. 2, Q 19200, p. 397.

39 Jack, *Peep over the Barleycorn*, pp. 74–75.

40 Quoted in Pretorius, *Life on Commando*, p. 141.

The ability of the Boers to fight from behind cover was especially useful during extended fire fights at close range.[41] In the early stages of the war, some British units attempted to use volleys during fire fights, but it was soon found that the slow, static nature of volley firing proved ineffective against dispersed and concealed enemies.[42] By contrast the Boers proved especially adept at 'snap shooting', leaning out from behind cover only long enough to acquire a target and fire, and then ducking out of sight once more. J.B. Atkins, a British journalist, witnessed snap shooting at the Battle of Hart's Hill, 23 February 1900, writing: 'Boer heads and elbows shot up and down; the defenders were aiming, firing, ducking.'[43] In the face of such conditions, the British were forced to adopt a far greater degree of independent firing themselves.[44] Unfortunately, pre-war training had done little to prepare the average soldier for this type of action, and, combined with the difficulties of atmosphere and the relative invisibility of many of the Boer positions, this made fire fights a difficult proposition. Major-General Sir William Gatacre noted the difference in fighting style:

> [The average British soldier] was rather slow in getting his aim, and he found he was unaccustomed to use his rifle without exposing himself, which at once brought a Mauser bullet in his direction ... The Boer, on the contrary, was particularly good at getting his bead on to the enemy's hat or mess tin quickly, and in getting covered again before men could aim and fire.[45]

It was within fire fight range that the majority of British officers felt the Boers had truly demonstrated their marksmanship skills. Major-General J.P. Brabazon argued: '[W]here they beat us so completely was that when we got onto kopjes at close quarters, say, a few hundred yards, a man could not put a finger up over a rock or ridge without being shot.'[46] Major-General A.H. Paget related his front line experience at the Battle of Modder River, noting that '[i]n these early fights [the Boers'] shooting was very accurate; every bullet had some mark, and there was no wild shooting at all, and when we got to the closer ranges, in places which were fire swept, everybody was hit.'[47] E.E. Carr recalled the difficulty of assaulting Boers in strong defensive positions, stating, 'I have seen men rolled

---

41 Opinions differed as to what exactly constituted 'close range' in the Anglo-Boer War. In the aftermath of the war, British regulations codified a range 600 yards or less as 'decisive' range for fire fights. See War Office, *Combined Training 1905* (London, 1905), p. 100.

42 For a graphic description of the difficulties of engaging concealed Boers with volleys, see Jack, *Peep over the Barelycorn*, pp. 71–72.

43 J.B. Atkins, *The Relief of Ladysmith* (London, 1900), p. 295.

44 Jack, *Peep over the Barleycorn*, p. 73.

45 *Elgin Commission*, vol. 2, Q 16772, p. 272.

46 Ibid., vol. 1, Q 6859, p. 294.

47 Ibid., vol. 2, Q 16441, p. 259.

over like rabbits and slaughtered, as the Inniskillings Fusiliers were at Pieter's Hill on the first attempt just before the relief of Ladysmith.'[48]

In intense fire fights, the skill of individual Boer marksmen could be striking. Colonel Forbes MacBean noted the presence of 'a certain percentage of men who are uncommonly good shots' in the average Boer firing line.[49] These elite marksmen were capable of causing disproportionate casualties. Henry Colvile noted that 'the Boers had a certain number of picked shots who did great damage', while A.H. Paget echoed the view, commenting that 'some of the shooting of the Boers was extraordinary'.[50] Even Archibald Hunter acknowledged the presence of crack shots amongst Boer forces, relating that '[t]here are certain shots who have earned their living as professional hunters, and from 200 yards to 300 yards [range] they are undoubtedly marvellous shots'.[51]

The skills of these marksmen were often attributed to frontier life and the popularity of game hunting.[52] However, game had been in decline throughout the 1880s and 1890s.[53] Furthermore, the growth of urban centres in the Transvaal and Orange Free State during the 1880s and 1890s meant that Boer forces contained a proportion of city-based volunteers who were unlikely to be natural riflemen.[54] Nevertheless, rifle culture remained a source of fascination in the Boer republics in the years prior to the war.[55] Howard Hillegas felt that rifle shooting was the 'chief amusement' in the Transvaal in the 1890s, writing that the 'demand for rifle ammunition was constant, and firing at marks may almost be said to have taken the place occupied by billiards in Europe'.[56] Furthermore, beginning in the early 1890s and intensifying in the aftermath of the botched Jameson Raid of 1895, the governments of the Boer republics put renewed emphasis on promoting rifle culture. Major-General Sir Frederick Maurice noted of this policy that '[e]very effort, in short, was made to preserve the old skill and interest in rifle-shooting, which it was feared would vanish with the vanishing elands and gemsbok. If the skill had diminished, the interest had not.'[57]

48 Ibid., vol. 2, Q 19198, p. 397. Colonel Carr appears to have confused the attacks at Hart's Hill (23 February 1900) and Pieter's Hill (27 February 1900). The Inniskillings Fusiliers suffered severe casualties at Hart's Hill but were not involved at Pieter's Hill. I am grateful to Ken Gillings for supplying this information.

49 *Elgin Commission*, vol. 2, Q 19593, p. 415.

50 Ibid., vol. 2, Q 16440, p. 259; Q 16989, p. 292.

51 Ibid., vol. 2, Q 14585, p. 138.

52 Ibid., vol. 2, Q 21950, p. 564.

53 NAUK, WO 33/154, Military Notes on the Dutch Republics of South Africa, p. 49.

54 Hillegas, *Boers in War*, pp. 19–20.

55 Maurice, *History of the War*, vol. 1, p. 80.

56 Hillegas, *Boers in War*, pp. 19–20.

57 Maurice, *History of the War*, p. 80.

Nevertheless, not all Boers were gifted marksmen and their shooting could sometimes be wild. However, the fact that the majority of burghers were equipped with modern magazine loading Mauser rifles helped to offset any disadvantages due to lack of individual accuracy. Less talented riflemen could make up for this deficiency through sheer volume of fire. As J.P. Brabazon noted: 'If you pump lead in a certain direction at a proper distance you must hit somebody.'[58] Charles Callwell saw the magazine rifle as the key element of Boer War tactics, noting that, due to its rate of fire,

> a mere handful of men, lying down under shelter, can bring such a hail of bullets to bear upon ground extending for a considerable distance to their front that hostile troops attempting to cross this will suffer appalling losses in doing so, even if they succeed in the venture.[59]

Facing such rapid fire could be a harrowing experience for soldiers in the front line. An officer of the 60th Rifles recorded his experience at the Battle of Talana Hill:

> I don't suppose I am ever likely to go through a more awful fire than broke out from the Boer line as we dashed forward. The ground in front of me was literally rising in dust from the bullets, and the din echoing between the hill and the wood below and among the rocks from the incessant fire of the Mausers seemed to blend with every other sound into a long drawn-out hideous roar ... the whole ground we had already covered was strewn with bodies.[60]

The modern rifles of the Boers offered additional advantages beyond rate of fire. The flat trajectory of the weapons made them more accurate and allowed the Boers to create deadly fire swept zones at battles such as Modder River and Magersfontein.[61] Indeed, at Magersfontein, the Boers had sited their main position at the base of a kopje, partially as means of taking advantage of the sweeping effect of flat trajectory fire.[62] In addition, the Mauser rifle benefited from the use of smokeless powder, meaning that there was no tell-tale puff of smoke to reveal a firer's location. This was a critical advantage and greatly enhanced the ability of the Boers to fight from behind cover. Charles Callwell considered it the decisive element of Boer marksmanship, arguing:

---

58 *Elgin Commission*, vol. 1, Q 6860, p. 294.

59 Charles Callwell, *The Tactics of Today* (Edinburgh, 1903), pp. 31–32.

60 Quoted in Leo Amery, *The Times History of the War in South Africa* (7 vols, London, 1902), vol. 2, p. 164.

61 The Battle of Magersfontein was fought on 11 December 1899.

62 G.R. Duxbury, *The Battle of Magersfontein 11th December 1899* (Johannesburg, 1995), p. 2.

> The disappearance of black powder has exerted a far more potent influence in moulding tactics into a new shape than the increased power and accuracy or the rapid fire of the modern rifle and gun. Concealment has been so greatly facilitated by this that it has gained a new and commanding importance. It was a standing grievance in South Africa that the Boers could only be heard and not seen.[63]

There had been some consideration of the effects of modern rifles within the British Army prior to the outbreak of the war, but such discussions had produced few tactical changes.[64] Interestingly, Sir John Ardagh, Director of Military Intelligence, argued prior to the conflict that the fact that the British were armed with smokeless, flat trajectory rifles would help to offset the dangers posed by natural Boer marksmanship, stating that modern weapons had 'much diminished the advantage offered by accuracy in judging distances'.[65] In fact, the advantages of modern rifles had the effect of greatly magnifying Boer strengths. Long range, flat trajectory rifles allowed the Boers to engage at great distances; the use of smokeless powder vastly enhanced the Boer's capacity for fighting from behind cover and improved individual accuracy; and the use of magazine loading allowed a far greater rate of fire to be maintained. Expert marksmen could benefit from the range and accuracy of their rifles, while less talented Boers could make up for lack of individual skill with sheer weight of fire. Despite wielding a weapon of similar quality, the British Army enjoyed few advantages by comparison. Lord Methuen offered a bleak assessment of the issue:

> The shooting of the Regular troops was conducted under exceptional difficulties on account of the clearness of the atmosphere and because the enemy offered no good target, but my opinion gained during my experience of the Tirah and the South African campaigns is that the shooting of our infantry is not worthy of the accuracy and the long range powers possessed by the present rifle.[66]

The combination of Boer rifle culture and modern magazine rifles lay at the core of many British tactical problems in the Second Anglo-Boer War. Frontal attacks against Boer positions frequently suffered heavy losses, and it took a considerable degree of in-theatre learning before the British Army was able to gain the upper hand on the battlefield.[67] The effectiveness of Boer firepower necessitated a profound reconsideration of assault tactics, with a

---

63 Callwell, *Tactics of Today*, p. 7.

64 Spiers, *Late Victorian Army*, p. 315.

65 NAUK, WO 33/154, Military Notes on the Dutch Republics of South Africa, p. 50.

66 *Elgin Commission*, Q 14188, p. 121.

67 For discussion of this aspect of the campaign, see Thomas Pakenham, *The Boer War* (London, 1979) and Stephen M. Miller, *Lord Methuen and the British Army: Failure and Redemption in South Africa* (London, 1999).

fresh emphasis on dispersed attack formations, prolonged artillery support and flanking movements.

III

In the aftermath of the conflict, the topic of Boer shooting was much discussed at the Royal Commission on the South African War. Twenty-one witnesses were questioned directly about Boer marksmanship and others spoke on the topic in general terms. Interestingly, several officers cast aspersions on the quality of Boer marksmanship. Colonel A.W. Thorneycroft and Major-General Sir H.M.L. Rundle both considered that Boer shooting had much declined from the First Anglo-Boer War, although both acknowledged that it still remained superior to that of their own soldiers.[68] Redvers Buller actually considered that British shooting was superior to that of the Boers.[69] Major-General Sir H.J.T. Hildyard thought that the marksmanship of his troops was comparable to that of the Boers, a view echoed by Forbes MacBean and Henry Colvile.[70] However, every critical witness qualified their statements on the topic. For example, Buller only considered British shooting to be superior if the British knew the range to the enemy position, a comparatively rare experience for much of the war.[71] Hildyard acknowledged his view was only an impression and 'was a very difficult thing to prove'.[72] MacBean admitted that he considered Boer fire to be of 'a fairly high average' and recognised the presence of dangerous sharpshooters amongst the commandos.[73] Colvile attempted to argue that the British *shooting* was as good as the Boers, but that the *hitting* was worse, due to the Boers' ubiquitous use of cover![74]

However, the majority of witnesses praised Boer marksmanship, albeit sometimes grudgingly. Major-General Sir Bruce Hamilton directly refuted Henry Colvile's evidence, arguing that Boer shooting was considerably superior.[75] When questioned by the commissioners as to the reason for the divergent views on the quality of Boer shooting, Hamilton responded perceptively: 'I think British officers are very anxious to stick up for the shooting of their men.'[76] A.H.

---

68 *Elgin Commission,* vol. 2, Q 12440, p. 19; Q 17879, p. 331.

69 Ibid., Q 14383, p. 212.

70 Ibid., Q 15982, p. 241; Q 16988, p. 292; Q 19593, p. 415.

71 Ibid., Q 15483, p. 212.

72 Ibid., Q 15982, p. 241.

73 Ibid., Q 19593, p. 415.

74 Ibid., Q 16988, p. 202.

75 Ibid., Q 17482, p. 314.

76 Ibid., Q 17479, p. 314.

Paget had much praise for Boer shooting, noting: 'I am going more not by what I saw when I had a higher command, but what I saw when I was in the fighting line myself ... I was in the fighting line and saw everything that was going on, and certainly the Boer shooting was very good indeed.'[77] However, senior officers often had praise for Boer marksmanship; Charles Warren and William Gatacre both considered it to have been superior to that of the British.[78] Lord Kitchener saw Boer rifle culture as the key element in Boer success:

> Our men were not as quick and accurate as their opponents in shooting rapidly, but they had not been trained for this during peace time, and could not, therefore be expected to excel in what the Boers had learned to practise from childhood.[79]

Lord Roberts was highly critical of British musketry in comparison to that of the Boers, arguing that the average British soldier:

> was the exact opposite of the Boer, especially in his want of knowledge of the ground and how to utilise it, and in his defective powers of observation. His shooting cannot be described as good ... there was no real marksmanship ... The shooting at short ranges ... was ineffective, and at long ranges the distance was seldom accurately estimated.[80]

The final report of the Royal Commission concluded that Boer marksmanship had been superior to that of the British, identifying the capacity of the Boers to fight from behind cover, their superior skill in judging distances and ability to hit fleeting targets as critical factors.[81]

The value of skilful marksmen wielding modern weaponry was clear to many veterans of the conflict. Alexander Thorneycroft considered it an 'essential point' from the war, arguing that '[w]hen you get to a decisive range, say 300 yards, if your men are first-class shots with good fire sights on their rifles for close shooting, you are at an enormous advantage'.[82] Ian Hamilton went even further, arguing that Britain should take inspiration from the Boer military system and adapt it to her own needs:

> I believe that an army composed of individuals each so highly trained as to be able to take full advantage of the terrain, and of his wonderful modern weapon,

77 Ibid., Q 16445, 16446, p. 259.

78 Ibid., Q 15660, p. 224; Q 16772, p. 272.

79 Ibid., vol. 1, Q 173, p. 7.

80 Ibid., vol. 2, Q 10442, p. 440.

81 Ibid., vol. 4, p. 48.

82 Ibid, vol. 2, Q 12435, p. 19.

> and each animated with a morale and trained to an efficiency which will make him capable of acting in battle on his own initiative, will break through, scatter, and demolish less efficient opposing forces, even if greatly superior in numbers.[83]

The British Army underwent considerable tactical reform in the aftermath of the conflict, with a particular focus on improving marksmanship.[84] In 1902, Lord Roberts stated that the first object in the training of a soldier was 'to make him a good shot'.[85] To this end, the old system of volley firing was abandoned and was replaced with training that aimed to make each soldier an effective individual rifleman. Musketry training was heavily based on the experience of the Second Anglo-Boer War. There was a concerted effort to mimic the skills of the Boers, with an emphasis on 'snap shooting', firing from behind cover and engaging fleeting targets at unknown distances.[86] The culmination of British marksmanship training was the 'Mad Minute', in which a soldier was required to fire 15 aimed rounds at a target at least three hundred yards distant within 60 seconds. This famous exercise was directly inspired by the effectiveness of sudden, intense bursts of fire in South Africa.[87]

Admiration of the rifle culture of the Boers prompted some authors to urge that attempts be made to inculcate a similar attitude towards guns within the British Empire.[88] Although this was impractical for the bulk of British civilian society, there was a marked change towards rifle training within the British Army. Writing in 1904, an anonymous officer noted with satisfaction: 'Greater interest is now shown by everybody, especially by the private soldier, and the keenness displayed by all ranks is as great as could be desired.'[89] Particular pride was attached to the completion of the 'Mad Minute' exercise, which was generally considered to be the true test of a marksman.[90] The award of coveted marksmanship badges and extra pay for soldiers who had reached the required standard further encouraged training and development. Individual training was supplemented by a wide variety of rifle competitions, many of which attracted

---

83 Ibid., Q 13941, p. 107.

84 For a fuller discussion of the reforms of this period, see Spencer Jones, *From Boer War to World War: Tactical Reform in the British Army 1902–1914* (Norman, 2012).

85 Quoted in 'K.', 'Suggestions for the Improvement of the Annual Course of Musketry', *United Service Magazine*, June 1904, p. 300.

86 War Office, *Musketry Regulations Part 1, 1909* (London, 1909), pp. 258–261.

87 Joint Services Command and Staff College Library, Report on a Conference of General Staff Officers at the Staff College, 2–11 January 1906, p. 118.

88 J. Peters, 'Teach the Boys to Shoot', *United Service Magazine*, March 1904, pp. 598–601.

89 'K.', 'Annual Course of Musketry', p. 300.

90 Richard Van Emden (ed.), *Tickled to Death to Go: Memories of a Cavalryman in the First World War* (Staplehurst, 1996), p. 24.

considerable participation.[91] Indeed, by 1913, some of the competitions were attracting so many entrants that they were in danger of becoming unmanageable.[92]

The impressive marksmanship skills of the British Army in 1914 owed a great deal to the training methods adopted in the aftermath of the South African conflict. A sergeant from the Lincolnshire Regiment recalled his battalion putting such skills to use at the Battle of Mons, 23 August 1914, with crack shots ordered to engage in independent firing to pick off German officers. The sergeant recalled of the tactic: 'That is another trick taught to us by Brother Boer, and our Germans did not like it at all.'[93]

The experience of facing Boer firepower left a deep and lasting impression on the British Army. The grudging respect that had developed during the war evolved into admiration in the aftermath of the conflict. As has been demonstrated, the British came to recognise that a variety of factors were responsible for the skills of the average Boer marksman, some of which, such as the clear atmosphere, were unique to the theatre of war. Nevertheless, the value of modern magazine rifles in the hands of skilled marksmen was a lesson that was taken to heart. Universal skills, such as firing from behind cover, snap shooting and rapid target acquisition went on to become the cornerstones of British musketry training in the aftermath of the conflict. Inspired by these changes, the British Army developed its own unique rifle culture during the pre-First World War period, with marksmen being recognised and rewarded for their skills. If imitation is the sincerest form of flattery, then the post-war training methods of the British Army paid a handsome compliment to the marksmanship of Boers.

---

91 The Smith-Dorrien Cup and the Evelyn Wood Cup were two particularly popular events.

92 NAUK, WO 279/32, Aldershot Command Papers, Comments on the Training Season 1913, p. 11.

93 Quoted in A. St John Adcock, *In The Firing Lines* (London, 1914), p. 25.

Chapter 13

Irish Paramilitarism and Gun Cultures, 1910–1921

Timothy Bowman

Between 1910 and 1921 a number of paramilitary forces were formed in Ireland. The Ulster Volunteer Force was formed by Unionists in January 1913, reaching a peak strength of around 100,000 by August 1914, and the Irish Volunteers were formed in the Autumn of the same year by advanced Nationalists, reaching their peak strength of around 191,000 in September 1914. After an acrimonious split between moderate and advanced Nationalists within their ranks in September 1914, the smaller group of around 10,000 men retained the name Irish Volunteers and were to form the basis of the Irish Republican Army, which emerged as a guerrilla force in 1919. The more moderate Irish National Volunteers declined rapidly after providing a large number of volunteers to the British Army, and many of the remaining members appear to have defected to the Irish Volunteers in 1917–1918. The Ulster Volunteer Force formed the basis of the 36th (Ulster) Division but otherwise soon went into decline on the outbreak of the Great War. Attempts were made to re-raise it, along with other Ulster Unionist paramilitary forces, in 1920, but it was shortly replaced by the Ulster Special Constabulary, a paramilitary and largely part-time force raised by the new Northern Ireland government. The small Irish Citizens' Army was a workers' militia, closely allied to the more extreme elements of the Irish Trades Union movement. Following the Easter Rising and the death of its leader, James Connolly, the force appears to have merged with the Irish Volunteers.

Of course, it would be wrong to see the gun as having been introduced into Irish politics in 1910. Both Unionists and Nationalists drew on longstanding traditions of physical force. A Protestant or Unionist volunteering tradition can be identified stretching back to the plantations of the late sixteenth and early seventeenth centuries. This was at its most politicised in the period from 1775–1792, when the largely Protestant Irish Volunteers essentially achieved a form of colonial self-government, but the Irish Yeomanry of 1796–1834 was also regarded as a highly politicised force.[1] The first and second Home Rule bills

---

[1] Allan Blackstock, *An Ascendancy Army: The Irish Yeomanry 1796–1834* (Dublin, 1998); D.W. Miller, 'Non-professional Soldiery, c. 1600–1800', in T. Bartlett and K. Jeffery (eds), *A Military History*

had witnessed drilling by Unionists on a relatively small scale.[2] For Nationalists, a belief in physical force went back at least as far as the United Irishmen of the 1790s, although some attempts were made to include the Irish Volunteers of the 1770s in a Nationalist volunteering tradition and encompassed the Young Irelanders of the mid-nineteenth century. The Fenian movement had reached its height in the 1860s as a mass movement, but survived into the twentieth century as a secret oath-bound society, and many of its members were involved in the establishment of the Irish Volunteers.[3]

Assessing the prevalence of a gun culture in Ireland in these years is no easy matter, given the limited source materials available. While IRA leaders such as Tom Barry and Ernie O'Malley produced popular, well written works, which have remained continuously in print, the UVF generated only three published memoirs: those by R.J. Adgey, Fred Crawford and Percy Crozier.[4] These UVF memoirs were little reviewed or noticed, with the exception of Crozier's work, and this was largely due to his provocative comments on his experiences in the Great War and Anglo-Irish War, not in the pre-war UVF; all have been long out of print. Ironically the largest of the paramilitary armies of this period, the Irish National Volunteers, gave rise to not a single published memoir.

The existing archival material concerning the UVF has a bias in favour of rural units, commanded by members of 'Big House' Unionism. This is because, in 1961, when Lord Brookeborough, the Prime Minister of Northern Ireland, was considering the forthcoming fiftieth anniversary of the Third Home Rule crisis, he wrote to a number of 'old-established Ulster families prominently identified with the anti-Home Rule movement', asking for material to be deposited in the Public Record Office of Northern Ireland (PRONI).[5] The Irish Volunteers of 1913 generated a reasonable amount of published works:

---

of Ireland (Cambridge, 1996); D.H. Smyth, 'The Volunteer Movement in Ulster: Background and Development, 1745–85', unpublished PhD thesis, Queen's University of Belfast, 1974; and Stephen Small, *Political Thought in Ireland 1776–1798: Republicanism, Patriotism and Radicalism* (Oxford, 2002), pp. 83–112.

2 Timothy Bowman, *Carson's Army: The Ulster Volunteer Force, 1910–22* (Manchester, 2007), pp. 15–18.

3 R.V. Comerford, *The Fenians in Context: Irish Politics and Society, 1848–1882* (Dublin, 1985); Richard Davis, *The Young Ireland Movement* (Dublin, 1987); and M.J. Kelly, *The Fenian Ideal and Irish Nationalism, 1882–1916* (Woodbridge, 2006).

4 Tom Barry, *Guerrilla Days in Ireland* (Dublin, 1949); Ernie O'Malley, *On Another Man's Wound* (Dublin, 1936); Ernie O'Malley, *The Singing Flame* (Dublin, 1978); R.J. Adgey, *Arming the Ulster Volunteers* (Belfast, n.d. [but *c.*1947]); Fred Crawford, *Guns for Ulster* (Belfast, 1947); and F.P. Crozier, *Impressions and Recollections* (London, 1930), pp. 142–181.

5 PRONI, *Report of the Deputy Keeper of the Records for the Years 1960–1965* (Belfast, 1968), p. 20.

memoirs and even a short history.[6] However, few archival sources survive concerning John Redmond's attempt to subsume the Irish Volunteers within the Irish Parliamentary Party and the organisation of the Irish National Volunteers (INV) from September 1914 to their demise in 1918.

The Bureau of Military History collected a wide range of witness statements in the late 1940s, 1950s and 1960s. However, while this provides us with a much better understanding of the views of the rank and file within the Irish Volunteers and IRA than we have for the UVF, it remains a problematic source. Those who were opposed to the Anglo-Irish Treaty were unlikely to provide statements, and no attempt was made to challenge the tendency of some interviewees to over-estimate their importance in the events of 1919–1921. Some, especially those who had pursued political careers, tried to sanitise or excuse their actions. In any case, those interviewed were overwhelmingly the IRA activists, those prepared to shoot members of the Crown forces and civilians, who made up a small proportion of the organisation as a whole.[7]

Other source material is problematic. Irish newspapers of the period were prone to heavy political bias, clearly identifying themselves as Nationalist or Unionist in sentiment, which heavily coloured their reportage of issues as factual as the numbers attending volunteer parades.[8] Royal Irish Constabulary (RIC) reports are also a problematic source, partly as most senior RIC personnel were Unionists, but also because, as the events of 1919–1921 dramatically proved, the RIC was not a particularly good investigative force.

In the early months of their formation, the UVF and Irish Volunteers were largely unarmed and it was to be 1914 until a firm 'gun culture' can be seen as emerging. Long before the romanticised picture of the armed 'freedom fighter' entered the Irish popular psyche, some militant rank-and-file members of these paramilitary forces demanded arms.[9] Alvin Jackson has noted the decision to arm the UVF was taken reluctantly and was an attempt to bolster morale, which was seen as wavering in early 1914. The, admittedly few, surviving UVF muster rolls and inspection reports certainly support this argument.[10] Reflecting the views of

---

6 Most of these were reprinted in F.X. Martin (ed.), *The Irish Volunteers, 1913–1915* (Dublin, 1963).

7 Ferghall McGarry bravely outlines the problems associated with the use of the witness statements in some detail in his work based on them: *Rebels: Voices from the Easter Rising* (Dublin, 2011), pp. xii–xx.

8 Michael Wheatley, *Nationalism and the Irish Party: Provincial Ireland, 1910–1916* (Oxford, 2005), pp. 18–20.

9 On the image of the 'freedom fighter' see Mark Connelly, *The IRA on Film and Television: A History* (Jefferson, 2012).

10 Alvin Jackson, 'Unionist Myths, 1912–1985', *Past and Present*, 136 (1992); PRONI, D1390/19/1, Charles Falls papers, John Sears's notebooks for 1913 and 1914 (Sears was the County Instructor for the Fermanagh UVF); D1132/6/17, R.T.G. Lowry papers concerning 'J' Company,

the rank and file, a deputation of senior UVF members from Co. Antrim wrote a 'statement on the arms question' for the attention of Sir Edward Carson, the Irish Unionist leader, in January 1914. In this statement, 'very great dissatisfaction was expressed at the inadequate supply of arms – even for instructional purposes; only 150 .303 Rifles and carbines and 50 Vetterlis had been issued for a force of 10,700 men, no ammunition being allowed for the latter rifle.' The Co. Antrim officers went on to note that their men were tiring of basic drill and that the ex-army officers promised from England had not arrived. They also stated that other counties had received more rifles and ammunition and demanded an explanation for this.[11] Not quite so militant, the Co. Down committee asked if UVF Headquarters would issue arms directly or provide funds to the county committees so that they could make their own arrangements.[12] Carson reacted to these concerns by holding a military conference of all UVF commanders, where he perhaps hoped to defuse tensions. However, Major Robert McCalmont, M.P., seems to have caught the mood of the meeting. 'In effect he said the men of Ulster were being asked to give everything for the defence of their land and were being "let down", because, although the leaders were telling the world they were armed and would fight to the bitter end, as they undoubtedly would, still they were not supplying the rank and file with the necessary arms and ammunition with which to fight.'[13] Thus it seems that the provision of arms became necessary for UVF Headquarters to maintain its authority over county committees and local commanders.

Similar concerns were voiced within the Irish Volunteers. These were partially expressed through the 'official publication', *The Irish Volunteer*, although it should be noted that from February to November 1914 this was operated as a financial venture by the editors of the *Enniscorthy Echo* with no control from the Irish Volunteer executive. As a result it can be seen, at times, to reflect little more than the views of its editor, Laurence de Lacy, an active member of both the Irish Volunteers and Irish Republican Brotherhood.[14] In the very first issue of *The Irish Volunteer*, the editorial provided a very curious attitude to firearms ownership, noting that '[e]very man who has a spark of the Volunteer spirit loves a rifle, and apart from their use as a weapon in war the possession of a rifle rouses the enthusiasm of most recruits'. He then went on to state: 'There is little doubt

Dungannon Battalion; and D1518/3/9, Hamilton papers concerning 'A' Company, 2nd North Antrim Regiment.

11 'Statement on the arms question laid before Sir Edward Carson and James Craig 20 January 1914, by a deputation of Co. Antrim volunteers, consisting of General Sir William Adair', PRONI, D.1238/108, cited in Patrick Buckland, *Irish Unionism 1885–1923: A Documentary History* (Belfast, 1972), p. 243.

12 PRONI, D.1327/4/1, Co. Down Committee Book, entry for 8 April 1914.

13 Crozier, *Impressions and Recollections*, p. 148.

14 Bulmer Hobson, *Ireland Yesterday and Tomorrow* (Tralee, 1968), p. 68.

[that] before rifles are again used in war a vastly improved pattern will have appeared on the market, and the difficulty in getting them will be no greater than was the difficulty of getting the Mauser or Lee-Enfield a couple of months ago.'[15] By June 1914, as the Irish Volunteers reached the 160,000 membership mark, the editorial was much more strident, stating simply: 'We demand arms', although an article on the next page, extolling the virtues of the pike in war, did tend to dilute this message.[16]

However, many of those who were members of the paramilitary forces did not embrace a gun culture. In the UVF there was not a simple dynamic of a conservative leadership, still wedded to parliamentary opposition, attempting to withhold arms from a militant rank and file. There is strong evidence of men being conscripted into the UVF in some areas, notably Dungannon, Co. Tyrone. Elsewhere, it seems likely that men felt under some obligation to their employers to enlist in the force where employers had raised and commanded UVF units. More tellingly the high levels of absenteeism witnessed in the UVF in late 1913 and early 1914 demonstrated a certain lack of commitment to the cause, and the turnout on test mobilisations was often surprisingly poor. The decision of the UVF high command to subdivide the force between a 25,000-strong Special Service Force for service anywhere in Ulster and a static defence force consisting of the remaining 75,000 Volunteers was a recognition that, late in 1913, there were different levels of experience and training, but also commitment, amongst the rank and file.[17]

The Irish Volunteers and IRA also had a large number of inactive members. In early 1918, when conscription into the British armed forces threatened, many men joined the Irish Volunteers to avoid, rather than undertake, any military training.[18] James Cahill reflected that, in Cavan, the 'threat of conscription caused the Company to expand to twenty or thirty times its former strength. With the passing of the conscription menace the Company dwindled almost to its former strength.' J.D. Cummins noted a similar, if not quite so pronounced, pattern in Charleville, Co. Cork, where the company grew from 14 to 85 men

15 *The Irish Volunteer*, 7 February 1914.

16 *The Irish Volunteer*, 27 June 1914.

17 PRONI, D/1132/6/17, R.T.G. Lowry papers, circular letter from Lord Northland to members of 4th Tyrone Regiment and returns re. trial mobilisation of No.1 Section, 'J' Company, 4th Tyrone Regiment; D.3743/3, enrolment register for Comber West Company; D.1414/30, Battalion orders for 1st (North) Tyrone Regiment; D.1238/88, O'Neill papers, memorandum by Hacket Pain, 7 May 1914; *Belfast Evening Telegraph*, 12 October 1913 and *Northern Whig*, 24 July 1914.

18 Joost Augusteijn, *From Public Defiance to Guerrilla Warfare: The Experience of Ordinary Volunteers in the Irish War of Independence, 1916–1921* (Dublin, 1996), p. 85, and Peter Hart, *The I.R.A. and Its Enemies: Violence and Community in Cork, 1916–1923* (Oxford, 1998), p. 239.

before falling back to its original strength when the crisis passed.[19] Similarly David Fitzpatrick, in his study of Co. Clare, noted that the 'number of active volunteers multiplied the instant fighting finished' after the Truce. This pattern was also evident in Co. Cork, where Peter Hart estimated that only around 230 of the 1,000 strong Bandon Battalion had been 'active' volunteers.[20]

Of course, a vibrant civilian gun culture existed in Edwardian Great Britain and this was also evident in Ireland. Joyce Lee Malcolm notes that firearms ownership was allied with the concept of individual liberty and that the 1870 firearms legislation, which introduced gun licences, was widely breached, largely as police forces saw it as an excise, not a criminal matter.[21] Firearms ownership was also increasing at a time when crime rates were falling, so contemporaries did not make the same direct connection between illegal firearms ownership and crime as police forces do today.[22] As late as June 1918, in an appeal to the President of the USA, advanced Nationalists noted that the British government was attempting to deny Irishmen the right to bear arms, something which was enshrined in the British Bill of Rights of 1688 and the American Constitution.[23]

Before 1920, there were few legal restraints on the carrying of firearms in the United Kingdom as a whole, though some restrictive Irish legislation existed. The Irish Attorney General noted that the Peace Preservation Act (Ireland) of 1881 was not renewed in 1906; thus, the only Act in force in Ireland in 1913 that interfered with the use or carrying of arms was the Gun Licence Act of 1870. He went on to note that 'the prohibition is only for the use or carrying of the gun and does not prevent the possession of firearms or subject the possessor to the licence duty. Moreover, this is a Revenue Act which must be enforced exclusively by the Excise Authorities.'[24] This important legal question over the powers of the RIC to act in excise cases was to drag on for months. Only in July 1914 was Sir John Simon, the English Attorney General, to state definitively that the RIC had the same powers as customs officers to seize arms, making the observation that the RIC were not slow to act over illegal distilling, which

19 National Archives of Ireland (NAI), Bureau of Military History (BMH), Witness Statement (WS) 503, James Cahill, pp. 1–2, and WS1039, John D. Cummins, pp. 2–3.

20 David Fitzpatrick, *Politics and Irish Life 1913–1921: Provincial Experience of War and Revolution* (Dublin, 1977), p. 230, and Hart, *I.R.A.*, pp. 227–228.

21 Joyce Lee Malcolm, *Guns and Violence: The English Experience* (Cambridge, MA, 2002), pp. 95–141.

22 V.A.C. Gattrell, 'The Decline of Theft and Violence in Victorian and Edwardian England' in V.A.C. Gatrell, B. Lenman and G. Parker (eds), *Crime and The Law: The Social History of Crime in Western Europe since 1500* (London, 1980).

23 Bulmer Hobson, *A Short History of the Irish Volunteers* (Dublin, 1918), pp. i–ii, preface by Eoin MacNeill, citing the address by the Mansion House Conference of 11 June 1918.

24 National Archives of the UK (NAUK), Kew, London, CO904/182, 'City of Belfast. Importation of Arms', by John F. Moriarty, 23 August 1913.

was also an excise matter.[25] The Irish Attorney General went on to suggest that those UVF members carrying arms could be tried under a statute of Edward III for going armed in public without lawful occasion. He also suggested that UVF members who had participated in the Baronscourt Camp and had openly carried arms in field training there could be tried for 'obstructing the public highway' for their activities near Ardstraw Bridge.[26]

UVF members seem to have been entirely clear about their legal rights. Lieutenant Colonel John Madden complained at length to the RIC County Inspector for Monaghan, stating:

> I wish to bring to your notice the way the police are behaving here. I understand they are trespassing through my place without any leave from me. Only yesterday the Sergeant in Scotchouse came in and watched the drill of a section of volunteers who were drilling in my pleasure ground, ¾ of a mile from the boundary of the place; concealed in a wood. This is an unwarrantable liberty, and I have to give you notice that I shall forcibly remove any police sergeant or constable entering my place in future ... I have also to complain that your Sergeant chose to come up into my yard during my absence in Dublin the other day, ostensibly to look at my motor car, but really to try + cross examine my chauffeur as to whether I had any arms in my home. I may say that naturally I have arms in my house of various sorts, + always have had, + I should like to know what law there is against my having as many as I choose.[27]

Similarly, the legality of importing arms into Ireland was a grey area, at least until December 1913, when the government issued two Royal Proclamations. One of these forbade the importation into Ireland of arms and ammunition, apart from those intended solely for sporting purposes, mining or any other 'unwarlike' purposes, while the second prohibited the carriage by sea of military arms and ammunition.[28] Otherwise, as the Solicitor General noted in April 1913, guns might be seized under the Customs Laws Consolidation Act 1876, the old Riot Act of the Irish Parliament and possibly the Prerogative Power of the Lord Lieutenant.[29]

---

25 NAUK, CO904/182, letter, Sir John Simon to Augustine Birrell, 31 July 1914.

26 NAUK, CO904/182, 'City of Belfast. Importation of Arms', by John F. Moriarty, 23 August 1913, and 'Armed party of the Ulster Volunteers manoeuvring on the public road near Baronscourt, 17/6/14'.

27 PRONI, D.3465/J/37/62, Madden papers, letter, John C. Madden to County Inspector E.M. Tyache, 24 February 1914.

28 *The Times*, 6 December 1913.

29 NAUK, CO904/182, 'Importation of Arms: Summary of Solicitor General's Opinion of 9th April, 1913', and Bodleian Library, Oxford, MS. Asquith 38, 'Ulster – Carriage of Rifles 1913, Attorney General's opinion'.

The legal authority under which the government made its largest seizure of 4,500 rifles destined for the UVF in Hammersmith in June 1913 was the Gun Barrel Proof Act of 1868. The UVF could have retrieved these weapons for a fine of £2 per rifle, as being of Italian origin and imported from Hamburg, they had not received the required quality control from the Gunmakers Company as required by the 1868 Act. However, this was not done, largely as the rifles seized had cost considerably less than £2 each. It is possible that this action was taken by the government under pressure from the Court of the Gunmakers Company, which had taken an interest in the foreign firearms being imported into Ulster as early as November 1911.[30] These legal anomalies ended with the outbreak of the First World War, since the Defence of the Realm Act (1914) and, following the formal end of the European hostilities, the Restoration of Order in Ireland Act (1920) imposed clear legal sanctions on those importing firearms into Ireland or carrying them without British government authority.

It appears that the RIC first became aware of guns being brought into Ulster in relatively large numbers by Unionists in December 1911, when, with some press attention, 24 Martini-Henry rifles and 1,000 rounds of ammunition were delivered to Orange Lodges in Co. Londonderry.[31] It seems that James Craig, M.P. and future Prime Minister of Northern Ireland, was first involved in gun-running at this time; Major Fred Crawford, later to be responsible for the large scale gun-running of April 1914, seems to have started his activities in July 1911.[32] Given the grey legal area in which the arming of Ulster Unionists was conducted it is not surprising to find that, at least initially, Unionists relied on established gunsmiths in Belfast to meet their requirements.

Most of the rifles issued to the UVF were brought into Ulster by Major Fred Crawford, a retired militia officer who was the UVF's Director of Ordnance. Initially, he worked with Sir William Bull M.P. and Bull's brother in law, Captain H.A. Budden, who set up a front firm, John Ferguson and Company in Hammersmith. They also operated through a legitimate firm of motor body builders, F.M. Foyer & Co. Ltd., again based in Hammersmith and owned by Bull's former chauffeur. Crawford's arrangements with Bull, Budden and Foyer were initially successful; it was through this Hammersmith operation that Crawford imported six Vickers machine guns and thousands of rifles into Ulster. Disaster struck for Crawford on 9 June 1913, when the Metropolitan

30 Details of the Hammersmith raid are taken from the manuscript notes in Sir William Bull's copy of McNeill, *Ulster's Stand for Union*, PRONI, D.3813/3. See also NAUK, CO904/28/2, letter, D.C. Lee, Clerk to the Gunmakers Company, to Secretary of State for War, 3 November 1911.

31 NAUK, CO904/28/1, report by District Inspector E.S. Cory, 11 December 1911; *Irish Times*, 7 December 1911, and *Northern Whig*, 1 December 1911.

32 PRONI, D.1415/B/38/1, Craigavon papers, Lady Craigavon's diary, entry for 8 December 1911; and D.1700/5/17/2C-E, Crawford papers, various letters, Crawford to Craig, July and August 1911.

Police, possibly acting under information received from Budden, seized 4,500 rifles, which they held under the 1868 Proof Act.[33]

Crawford's largest gun-running success was the major operation of April 1914, which culminated in the landing of a large number of rifles at Larne, Bangor and Donaghadee. The events of this audacious enterprise have been recounted in detail elsewhere and need only be dealt with briefly here.[34] In essence, Crawford purchased 20,000 rifles and 2,000,000 rounds of ammunition in Hamburg in February 1914 and embarked them on a Norwegian vessel, the SS *Fanny* on 2 April 1914. Following a circuitous route and trans-shipment of the rifles and ammunition on to a coal vessel, the SS *Clyde Valley* on 19–20 April, the cargo was unloaded in the three Ulster ports on the 24–25 April 1914. The importance of this large scale gun-running was partly that it showed what the UVF could achieve in a military operation. The UVF was mobilised throughout Ulster, with local units seizing Larne, Bangor and Donaghadee, quickly overawing the local police, coastguards and customs officials. Telegram and telephone systems were short-circuited, which meant that no police or army reinforcements could be summoned. This was a crucial factor, given that a battalion of British troops was based in Holywood, a mere seven miles from Bangor. In addition a careful deception plan was effectively executed by 2,500 UVF members under the command of Colonel Couchman. This group met the SS *Balmerino* when she docked in Belfast and drew RIC and customs attention to this vessel, which in fact carried no arms at all. Perhaps the major success of this operation was that only one death occurred during it, when a coastguard, H.E. Painter, died of a heart attack at Donaghadee, having cycled furiously to wake his commanding officer.

However, the Larne gun-running was not flawless. There was a shortage of transport at Larne, with General Adair, the UVF commander in Co. Antrim, asking for any available transport (including horse drawn wagons and hand carts) to be sent to Larne; as a result, it is not clear how far the guns were transported from Larne, Bangor and Donaghadee in the crucial hours following their unloading. Many of the cars and lorries entering and leaving the harbours did nothing to disguise their number plates, which meant that the RIC quickly ascertained who had been involved and the likely destinations of the rifles landed. It also seems that Crawford had not recruited enough sailors to properly man the *Clyde Valley* and that crucial harbour facilities had not been secured (in Bangor water had to be fed to the *Clyde Valley* by the district council's fire cart). This all meant that the unloading proceeded much more slowly than had been

---

33 All details from PRONI, D.1415/B/34, Craigavon papers, Fred Crawford, 'The Arming of Ulster', pp. 5–35.

34 A.T.Q. Stewart, *The Ulster Crisis: Resistance to Home Rule, 1912–1914* (London, 1967), pp. 105–249.

planned, with the unloading at Bangor and Donaghadee, which should have been carried out under cover of darkness, not completed until 8.30 a.m.[35]

The distribution of arms throughout Ulster could easily have been prevented by the RIC, as one senior police officer stated in his memoirs.

> The gun-running is always said to have been carried out in such a masterly fashion that the RIC were caught napping, and all the arms distributed before police were aware as to what was happening. This was not so in my case at any rate. Sometime during the day in question I heard that arms were to be brought that evening to the house of Mr. J. Porter-Porter, of Belleisles. I did not know what to do and could not get instructions. It would have been very easy to seize the arms as they all seemed to be in packages and, of course, useless for immediate use. I brought just a few men to Mr. Porter-Porter's demesne. When I got to the front gate I knew the information was correct, as he and his wife were both there. I sat on the other side of the road and a jocose conversation was kept up. After a time along came the cars with their packages in the back, and in they went to Belleisle. I knew a number of the drivers and some of them waved as they passed. The arms were delivered and there was no interference at any time.[36]

The Larne gun-running was a major propaganda coup in the British and Irish press, which was a major preoccupation of the Irish Unionist leadership and one of the reasons why the risk was taken of landing 20,000 rifles in one operation. This was entirely contrary to the wishes of Major F.H. Crawford, who wrote: '[t]he Unionist papers ought to be squared so that they will not refer to it at all, at least for some days after. This will give you a better chance of getting redistributed later.'[37]

Other UVF gun-running operations were carried out through 1912–1914 by R.J. Adgey, who operated a shop, which appears to have mainly been a pawnbrokers, but which also dealt in second-hand firearms. As such he had been involved in selling arms to individual Unionists from the beginning of the Ulster Crisis. The extent of Adgey's involvement in supplying the UVF is unclear, but he claimed to have imported 'many' rifles into Ulster disguised as hardware or secreted in bleaching powder barrels. In November 1914, the RIC understood him to have imported 1,188 Martini-Enfield rifles in that month alone.[38] W.P. Johnston, who was interviewed by A.T.Q. Stewart, claimed to have been involved in importing hundreds of rifles into Ulster from Manchester, under the cover of

35 Ibid., pp. 176–212; *Belfast Evening Telegraph*, 25 April 1914; McNeill, *Ulster's Stand for Union*; and PRONI, D.1415/B/34, Craigavon papers, Fred Crawford, 'The Arming of Ulster'.

36 PRONI, D/3160/1, John Regan, 'Memoirs of service in the RIC and RUC', p. 76.

37 PRONI, D.1700/5/17/16A, Crawford papers, letter, Crawford to Spender, 21 April 1914.

38 Adgey, *Arming the Ulster Volunteers*, pp. 2–25, and NAUK, CO904/28/2, report by Sergeant J. Edwards, 6 November 1914.

his family's textile printing business.[39] It also seems likely, based on reports of the seizures of arms, that five other individuals organised gun-running, though they may simply have been aiding Crawford and Adgey's in their activities.[40] The UVF appear to have been attempting to obtain rifles from sympathetic British Army personnel.[41]

The quality of guns that the Irish paramilitary forces possessed varied enormously. Charles Townshend has noted that, for the UVF, this would have created a 'logistical nightmare' in any conflict, and this seems to have been the case, given the unwillingness of UVF units to agree to any sensible redistribution of arms.[42] By mid-1914, the UVF appears to have been equipped with a variety of Lee-Metford, Martini-Henry, Mauser, Steyr and Vetterli rifles, all taking different types of ammunition. Most UVF rifles were of the single-shot variety, and a large minority were Vetterli rifles, which were purchased from the bargain basement of the international arms market, having been withdrawn from service in the Italian Army in 1887.[43] Brigadier General Count Gleichen was utterly dismissive of the military value of these Italian rifles noting that 'they were not good, but weedy + weak + only cost 5 francs apiece, including belt and bayonet!'[44] It was probably the poor quality of these rifles that made redistribution of arms throughout the UVF, as suggested by Colonel Hackett Pain, impossible, as no County Committee wanted to risk their units being re-equipped solely with these Italian rifles, while their more highly prized British, German and Austrian rifles were sent elsewhere.

The distribution of UVF arms is interesting, suggesting that political rather than military decisions were important in deciding how to arm particular units. It is noticeable that the 'frontier' counties did not receive the best equipment, as one would have expected, with a disproportionately high number of the despised Italian rifles present in Counties Armagh, Fermanagh, Londonderry and Monaghan. It is also noticeable that Co. Antrim received very few of the Italian rifles, despite the fact that the county was relatively secure, in terms both of being well behind any likely Unionist 'front line' and having a small Catholic minority. The figures for Belfast are interesting, too, for while the highest number of rifles were in the city, along with the entire compliment of machine

39 Stewart, *Ulster Crisis*, pp. 99–100, and PRONI, D2966/138/1, 'Report of Gun-running to Ulster 1913–1914 as told by W. P. Johnston'.

40 *The Fermanagh Times*, 5 June 1913; *Northern Whig*, 18 June 1913, and 7 July 1913.

41 *Northern Whig*, 23 July 1914.

42 Charles Townshend, *Political Violence in Ireland: Government and Resistance since 1848* (Oxford, 1988), p. 255.

43 John Whittam, *The Politics of the Italian Army, 1861–1918* (London, 1977), p. 194.

44 Nuffield College, Mottistone papers, 22/f.193–4, Report by Brigadier General Count Gleichen, 'The Ulster Volunteer Force', 14 March 1914; and Count Gleichen, *A Guardsman's Memories* (London, 1932), p. 366.

guns, the numbers of rifles were disproportionately low, given the large numbers of UVF personnel in the city and the fact that the Special Service Force was centred on the Belfast regiments.

The number of firearms that the UVF possessed is impossible to calculate. As noted above, firearms control was very lax and revolver ownership was certainly widespread.[45] A.T.Q. Stewart has stated that in July 1914 the force had 37,048 rifles.[46] This is a curiously precise figure, given the efforts made by local UVF commanders to obtain (and, of course, retain) rifles, without reference to UVF Headquarters. Alvin Jackson has come to an estimate of around 40,000 rifles in UVF hands by July 1914, while Josephine Howie suggested a figure of 60,000 rifles by early May 1914.[47] A police report of February 1917 (compiled at a time when the UVF were co-operating with the RIC) suggested that the UVF then had access to 53,130 rifles, though this figure is probably an underestimate, given the unwillingness of UVF members to hand in rifles to central armouries.[48] Howie's figure of 60,000 would seem credible, including rifles of all calibres and vintages. Certainly the UVF Special Service Force of 25,000 men could have been armed with modern rifles and probably all 100,000 members of the force could have been armed with firearms of some sort, if revolvers and shotguns are included.

Irish Volunteers had access to fewer rifles overall, perhaps something in the region of 6,000, and the Irish Citizen Army's arsenal seems to have consisted of rifles stolen from the Irish Volunteers during the confusion of the Howth gun-running. The German rifles, old but effective, landed at Howth in July 1914 were probably the best available to the Irish Volunteers. There was some debate over where they should be sent, with Redmondites on the Irish Volunteer Committee wanting them sent to units in Ulster, which logically would be in the forefront of any civil war. However, the other members of the Committee rejected this, and the prized Howth guns seem to have been distributed largely to units in Dublin. However, even in Dublin, some Irish Volunteer units received little. Liam Archer, a future Lieutenant General in the Irish Army, remembered attending an Irish Volunteer meeting in mid-1914 at which he asked if rifles would be provided. The 'answer was that it was not intended the movement should be armed but that the *Daily Sketch* "side" would be taken care

45 NAUK, CO904/28/1, Report by District Inspector P. McHugh, Londonderry, 27 August 1913, in file entitled, '1886–1913 Arms Importation and Distribution'.

46 Stewart, *Ulster Crisis*, p. 248.

47 Alvin Jackson, 'British Ireland: What if the Home Rule Had Been Enacted in 1912?', in Niall Ferguson (ed.), *Virtual History: Alternatives and Counterfactuals* (London, 1998), p. 220, and Josephine Howie, 'Militarising a society: The Ulster Volunteer Force, 1913–14' in Y. Alexander and A. O'Day (eds), *Ireland's Terrorist Dilemma* (Dordrecht, 1986), p. 221.

48 NAUK, CO904/29/2, 'Return of rifles in possession of Ulster Volunteer Force', 28 February 1917.

of. Sometime later a consignment of Italian rifles reached the Company. They were pronounced to be "duds" by some members of the Company who had British Army experience.'[49] In August 1914 the Belfast Regiment of the Irish Volunteers, which had as many as 4,000 members, had access to just 100 Italian rifles with no ammunition, one Lee-Metford with 100 rounds and six miniature .22 rifles and an air rifle for target practice.[50] Irish Volunteer units in rural Ireland were left with a similarly unimpressive mixed bag of firearms. In Dunboyne, Co. Meath, the Irish Volunteers' only firearms were five big game rifles stolen from the house of Mr M. Carthy and a .22 rifle for practice. In Tipperary town the Irish Volunteer Company had about 50 Martini-Henry rifles and 'some' Italian rifles, without any ammunition. Meanwhile in Drumcollogher, Co. Limerick, the Volunteers had no firearms and paraded with wooden dummy rifles.[51] While UVF rifles appear to have been issued free to members or sold to them at a purely nominal sum, some Irish Volunteers paid high prices for their rifles. H.T. Banks remembered purchasing his Long Lee-Enfield for £4 5s. 0d. on an 'easy payment system' through the 1st Dublin Battalion; a similar system appears to have operated in Co. Kerry.[52]

For the Irish Volunteers as well as the UVF, gun-running had a propaganda as well as a strictly military objective. Conscious that Irish Volunteer funds were much more limited than those of the UVF, Bulmer Hobson reflected,

> On thinking the matter over, I decided that 1,500 rifles would not go very far towards solving our problem, but that if we could bring them in in a sufficiently spectacular manner, we should probably solve our financial problem and the problem of arming the Volunteers as well. With this in mind, I decided to bring the guns in daylight, in the most open manner and as near to Dublin as possible.[53]

As a result of Hobson's planning, on 26 July 1914, 900 rifles and 26,000 rounds of ammunition were brought into Howth by yacht in broad daylight. These rifles were obsolete Mausers, purchased in Antwerp and withdrawn from service in the German Army in 1891. This was followed on 1 August 1914 by the landing of 600 rifles and 19,000 rounds at Kilcoole, Co. Wicklow. The actual landing of the rifles at Howth went ahead as planned, but the transportation of them to Dublin led to confrontation with the Dublin Metropolitan Police and troops. Deputy Commissioner W.V. Harrell decided to take firm action and,

---

49 R.M. Fox, *The History of the Irish Citizen Army* (Dublin, 1943), p. 74; Hobson, *Ireland Yesterday and Tomorrow*, pp. 51–52; and NAI, BMH, WS819, Liam Archer.

50 Marnie Hay, *Bulmer Hobson and the Nationalist Movement in 20th Century Ireland* (Manchester, 2009), p. 143; NAI, BMH, WS 971, George F. H. Berkeley, and WS 229, Frank Booth.

51 NAI, BMH, WS12, Peader Bracken, WS517, Maurice Crowe, and WS1213, Timothy O'Shea.

52 NAI, BMH, WS1637, Henry T. Banks, and WS146, Diarmuid Crean, p. 3.

53 Hobson, *Ireland Yesterday and Tomorrow*, p. 60.

with a force of around 80 police and 100 troops of the King's Own Scottish Borderers, attempted to seize the rifles. Eighteen constables of the Dublin Metropolitan Police refused to act, perhaps out of sympathy for the Irish Volunteer cause or perhaps showing a keener awareness of the legal niceties than Harrell. However, the police and military were dramatically 'talked to a halt' by Thomas MacDonagh and Darrell Figgis and, after a failed attempt by the police to seize the rifles, the Volunteers were able to disperse with them. Hobson noted:

> [w]hile I expected that the authorities in Dublin Castle would attempt to prevent the landing of rifles, I knew that the Liberal Government in England, having already remained inactive on the occasion of the Carsonite gun-running at Larne, would find it very embarrassing to take active measures against us. I rather suspected, what was afterwards established as a fact, that the local police or soldiers were acting without orders from their superiors.[54]

This of course ended in tragedy, as the troops, returning to their barracks via Bachelor's Walk, were harassed by a crowd and, following a misconstrued hand signal by their commanding officer, Major Haig, they opened fire on the crowd, leaving three dead and at least 35 injured. Following the Royal Commission into the incident, Harrell was discharged and the Chief Commissioner, Sir John Ross, resigned in protest.[55]

With the outbreak of the First World War the most enthusiastic paramilitaries – Unionist, moderate Nationalist and, in some cases, future IRA leaders like Tom Barry – enlisted in the British Army; perhaps 30,000 from the UVF and 32,000 from the INV. This left a bewildered group of temporary commanding officers in both the UVF and INV attempting to work out who had actually possessed a rifle and where these had been stored. The INV, tentatively being established by John Redmond as a home defence force, did still purchase and manage to import around 200 rifles in the early months of the Great War with the full knowledge of the Dublin Castle administration. Ben Novick estimates that by March 1915 the INV had 9,000 rifles.[56] The war curtailed Tom Kettle's attempts to purchase more rifles for the INV in Belgium, and Redmond's attempts to have the force recognised as an adjunct of the Territorial Force and armed by the British government came to nought. As a result the INV went into sharp decline, its last major parade being held in Dublin in Easter 1915.[57] The UVF was also recast as a home defence force, with some contingency plans for

[54] Bulmer Hobson, 'Gun-running at Howth and Kilcoole' in Martin, *Irish Volunteers*, p. 37.

[55] Townshend, *Political Violence in Ireland*, pp. 274–276.

[56] Ben Novick, 'The Arming of Ireland: Gun running and the Great War 1914–16', in A. Gregory and S. Paseta (eds), *Ireland and the Great War: 'A War to Unite Us All?'* (Manchester, 2002), pp. 94–107.

[57] J.B. Lyons, *The Enigma of Tom Kettle: Irish Patriot, Essayist, Poet, British Soldier, 1880–1916* (Dublin, 1983), pp. 249–252; Charles Hannon, 'The Irish Volunteers and the Concepts of Military

the defence of the Ulster coastline against threatened German raids being drawn up. However, there were fears that the incorporation of the UVF in the UK-wide Volunteer Training Corps would put it under firm War Office control, and, rather than see this, the Irish Unionist leadership was prepared to let the force decay. The extent of this decrepitude is shown by the fact that during the Easter Rising of 1916 the UVF, once boasting 100,000 members, could mobilise just 1,300 men to support British forces. In 1918 there were serious concerns about Irish Volunteers raiding houses for UVF arms, and many, especially in exposed rural areas, were placed into British Army arsenals. Brigadier General Ambrose Ricardo, the senior UVF officer in Co. Tyrone, noted his concerns about this process: '[m]y advice was a voluntary surrender, to be stored in Omagh Military Barracks; any individual to retain his rifle if he was keen to do so, but at his own risk and against the advice of local leaders. As usual the local people receive no lead from Belfast.'[58]

The separatist Irish Volunteers carefully collected and preserved the rifles landed at Howth and Kilcoole. These appear to have formed a significant part of the arsenal at the disposal of those who rebelled in Easter 1916, being seized by the British military when the rebels surrendered.[59] Of course, an attempt was made by the Germans to transport 20,000 Russian rifles to Tralee Bay on board the *SMS Libau*, which masqueraded as the Norwegian steamer, the *Aud*. It was envisaged that these would be used to arm Irish Volunteers in the south west, who would then be able to act in conjunction with those who would rebel in Dublin. However, the *Aud* was scuttled by her captain when detained by a Royal Navy patrol.[60]

From 1917 the Irish Volunteers (after 1919 known as the IRA) turned to two major sources of supply for their arms. Firstly, rifles were seized; in total Charles Townshend has calculated that there were 3,218 raids for arms carried out between January 1919 and the Truce of 1921.[61] Most of these seizures were on a very small scale and were normally nothing more than shotguns stolen from local farmers. However, some rifles were stolen from existing INV or UVF stocks, something made all the easier by the widespread defection of INV personnel to the Irish Volunteers in 1918.[62] Indeed, some IRA units even stole rifles from

Service and Defence, 1913–24', unpublished PhD thesis, University College Dublin, 1989, pp. 82–83; Wheatley, *Nationalism and the Irish Party*, pp. 208–212, 235.

58 Bowman, *Carson's Army*, pp. 167–170; and PRONI, D627/436/41, letter, Ricardo to H. de F. Montgomery, 1 September 1918.

59 NAUK, WO32/9574, 'Appreciation of the situation in Ireland by the General Officer Commanding and a return of captured arms, 1916'.

60 M.T. Foy and Brian Barton, *The Easter Rising* (Stroud, 2011), pp. 70–71, and Charles Townshend, *Easter 1916: The Irish Rebellion* (London, 2005), pp. 126–131.

61 Charles Townshend, *The British Campaign in Ireland, 1919–21* (Oxford, 1975), p. 214.

62 NAI, BMH, WS517, Maurice Crowe, and WS647, Edward Boyle.

each other; one member of the force in Callan, Co. Kilkenny, remembered plans to seize the 50 rifles of the Kilkenny City volunteers as they were felt to be inactive.[63] Weapons were seized, or, on some rare occasions, purchased, from Crown forces. This was generally on a very small scale, for example when the local IRA stole four Webley Revolvers and 150 rounds of ammunition from Ulster Special Constabulary stores in Glenanne or when the IRA raided the RIC barracks in Trim in October 1920, making off with 12 rifles and 1,000 rounds of ammunition.[64] However, there were some major successes, such as the theft of 200 rifles from the 11th Royal Dublin Fusiliers in Longford in November 1917 and the dramatic seizure of 75 rifles, 72 bayonets and 4,000 rounds of ammunition at Collinstown aerodrome, near Dublin, in March 1919.[65]

Finally, some firearms were successfully imported from overseas, but, in attempting this, the IRA was hindered not just by the vigilance of British forces and U.S. customs, but also by fierce internal rivalries, notably that between Michael Collins and Cathal Brugha, fraud and determined competition from various European paramilitary organisations. Curiously attempts to tap sympathetic Irish-American sponsors and funds seem to have focused on obtaining Thompson sub-machine guns, oblivious to the fact that these weapons were of little use, unless supplied with very considerable quantities of ammunition. Eventually, of the 495 Thompson sub-machine guns purchased in the USA, 49 appear to have actually been used by the IRA in action. Peter Hart estimated that old established smuggling routes through Liverpool, initially organised by the Irish Republican Brotherhood, delivered 289 handguns, 53 rifles and 24,141 rounds of ammunition to the IRA between 1919 and the Truce of 1921. More elaborate gun-running ventures delivered two small boatloads of old rifles, obtained in Germany, which arrived after the Truce.[66] Despite these raids and gun-running, IRA stocks remained very low. Tom Barry remembered that in the 3rd West Cork Brigade, approximately 3,000 strong, and one of the better equipped formations, '[e]ven in the middle of 1920, the whole Brigade armament was only thirty-five serviceable rifles, twenty automatics or revolvers, about thirty rounds of ammunition per rifle, and ten rounds for each automatic or revolver'. The South Mayo Brigade claimed to have just five Mauser rifles, with very little ammunition in April 1921.[67]

In conclusion, it is evident that there was a strong paramilitary and also civilian gun culture in Ireland in these years. This is important in a wider

63 NAI, BMH, WS980, Edward J. Aylward, p. 23.

64 NAI, BMH, WS647, Edward Boyle; and *Meath Chronicle*, 2 October 1920.

65 NAUK, WO32/9507, 'Measures to be taken with regard to unauthorized drilling of parties in Ireland'; and Townshend, *British Campaign*, p. 19.

66 Peter Hart, *The I.R.A. at War 1916–1923* (Oxford, 2003), pp. 178–193; Peter Hart, *Mick: The Real Michael Collins* (London, 2005), pp. 260–261; and Townshend, *British Campaign*, pp. 180–181.

67 Barry, *Guerrilla Days in Ireland*, p. 9, and Augusteijn, *From Public Defiance*, pp. 144–145.

European context on two counts. Firstly, the late Professor F.X. Martin once raised the question: 'Was the Ulster Volunteer Force the first fascist army in modern times?'[68] Ideologically, of course, the UVF had little in common with Fascism – no new economic programme, no dreams of regaining lost provinces and essentially endorsing the policies of the Unionist Party. It would be very misguided to categorise UVF attitudes to Catholics as similar to Nazi attitudes to Jews and this, clearly, was not the comparison which Martin had in mind. However, in terms of Fascism as a mass movement characterised by impressive parades and rallies, there is clearly a resonance with the UVF. Secondly, the paramilitarism that had been largely confined to Ireland before the First World War became a European phenomenon immediately after the war.[69] The conditions that bred paramilitarism in Europe in 1919 were very different from those that gave birth to paramilitarism in Ireland in 1913, but the UVF and Irish Volunteers can be seen as the predecessors of the many private armies that sprung up in Europe after 1918.

In Ireland, between 1913 and 1921, not all paramilitaries enthusiastically endorsed a gun culture. Edward Carson and John Redmond were both somewhat reluctant paramilitary leaders, and large numbers of the UVF and IRA were inactive. Liberal firearms legislation meant that until 1914 arms importation into Ireland was relatively easy, and rapid developments in firearms technology meant that many obsolete rifles were available at very low prices. The Larne and Howth gun-running episodes were designed both to bring in significant numbers of arms and to provide a propaganda coup for the UVF and Irish Volunteers, respectively. After 1914, the importation of firearms into Ireland became much more difficult, and the IRA of 1919–1921 had to rely heavily on weapons seized from the Crown forces. Indeed, this chapter illuminates a paradox, in that it was the worst armed paramilitary force of this period that proved to be the most militant. Within the paramilitary forces, the distribution of firearms had often more to do with political patronage than military needs.

---

68 F.X. Martin, '1916: Myth, Fact and Mystery', *Studia Hibernica*, 7 (1967), p. 56.

69 As an introduction to paramilitarism in Europe immediately after the First World War see Richard Bessel, *Germany after the First World War* (Oxford, 1993); Julia Eichenberg, 'The Dark Side of Independence: Paramilitary Violence in Ireland and Poland after the First World War', *Contemporary European History*, 19 (2010); Robert Gerwarth and John Horne, *War in Peace: Paramilitary Violence in Europe after the Great War* (Oxford, 2012); and Timothy Wilson, *Frontiers of Violence: Conflict and Identity in Ulster and Upper Silesia, 1918–1922* (Oxford, 2010).

Bibliography

Only secondary sources are listed below. For both published and unpublished primary sources, readers are referred to the footnotes in each individual chapter.

Abernethy, Thomas Perkins, *Western Lands and the American Revolution* (New York: University of Virginia Press, 1937)

Adams, Donald R. Jr, 'Prices and Wages in Maryland, 1750–1850', *Journal of Economic History*, 46/3 (1986): 625–645

Adams, Kevin, *Class and Race in the Frontier Army: Military Life in the West, 1870–1890* (Norman: University of Oklahoma Press, 2009)

Alpers, Edward A., *Ivory and Slaves in East Central Africa: Changing Patterns of International Trade to the Later Nineteenth Century* (London: Heinemann, 1975)

Al-Sayegh, Fatma, 'Merchants' Role in a Changing Society: The Case of Dubai, 1900–1990', *Middle Eastern Studies*, 34/1 (1998): 87–102

Ambrose Brown, James, *The War of a Hundred Days: Springboks in Somalia and Abyssinia, 1940–41* (Johannesburg: Ashanti, 1990)

Anderson, David, and David Killingray, 'Consent, Coercion and Colonial Control: Policing the Empire, 1830–1940', in D. Anderson and D. Killingray (eds), *Policing the Empire: Government, Authority and Control, 1830–1940* (Manchester: Manchester University Press, 1991)

Anderson, David, and Richard H. Grove, (eds), *Conservation in Africa: People, Policies and Practices* (Cambridge: Cambridge University Press, 1987)

Arnold, David, 'The Armed Police and Colonial Rule in South India, 1914–1947', *Modern Asian Studies*, 11/1 (1977): 101–125

Aron, Stephen, *How the West Was Lost: The Transformation of Kentucky from Daniel Boone to Henry Clay* (Baltimore: Johns Hopkins University Press, 1996)

Aron, Stephen, '"Rights in the Woods" on the Trans-Appalachian Frontier', in A.R.L. Cayton and F.J. Teute (eds), *Contact Points: American Frontiers from the Mohawk Valley to the Mississippi, 1750–1830* (Chapel Hill: University of North Carolina Press, 1998)

Assael, Brenda, *The Circus and Victorian Society* (Charlottesville and London: University of Virginia Press 2005)

Atmore Anthony, et al., 'Firearms in South Central Africa', *Journal of African History*, 12/4 (1971): 545–556

Atta, John R. Van, '"A Lawless Rabble": Henry Clay and the Cultural Politics of Squatters' Rights, 1832–1841', *Journal of the Early Republic*, 28/3 (2008): 337–378

Attridge, Steve, *Nationalism, Imperialism and Identity in Late Victorian Culture: Civil and Military Worlds* (Basingstoke and New York: Palgrave Macmillan, 2003)

Augusteijn, Joost, *From Public Defiance to Guerrilla Warfare: The Experience of Ordinary Volunteers in the Irish War of Independence, 1916–1921* (Dublin: Irish Academic Press, 1996)

Barker, John, 'Where the Missionary Frontier Ran ahead of Empire', in N. Etherington (ed.), *Missions and Empire* (Oxford: Oxford University Press, 2005)

Barnes, Andrew, 'Catholic Evangelizing in One Colonial Mission: The Institutional Evolution of Jos Prefecture, Nigeria, 1907–1954', *Catholic Historical Review*, 84/2 (1998): 240–262

Barnes, John A., *Politics in a Changing Society: A Political History of the Fort Jameson Ngoni* (London: Oxford University Press, 1954)

Barr, Alwyn, 'The Black Militia of the New South: Texas as a Case Study', *Journal of Negro History*, 63/3 (1978): 209–219.

Beachey, R.W., 'The Arms Trade in East Africa in the Late Nineteenth Century', *Journal of African History*, 3/3 (1962): 451–467

Beattie, J.H.M, 'A Note on the Connexion between Spirit Mediumship and Hunting in Bunyoro, with Special Reference to Possession by Animal Ghosts', *Man*, 63 (1963): 188–189

Beckett, Ian F.W., *Riflemen Form: A Study of the Rifle Volunteer Movement, 1859–1908* (Aldershot: Ogilby Trusts, 1982)

Beckett, Ian F.W., 'The Indian Expeditionary Force on Malta and Cyprus, 1878', *Soldiers of the Queen*, 76 (1994): 6–11

Beckett, Ian F.W., 'The South African War and the Late Victorian Army', in P. Dennis and J. Grey (eds), *The Boer War: Army, Nation and Empire* (Canberra: Army History Unit, 2000)

Beckett, Ian F.W., *Isandlwana* (London: Brasseys, 2003)

Bederman, Gail, *Manliness and Civilization: A Cultural History of Gender and Race in the United States, 1880–1917* (Chicago: University of Chicago Press, 1995)

Beidelman, Thomas, 'Contradictions between the Sacred and the Secular Life: The Church Missionary Society in Ukaguru, Tanzania, East Africa, 1876–1914', *Comparative Studies in Society and History*, 23/1 (1981): 73–95

Beinart, William, and Lotte Hughes, *Environment and Empire* (Oxford: Oxford University Press 2007)

Beinart, William, and Peter Coates, *Environment and History: The Taming of Nature in the USA and South Africa* (London: Routledge, 1995)

Bellesiles, Michael, *Arming America: The Origins of a National Gun Culture* (New York: Alfred A. Knopf, 2000)

Benneyworth, Garth, 'Armed and Trained: Nelson Mandela's 1962 Military Mission as Commander in Chief of Umkhonto we Sizwe and Provenance for his Buried Makarov Pistol', *South African Historical Journal*, 63/1 (2011): 78–101

Bessel, Richard, *Germany after the First World War* (Oxford: Clarendon 1993)

Black, Jeremy, *Rethinking Military History* (London and New York: Routledge, 2004)

Blackstock, Allan, *An Ascendancy Army: The Irish Yeomanry, 1796–1834* (Dublin: Four Courts Press, 1998)

Blease, W. Lyon, *Suvorof* (London: Constable, 1920)

Bowman, Timothy, *Carson's Army: The Ulster Volunteer Force, 1910–22* (Manchester: Manchester University Press, 2007)

Braund, Kathryn E. Holland, *Deerskins and Duffels: The Creek Indian Trade with Anglo-America, 1685–1815* (Lincoln: University of Nebraska Press, 1993)

Brearley, H.C, 'The Negro and Homicide', *Social Forces*, 9/2 (1930): 247–253

Breen, T.H., '"Baubles of Britain": The American and Consumer Revolutions of the Eighteenth Century', *Past and Present*, 119 (1988): 73–104

Brelsford, W.V., *The Story of the Northern Rhodesia Regiment* (Bromley: Galago, 1990; 1st edn, 1954)

Brown, M.L., *Firearms in Colonial America: The Impact on History and Technology, 1492–1792* (Washington DC: Smithsonian Institution Scholarly Press, 1980)

Brown, Meredith, *Frontiersman: Daniel Boone and the Making of America* (Baton Rouge: Louisiana State University Press, 2008)

Brown, Richard M., *Strain of Violence: Historical Studies of American Violence and Vigilantism* (New York: Alfred A. Knopf, 1975)

Bruner, Jason, 'The Cambridge Seven, Late Victorian Culture, and the Chinese Frontier', *Social Sciences and Missions* (forthcoming)

Buckley, Jerome H., *The Victorian Temper: A Study in Literary Culture* (New York: Random House, 1964; 1st edn, 1951)

Burrell, Robert M., 'Arms and Afghans in the Makrān: An Episode in Anglo-Persian Relations, 1905–1912', *Bulletin of the School of Oriental and African Studies*, 49/1 (1986): 8–24

Calloway, Colin, *New Worlds for All: Indians, Europeans and the Remaking of Early America* (Baltimore: Johns Hopkins University Press, 1997)

Capeci, Dominic J. Jr, and Jack C. Knight, 'Reckoning with Violence: W.E.B. Du Bois and the 1906 Atlanta Race Riot', *Journal of Southern History*, 62/4 (1996): 727–766

Carnes, Mark C., and Clyde Griffen, *Meanings for Manhood: Constructions of Masculinity in Victorian America* (Chicago: University of Chicago Press, 1990)

Carr, Raymond, 'Country Sports', in G. Mingay (ed.), *The Victorian Countryside* (2 vols, London: Routledge and Kegan Paul, 1981)

Carruthers, Jane, 'Tracking in Game Trails: Looking afresh at the Politics of Environmental History in South Africa', *Environmental History*, 11/4 (2006): 804–829

Caulk, R.A., 'Firearms and Princely Power in Ethiopia in the Nineteenth Century', *Journal of African History*, 13/4 (1972): 609–630

Cawelti, John, *The Six Gun Mystique* (Bowling Green: Bowling Green State University Popular Press, 1999; 1st edn, 1971)

Cerino Badone, Giovanni, *Il bianco dei loro occhi. Storia della potenza di fuoco nelle guerre europee 1500–1800* (Milan: Edizioni Libreria Militare, in press)

Çetinsaya, Gökhan, 'The Ottoman view of the British Presence in Iraq and the Gulf: The Era of Abdulhamid II', *Middle Eastern Studies*, 39/2 (2003): 194–203

Chandler, David G., *The Campaigns of Napoleon* (New York: Macmillan, 1966)

Chetty, Suryakanthie, 'Gender under Fire: Interrogating War in South Africa, 1939–1945', unpublished MA dissertation, University of Natal, 2001

Chew, Emrys, 'Militarized Cultures in Collision: The Arms Trade and War in the Indian Ocean during the Nineteenth Century', *RUSI Journal*, 148/5 (2003): 90–96

Churchill, Robert H, 'Gun Regulation, the Police Power, and the Right to Keep Arms in Early America: The Legal Context of the Second Amendment', *Law and History Review*, 25/1 (2007): 139–75

Comerford, R.V., *The Fenians in Context: Irish Politics and Society, 1848–82* (Dublin: Wolfhound Press, 1985)

Connelly, Mark, *The IRA on Film and Television: A History* (Jefferson: McFarland & Co., 2012)

Cooke, James, 'Anglo-French Diplomacy and the Contraband Arms Trade in Colonial Africa', *African Studies Review*, 17/1 (1974): 27–41

Cornell, Saul, 'Early American Gun Regulation and the Second Amendment: A Closer Look at the Evidence', *Law and History Review*, 25/1 (2007): 197–204

Corum, James S., *The Roots of Blitzkrieg: Hans von Seeckt and German Military Reform* (Lawrence: University Press of Kansas, 1992)

Cottrol, Robert J., and Raymond T. Diamond, 'Never Intended to Be Applied to the White Population: Firearms Regulation and Racial Disparity. The Redeemed South's Legacy to a National Jurisprudence', *Chicago-Kent Law Review*, 70 (1994–1995): 1307–1335

Coupland, Reginald, *Zulu Battle Piece: Isandhlwana* (London: Collins & Co, 1948)

Cowling, Noëlle (ed.), 'Historical Survey of the Non-European Army Services outside of the Union of South Africa (Part II)', *Scientia Militaria*, 24/2 (1994): 26–48

Cramer, Clayton E., 'The Racist Roots of Gun Control', *Kansas Journal of Law and Public Policy*, 17 (1994–1995): 17–33

Cunningham, Hugh, *The Volunteer Force: A Social and Political History, 1859–1908* (London: Croom Helm, 1975)

Cunningham, Hugh, *Leisure in the Industrial Revolution* (London: Croom Helm, 1980)

Cutcliffe, Stephen H, 'Indians, Furs and Empires: The Changing Policies of New York and Pennsylvania, 1674–1768', unpublished PhD thesis, Lehigh University, 1976

Davis, Richard, *The Young Ireland Movement* (Dublin: Gill and Macmillan, 1987)

DeConde, Alexander, *Gun Violence in America: The Struggle for Control* (Boston: Northeastern University Press, 2001)

Del Boca, Angelo, *Gli italiani in Libia. Vol. 1: Tripoli bel suol d'amore, 1860–1920* (Milano: Mondadori, 1993)

Del Fra, Lino, *Sciara Sciat: Genocidio nell'oasi. L'esercito italiano a Tripoli* (Rome: ManifestoLibri, 2011)

Drønen, Tomas S., *Communication and Conversion in Northern Cameroon: The Dii People and Norwegian Missionaries, 1934–1960* (Leiden: Brill, 2009)

Dunlap, Thomas, *Nature and the English Diaspora: Environment and History in the United States, Canada, Australia and New Zealand* (Cambridge: Cambridge University Press, 1999)

Dunlop, John, *The Development of the British Army 1899–1914* (London: Methuen, 1938)

Dunn, R.E., 'Bu Himara's European Connection: The Commercial Relations of a Moroccan Warlord', *Journal of African History*, 21/2 (1980): 235–253

Dunn, Walter S., *The New Imperial Economy: The British Army and the American Frontier, 1764–1768* (Westport: Praeger, 2001)

Durie, Alastair J., '"Unconscious Benefactors": Grouse-Shooting in Scotland, 1780–1914', *International Journal of the History of Sport*, 15/3 (1998): 57–73

Durie, Alastair J., 'Game Shooting: an Elite Sport, *c*.1870–1980', *Sport in History*, 28/3 (2008): 431–449

Duxbury, G.R., *The Battle of Magersfontein, 11th December 1899* (Johannesburg: National Museum of Military History, 1995)

Dykstra, Robert R., *The Cattle Towns* (New York: Alfred A. Knopf, 1968)

Dykstra, Robert R., 'Body Counts and Murder Rates: The Contested Statistics of Western Violence', *Reviews in American History*, 31/4 (2003): 554–563

Echevarria, Antulio, 'The "Cult of the Offensive" Revisited: Confronting Technological Change before the Great War', *Journal of Strategic Studies*, 25/1 (2002): 199–214

Eichenberg, Julia, 'The Dark Side of Independence: Paramilitary Violence in Ireland and Poland after the First World War', *Contemporary European History*, 19/3 (2010): 231–248

Engelke, Matthew, 'Discontinuity and the Discourse of Conversion', *Journal of Religion in Africa*, 34/1–2 (2004): 82–109

Engelke, Matthew, *A Problem of Presence: Beyond Scripture in an African Church* (Los Angeles: University of California Press, 2007)

Etherington, Norman, *The Great Treks: The Transformation of Southern Africa, 1815–1854* (London and New York: Pearson, 2001)

Faragher, John M., *Daniel Boone: The Life and Legend of an American Pioneer* (New York: Holt, 1993)

Fisher, John, '"The Safety of our Indian Empire": Lord Curzon and British Predominance in the Arabian Peninsula, 1919', *Middle Eastern Studies*, 33/3 (1997): 494–520

Fisher, John, 'Property Rights in Pheasants: Landlords, Farmers and the Game Laws, 1860–80', *Rural History*, 11 (2000): 165–180

Fitzpatrick, David, *Politics and Irish Life 1913–1921: Provincial Experience of War and Revolution* (Dublin: Gill and Macmillan, 1977)

Ford, Matthew, 'The British Army and the Politics of Rifle Development', unpublished PhD thesis, King's College London, 2008

Foy, Michael T., and Brian Barton, *The Easter Rising* (Stroud: The History Press, 2011; 1st edn, 1999)

Friend, Craig T., *Kentucke's Frontiers* (Bloomington: Indiana University Press, 2010)

Gann, Lewis, H., *A History of Northern Rhodesia: Early Days to 1953* (London: Chatto & Windus 1964)

Gat, Azar, *A History of Military Thought: From the Enlightenment to the Cold War* (Oxford: Oxford University Press, 2001)

Gattrell, V.A.C., 'The Decline of Theft and Violence in Victorian and Edwardian England' in V.A.C. Gatrell, B. Lenman and G. Parker (eds), *Crime and The Law: The Social History of Crime in Western Europe since 1500* (London: Europa Publications, 1980)

Gerwarth, Robert, and John Horne, *War in Peace: Paramilitary Violence in Europe after the Great War* (Oxford: Oxford University Press, 2012)

Giddings, Paula J., *Ida: A Sword among Lions. Ida B. Wells and the Campaign against Lynching* (New York: Amistad, 2008)

Giglio, Vittorio, and Angelo Ravenni, *Le guerre coloniali d'Italia* (Milan: Vallardi, 1942)

Giliomee, Hermann, and Bernard Mbenga (eds), *New History of South Africa* (Cape Town: Tafelberg, 2007)

Gillespie, Greg, *Hunting for Empire: Narratives of Sport in Rupert's Land, 1840–1870* (Vancouver: UBC Press, 2007)

Gleeson, Ian, *The Unknown Force: Black, Coloured and Indian Soldiers through Two World Wars* (Johannesburg: Ashanti, 1994)

Godshalk, David F., *Visions: The 1906 Atlanta Race Riot and the Reshaping of Race Relations* (Greensboro, NC: University of North Carolina Press, 2005)

Goldberg, Jacob, 'The 1913 Saudi Occupation of Hasa Reconsidered', *Middle Eastern Studies*, 18/1 (1982): 21–29

Goldberg, Jacob, 'Captain Shakespear and Ibn Saud: A Balanced Reappraisal', *Middle Eastern Studies*, 22/1 (1986): 74–88

Goldman, Leslie F., 'The Second Amendment, the Slaughter-House Cases (1873), and United States v. Cruikshank (1876)', *Albany Government Law Review*, 1/2 (2008): 365–418

Gon, Philip, *The Road to Isandhlwana* (Johannesburg: A.D. Donker, 1979)

Gooch, John, 'The Weary Titan: Strategy and Policy in Great Britain, 1890–1918', in W. Murray, M. Knox and A. (eds), *The Making of Strategy: Rulers, States and War* (Cambridge: Cambridge University Press, 1994)

Gorn, Elliott J., '"Gouge and Bite, Pull Hair and Scratch": The Social Significance of Fighting in the Southern Backcountry', *Journal of American History*, 90/1 (1985): 18–43

Grant, Jonathan, *Rulers, Guns and Money: The Global Arms Trade in the Age of Imperialism* (Cambridge, MA: Harvard University Press, 2007)

Greaves, Adrian (ed.), *Redcoats and Zulus* (Barnsley: Pen & Sword Military, 2004)

Greaves, Adrian, *Forgotten Battles of the Zulu War* (Barnsley: Pen & Sword Military, 2012)

Grundlingh, Albert, '"The King's Afrikaners"? Enlistment and Ethnic Identity in the Union of South Africa's Defence Force during the Second World War, 1939–45', *Journal of African History*, 40/3 (1999): 351–365

Grundlingh, Louis, 'The Recruitment of South African Blacks for Participation in the Second World War', in D. Killingray and R. Rathbone (eds), *Africa and the Second World War* (Basingstoke: Macmillan, 1986)

Grundlingh, Louis, '"Non-Europeans Should Be Kept away from the Temptations of Towns": Controlling Black South African Soldiers during the Second World War', *International Journal of African Historical Studies*, 25/3 (1992): 539–60

Gudmundsson, Bruce I., *Stormtroop Tactics: Innovation in the German Army, 1914–18* (New York and London: Praeger 1989)

Gundersen, Joan R., 'Independence, Citizenship, and the American Revolution', *Signs*, 13/1 (1987): 59–77

Guy, Jeff, 'A Note on Firearms in the Zulu Kingdom with Special Reference to the Anglo-Zulu War, 1879', *Journal of African History*, 12/4 (1971): 557–570.

Hall, Sheldon, *Zulu: With Some Guts Behind It. The Making of the Epic Movie* (Sheffield: Tomahawk Press, 2006)

Hancock, W. Keith, *Smuts. Vol. 1: The Sanguine Years, 1870–1919* (Cambridge: Cambridge University Press, 1962)

Hannon, Charles, 'The Irish Volunteers and the Concepts of Military Service and Defence, 1913–24', unpublished PhD thesis, University College Dublin, 1989

Hannon, Neal, and Richard Taylor, *Virginia's Western War 1775–1786* (Mechanicsburg: Stackpole Books, 2002)

Harries, Patrick, *Butterflies and Barbarians: Swiss Missionaries and Systems of Knowledge in South-East Africa* (Oxford: James Currey, 2007)

Harrison, Brian, 'Animals and the State in Nineteenth-Century England', *English Historical Review*, 88/349 (1973): 786–820

Hart, Peter, *The I.R.A. and Its Enemies: Violence and Community in Cork, 1916–1923* (Oxford: Clarendon, 1998)

Hart, Peter, *The I.R.A. at War 1916–1923* (Oxford: Oxford University Press, 2003)

Hart, Peter, *Mick: The Real Michael Collins* (London: Macmillan, 2005)

Harvey, Karen, 'The History of Masculinity, circa 1650–1800', *Journal of British Studies*, 44/2 (2005): 296–311

Hatton, Ed, '"He Murdered Her Because He Loved Her": Passion, Masculinity, and Intimate Homicide in Antebellum America', in C. Daniels and M.V. Kennedy (eds), *Over the Threshold: Intimate Violence in Early America* (New York: Routledge, 1999)

Hay, Marnie, *Bulmer Hobson and the Nationalist Movement in Twentieth-Century Ireland* (Manchester: Manchester University Press, 2009)

Headrick, Daniel R., *The Tools of Empire: Technology and European Imperialism in the Nineteenth Century* (Oxford: Oxford University Press, 1981)

Headrick, Daniel R., *Power over Peoples: Technology, Environments, and Western Imperialism, 1400 to the Present* (Princeton: Princeton University Press, 2010)

Henron, Jan, *Britain's Forgotten Wars: Colonial Campaigns of the 19th Century* (Stroud: Sutton Publishing, 2002)

Herman, Daniel, *Hunting and the American Imagination* (Washington, DC: Smithsonian Institution Press, 2003)

Hess, Robert, 'The "Mad Mullah" and Northern Somalia', *Journal of African History*, 5/3 (1964): 415–433

Hichberger, Joan W.M., *Images of the Army: The Military in British Art, 1815–1914* (Manchester: Manchester University Press, 1988)

Horton, Robin, 'African Conversion', *Africa*, 41/2 (1971): 85–108

Horton, Robin, *Patterns of Thought in Africa and the West: Essays on Magic, Religion and Science* (Cambridge: Cambridge University Press, 1993)

Huggins, Mike, *The Victorians and Sport* (London: Hambledon, 2004)

Iliffe, John, *Honour in African History* (Cambridge: Cambridge University Press, 2005)

Isaac, Rhys, *The Transformation of Virginia, 1740–1790* (New York: W.W. Norton, 1988)

Isaacman, Allen F. and Barbara S., *Slavery and Beyond: The Making of Men and Chikunda Ethnic Identities in the Unstable World of South-Central Africa, 1750–1920* (Portsmouth, NH: Heinemann, 2004)

Jackson, Alvin, 'Unionist Myths, 1912–1985', *Past and Present*, 136 (1992).

Jackson, Alvin, 'British Ireland: What if the Home Rule Had Been Enacted in 1912?', in N. Ferguson (ed.), *Virtual History: Alternatives and Counterfactuals* (London: Macmillan, 1998)

Jackson, F.W.D., 'Isandhlwana, 1879: The Sources Re-examined', *Journal of the Society for Army Historical Research*, 43 (1965): 30–43, 113–132, 169–183

Jackson, F.W.D., *Hill of the Sphinx: The Battle of Isandlwana* (London: Westerners Publications, 2002)

Jackson, Lorna, 'Patriotism or Pleasure? The Nineteenth Century Volunteer Force as a Vehicle for Rural Working-Class Male Sport', *The Sports Historian*, 19/1 (1999): 125–139

Jacoby, Karl, *Crimes against Nature: Squatters, Poachers, Thieves and the Hidden History of American Conservation* (Berkeley: University of California Press, 2001)

Joly, Vincent, *Guerres d'Afrique: 130 ans de guerres coloniales. L'expérience française* (Rennes: Presses Universitaires de Rennes, 2009)

Johnson, Robert A., '"Russian at the Gates of India"? Planning the Defence of India, 1885–1900', *Journal of Military History*, 67/3 (2003): 697–743

Jones, Huw M., *The Boiling Cauldron: Utrecht District and the Anglo-Zulu War, 1879* (Bisley: Shermershill Press, 2006)

Jones, Spencer, *From Boer War to World War: Tactical Reform in the British Army 1902–1914* (Norman: University of Oklahoma Press, 2012)

Joyce, Simon, *Capital Offenses: Geographies of Class and Crime in Victorian London* (Charlottesville: University of Virginia Press, 2003)

Kaylani, Nabil M., 'Politics and Religion in 'Umān: An Historical Overview', *International Journal of Middle East Studies*, 10/4 (1979): 567–579

Keith, LeeAnna, *The Colfax Massacre: The Untold Story of Black Power, White Power and the Death of Reconstruction* (New York: Oxford University Press, 2008)

Kellett, Mark A., 'The Power of Princely Patronage: Pigeon-Shooting in Victorian Britain', *International Journal of the History of Sport*, 11/1 (1994): 63–85

Kelly, Matthew J., *The Fenian Ideal and Irish Nationalism, 1882–1916* (Woodbridge: Boydell and Brewer, 2006)

Kelly, William P., *Plotting America's Past: Fenimore Cooper and the Leatherstocking Tales* (Carbondale: Southern Illinois University Press, 1983)

Keppel-Jones, Arthur, *Rhodes and Rhodesia: The White Conquest of Zimbabwe, 1884–1902* (Kingston: McGill-Queen's University Press, 1983)

Killingray, David, *Fighting for Britain: African Soldiers in the Second World War* (Woodbridge: James Currey, 2010)

Kimmel, Michael S., *Manhood in America: A Cultural History* (Oxford: Oxford University Press, 2006)

Klein, Harry, *Springbok Record* (Johannesburg: SA Legion, 1946)

Klein, Kerwin L., 'Reclaiming the "F" Word, or Being and Becoming Postmodern', *Pacific Historical Review*, 65/2 (1996): 179–215

Knight, Ian, *Zulu: Isandlwana and Rorke's Drift* (London: Windrow & Greene, 1992)

Knight, Ian, *The Anatomy of the Zulu Army from Shaka to Cetshwayo, 1818–1879* (London: Greenhill Books, 1995)

Knight, Ian, 'Ammunition at Isandlwana: A Reply', *Journal of the Society for Army Historical Research*, 73 (1995): 237–250

Knight, Ian, *Go to your God like a Soldier* (London: Greenhill Books, 1996)

Knight, Ian, 'Old Steady Shots: The Martini-Henry Rifles, Rates of Fire and Effectiveness in the Anglo-Zulu War', *Journal of the Anglo Zulu War Historical Society*, 11 (2002): 1–5

Knight, Ian, *Zulu Rising: The Epic Story of iSandlwana and Rorke's Drift* (London: Macmillan, 2010)

Knight, Ian, and Ian Castle, *Zulu War* (Oxford: Osprey, 2004)

Knox, MacGregor, 'Mass Politics and Nationalism as Military Revolution: The French Revolution and after', in M. Knox and W. Murray (eds), *The Dynamics of Military Revolution, 1300–2050* (Cambridge: Cambridge University Press, 2001)

Kozuskanich, Nathan R., 'Pennsylvania, the Militia, and the Second Amendment', *Pennsylvania Magazine of History and Biography*, 133/3 (2009): 119–147

Kwint, Marius, 'The Legitimization of the Circus in Late Georgian England', *Past and Present*, 174 (2002): 72–115

Kynoch, Gary, 'Your Petitioners Are in Mortal Terror: The Violent World of Chinese Mineworkers in South Africa, 1904–10', *Journal of Southern African Studies*, 31/3 (2005): 531–546

Laband, John, *Kingdom in Crisis: The Zulu Response to the British Invasion of 1879* (Manchester: Manchester University Press, 1992)

Laband, John, *Rope of Sand: The Rise and Fall of the Zulu Kingdom in the Nineteenth Century* (Johannesburg: Jonathan Ball, 1995)

Laband, John, '"Bloodstained Grandeur": Colonial and Imperial Stereotypes of Zulu Warriors and Zulu Warfare', in B. Carton, J. Laband and J. Sithole (eds), *Zulu Identities: Being Zulu, Past and Present* (Pietermaritzburg: University of KwaZulu-Natal Press, 2008)

Laband, John, 'The War-Readiness and Military Effectiveness of the Zulu Forces in the 1879 Anglo-Zulu War', *Natalia*, 39 (2009): 37–46

Laband, John, *Historical Dictionary of the Zulu Wars* (Lanham: The Scarecrow Press, 2009)

Lane, Charles, *The Day Freedom Died: The Colfax Massacre, the Supreme Court, and the Betrayal of Reconstruction* (New York: Holt, 2008)

Langworthy, Harry W., 'Introduction: Carl Wiese and Zambezia', in Carl Wiese, *Expedition in East-Central Africa, 1888–1891: A Report*, ed. H.W. Langworthy (Norman: University of Oklahoma Press, 1983)

Laver, Harry S, 'Rethinking the Social Role of the Militia: Community-Building in Antebellum Kentucky', *Journal of Southern History*, 68/4 (2002): 777–816

Li, Chien-hui, 'A Union of Christianity, Humanity, and Philanthropy: The Christian Tradition and the Prevention of Cruelty to Animals in Nineteenth-Century England', *Society and Animals*, 8/3 (2000): 265–285

Limerick, Patricia N., *Legacy of Conquest: The Unbroken Past of the American West* (New York: Norton, 1987)

Loewenstein, Andrew, '"The Veiled Protectorate of Kowait": Liberalized Imperialism and British Efforts to Influence Kuwaiti Domestic Policy during the Reign of Sheikh Ahmad al-Jaber, 1938–1950', *Middle Eastern Studies*, 36/2 (2000): 103–123

Longworth, Philip, *The Art of Victory: The Life and Achievements of Generalissimo Suvorov, 1729–1800* (London: Constable, 1965)

Loo, Tina, 'Of Moose and Men: Hunting for Masculinities in British Columbia, 1880–1939', *Western Historical Quarterly*, 32 (2002): 296–319

Lowerson, John, *Sport and the English Middle Classes, 1870–1914* (Manchester: Manchester University Press, 1993)

Lynn, John A., *The Bayonets of the Republic: Motivation and Tactics in the Army of Revolutionary France, 1791–94* (Urbana: University of Illinois Press, 1984)

Lyons, J.B., *The Enigma of Tom Kettle: Irish Patriot, Essayist, Poet, British Soldier, 1880–1916* (Dublin: Glendale Press, 1983)

MacKenzie, John M. (ed.), *Imperialism and Popular Culture* (Manchester: Manchester University Press, 1986)

MacKenzie, John M., 'The Imperial Pioneer and Hunter and the British Masculine Stereotype in Late Victorian and Edwardian Times', in J.A. Mangan and J. Walvin (eds), *Manliness and Morality: Middle-Class Masculinity in Britain and America, 1800–1940* (Manchester: Manchester University Press, 1987)

MacKenzie, John M., *The Empire of Nature: Hunting, Conservation and British Imperialism* (Manchester: Manchester University Press, 1988)

MacKenzie, John M. (ed.), *Popular Imperialism and the Military: 1850–1950* (Manchester: Manchester University Press, 1992)

Macola, Giacomo, 'Reassessing the Significance of Firearms in Central Africa: The Case of North-Western Zambia to the 1920s', *Journal of African History*, 51/3 (2010): 301–321

Majundar, Boria, 'Tom Brown Goes Global: The "Brown" Ethic in Colonial and Post-Colonial India', *International Journal of the History of Sport*, 23/5 (2006): 805–820

Malcolm, Joyce Lee, *To Keep and Bear Arms: The Origins of an Anglo-American Right* (Cambridge, MA: Harvard University Press, 1994)

Malcolm, Joyce Lee, *Guns and Violence: The English Experience* (Cambridge, MA: Harvard University Press, 2002)

Malgeri, Francesco, *La guerra libica (1911–1912)* (Rome: Edizioni di Storia e Letteratura, 1970)

Mangan, J.A., 'Britain's Chief Spiritual Export: Imperial Sport as Moral Metaphor, Political Symbol and Cultural Bond', in J.A. Mangan (ed), *The Cultural Bond: Sport, Empire and Society* (London: Frank Cass, 1992)

Mangan, J.A., 'Christ and the Imperial Playing Fields: Thomas Hughes's Ideological Heirs in Empire', *International Journal of the History of Sport*, 23/5 (2006): 777–804

Mangan, J.A., and Callum McKenzie, '"Duty unto Death" – the Sacrificial Warrior: English Middle Class Masculinity and Militarism in the Age of the New Imperialism', *International Journal of the History of Sport*, 25/9 (2008): 1080–1105

Mangan, J.A., and Callum McKenzie, 'Martial Conditioning, Military Exemplars and Moral Certainties: Imperial Hunting as Preparation for War', *International Journal of the History of Sport*, 25/9 (2008): 1132–67

Mangan, J.A., and Callum McKenzie, 'Imperial Masculinity Institutionalized: The Shikar Club', *International Journal of the History of Sport*, 25/9 (2008): 1218–1242

Marcus, Harold, 'The Embargo on Arms Sales to Ethiopia, 1916–1930', *International Journal of African Historical Studies*, 16/2 (1983): 263–279

Marjomaa, Risto, 'The Martial Spirit: Yao Soldiers in British Service in Nyasaland (Malawi), 1895–1939', *Journal of African History*, 44/3 (2003): 413–432

Martin, F.X, '1916: Myth, Fact and Mystery', *Studia Hibernica*, 7 (1967): 7–126

Martin, Henry J., and Neil D. Orpen, *South Africa at War: Military and Industrial Organisation and Operations in Connection with the Conduct of the War, 1939–1945* (Cape Town: Purnell, 1979)

Martin, John, 'Pigeon Shooting', in T. Collins, J. Martin and W. Vamplew (eds), *Encyclopedia of Traditional British Rural Sports* (London: Routledge, 2005)

Martin, Samuel, 'British Images of the Zulu, ca. 1820–1879', unpublished PhD thesis, University of Cambridge, 1982

Marx, Leo, *Machine in the Garden: Technology and the Pastoral Ideal in America* (Oxford: Oxford University Press, 2000; 1st edn, 1964)

McCracken, John, *Politics and Christianity in Malawi, 1875–1940: The Impact of the Livingstonia Mission in the Northern Province* (Cambridge: Cambridge University Press, 1977)

McCracken, John, 'Coercion and Control in Nyasaland: Aspects of the History of a Colonial Police Force', *Journal of African History*, 27/1 (1986): 127–147

McGarry, Ferghall, *Rebels: Voices from the Easter Rising* (Dublin: Penguin Ireland, 2011)

McGrath, Roger D., *Gunfighters, Highwaymen, and Vigilantes: Violence on the Frontier* (Berkeley: University of California Press, 1984)

McKenzie, Callum, 'The British Big-Game Hunting Tradition, Masculinity and Fraternalism with Particular Reference to "The Shikar Club"', *The Sports Historian*, 20/1 (2000): 70–96

McMeekin, Sean, *The Berlin–Baghdad Express: The Ottoman Empire and Germany's Bid for World Power, 1898–1918* (London: Penguin, 2010)

McMurry, Linda O., *To Keep the Waters Troubled: The Life of Ida B. Wells* (Oxford: Oxford University Press, 1998)

Melosi, Martin V., 'Equity, Eco-Racism and Environmental History', *Environmental History Review*, 19/3 (1995): 1–16

Menning, Bruce W., 'Train Hard, Fight Easy: The Legacy of A.V. Suvorov and His "Art of Victory"', *Air University Review* (November–December 1986): 79–88

Meyer, Birgit, '"If You Are a Devil, You Are a Witch and, If You Are a Witch, You Are a Devil." The Integration of "Pagan" Ideas into the Conceptual Universe of Ewe Christians in Southeastern Ghana', *Journal of Religion in Africa*, 22/2 (1992): 98–132

Middlekauff, Robert, *The Glorious Cause: The American Revolution, 1763–1789* (Oxford: Oxford University Press, 1982)

Miers, Sue, 'Notes on the Arms Trade and Government Policy in Southern Africa between 1870 and 1890', *Journal of African History*, 12/4 (1971): 571–577

Miller, David W., 'Non-professional Soldiery, c. 1600–1800', in T. Bartlett and K. Jeffery (eds), *A Military History of Ireland* (Cambridge: Cambridge University Press, 1996)

Miller, Henry J., 'John Leech and the Shaping of the Victorian Cartoon: The Context of Respectability', *Victorian Periodicals Review*, 42/3 (2009): 267–291

Miller, Stephen M., *Lord Methuen and the British Army: Failure and Redemption in South Africa* (London: Frank Cass, 1999)

Mkutu, Kennedy A., *Guns and Governance in the Rift Valley: Pastoralist Conflict and Small Arms* (Oxford: James Currey, 2008)

Moreman, T.R., 'The Arms Trade and the North-West Frontier Pathan Tribes, 1890–1914', *Journal of Imperial and Commonwealth History*, 22/2 (1994): 187–216

Morris, Donald R., *The Washing of the Spears, The Rise and Fall of the Zulu Nation*, (London: Pimlico, 1994; 1st edn, 1965)

Munsche, P.B., *Gentlemen and Poachers: The English Game Laws, 1671–1831* (Cambridge: Cambridge University Press, 1981)

Myrdal, Gunnar, *An American Dilemma: The Negro Problem and American Democracy* (London: Harper and Brothers Publishers, 1944)

Nash, Roderick, *Wilderness and the American Mind* (New Haven: Yale University Press, 1967)

Nasson, Bill, 'A Great Divide: South African Responses to the Great War, 1914–1918', *War and Society*, 12/ 1 (1994): 47–64

Nasson, Bill, 'War Opinion in South Africa, 1914', *Journal of Imperial and Commonwealth History*, 23/2 (1995): 248–278

Nasson, Bill, *The South African War* (London: Hodder Arnold, 1999)

Nasson, Bill, 'Why They Fought: Black Cape Colonists and Imperial Wars, 1899–1918', *International Journal of African Historical Studies*, 37/1 (2004): 55–70

Nasson, Bill, *Springboks on the Somme: South Africa in the Great War, 1914–1918* (Johannesburg: Penguin 2007)

Nasson, Bill, *South Africa at War, 1939–1945* (Johannesburg: Jacana, 2012)

Ndee, Hamad S., 'Western Influences on Sport in Tanzania: British Middle-Class Educationalists, Missionaries and the Diffusion of Adapted Athleticism', *International Journal of the History of Sport*, 27/5 (2010): 905–936

Neumann, R.P., 'Dukes, Earls, and Ersatz Edens: Aristocratic Nature Preservationists in Colonial Africa', *Environment and Planning D: Society and Space*, 14/1 (1996): 79–98

Novick, Ben, 'The Arming of Ireland: Gun Running and the Great War 1914–16', in A. Gregory and S. Paseta (eds), *Ireland and the Great War: 'A War to Unite Us All?'* (Manchester: Manchester University Press, 2002)

Pakenham, Thomas, *The Boer War* (London: Weidenfeld and Nicolson, 1979)

Pankhurst, Richard, 'Guns in Ethiopia', *Transition*, 20 (1965): 26–33

Parsons, Timothy H., '"Wakamba Warriors Are Soldiers of the Queen": The Evolution of the Kamba as a Martial Race, 1890–1970', *Ethnohistory*, 46/4 (1999): 671–701

Parsons, Timothy H., *The African Rank-and-File: Social Implications of Colonial Military Service in the King's African Rifles, 1902–1964* (Portsmouth, NH: Heinemann, 1999)

Parry, Jonathan, *Democracy and Religion* (Cambridge: Cambridge University Press, 1983)

Paul, Catriona, '"The Horsemen Had the Start": Horse Ownership and Advantage in Early Kentucky, 1770–1830', unpublished PhD thesis, University of Dundee, 2012

Peel, John D.Y., 'Conversion and Tradition in Two African Societies: Ijebu and Buganda', *Past and Present*, 77 (1977): 108–141

Peers, Douglas M., '"Those Noble Exemplars of True Military Tradition": Construction of the Indian Army in Mid-Victorian Press', *Modern Asian Studies*, 31/1 (1997): 109–142

Phimister, Ian, 'Union of South Africa', in M.R.D. Foot and I.C.B. Dear (eds), *The Oxford Companion to World War II* (Oxford: Oxford University Press, 2005)

Pretorius, Fransjohan, *Life on Commando during the Anglo-Boer War 1899–1902* (Cape Town: Human and Rosseau, 1999)

Pretorius, J.D., 'Ideology and Identities: Printed Graphic Propaganda of the Communist Party of South Africa, 1921–1950', unpublished DLit. et Phil. thesis, University of Johannesburg, 2011

Prins, Gwyn, *The Hidden Hippopotamus: Reappraisal in African History. The Early Colonial Experience in Western Zambia* (Cambridge: Cambridge University Press, 1980)

Quarles, Benjamin, 'The Colonial Militia and Negro Manpower', *Mississippi Valley Historical Review*, 45/4 (1959): 643–652

Ram, K.V., 'British Government, Finance Capitalists and the French Jibuti-Addis Ababa Railway, 1898–1913', *Journal of Imperial and Commonwealth History*, 9/2 (1981): 146–168

Rau, William E., 'Mpezeni's Ngoni of Eastern Zambia, 1870–1920', unpublished PhD thesis, UCLA, 1974

Read, Margaret, 'Tradition and Prestige among the Ngoni', *Africa*, 9/4 (1936): 453–484

Read, Margaret, *The Ngoni of Nyasaland* (London: Frank Cass, 1970; 1st edn, 1956)

Reid, Darren, 'Walking the Line of Fire: Violence, Society and the War for the Kentucky and Trans-Appalachian Frontier, 1774–1795', unpublished PhD thesis, University of Dundee, 2011

Reiger, John, *American Sportsmen and the Origins of Conservation* (Corvallis: Oregon State University Press, 2001; 1st edn, 1975)

Richardson, Joel M., 'Florida Black Codes', *Florida Historical Quarterly*, 47/4 (1969): 365–379

Richter, Daniel K., *Facing East from Indian Country: A Native History of Early America* (Cambridge MA: Harvard University Press, 2001)

Ritvo, Harriet, *The Animal Estate: The English and Other Creatures in the Victorian Age* (Cambridge, MA: Harvard University Press, 1987)

Roberts, Andrew D., 'Firearms in North-Eastern Zambia before 1900', *Transafrican Journal of History*, 1/2 (1971): 3–21

Roberts, Andrew D., *A History of Zambia* (London: Heinemann, 1976)

Robin, Libby, and Tom Griffiths (eds), *Ecology and Empire: Environmental History of Settler Societies* (Edinburgh: Keele University Press, 1998)

Rohrbough, Malcolm J., *Trans-Appalachian Frontier: People, Societies, and Institutions, 1775–1850* (Bloomington: Indiana University Press, 2008; 1st edn, 1978)

Rorabaugh, W.J., 'The Political Duel in the Early Republic: Burr v. Hamilton', *Journal of the Early Republic*, 15/1 (1995): 1–23

Ross, Steven T., *From Flintlock to Rifle: Infantry Tactics, 1740–1866* (London: Frank Cass, 1996)

Rotundo, E. Anthony, *American Manhood: Transformations in Masculinity from the Revolution to the Modern Era* (New York: BasicBooks, 1993)

Sahlins, Marshall, *Islands of History* (Chicago: University of Chicago Press, 1985)

Samuels, Martin, *Doctrine and Dogma: German and British Infantry Tactics in the First Word War* (New York: Greenwood Press, 1992)

Sanders, Peter, *'Throwing Down White Man': Cape Rule and Misrule in Colonial Lesotho, 1871–1884* (Pontypool: Merlin, 2011)

Sanneh, Lamin, *Translating the Message: The Missionary Impact on Culture* (Maryknoll: Orbis Books, 1989)

Schindler, John R., *Isonzo: The Forgotten Sacrifice of the Great War* (Westport: Praeger, 2001)

Shalhope, Robert E., 'The Ideological Origins of the Second Amendment', *Journal of American History*, 69 (1982): 599–614

Shalhope, Robert E., and Lawrence D. Cress, 'The Second Amendment and the Right to Bear Arms: An Exchange', *Journal of American History*, 71 (1984): 587–93

Shear, Keith, '"Taken as Boys": The Politics of Black Police Employment and Experience in Early Twentieth-Century South Africa', in L. Lindsay and S. Miescher (eds), *Men and Masculinities in Modern Africa* (Portsmouth, NH: Heinemann, 2003)

Shoemaker, Robert, 'Male Honour and the Decline of Public Violence in Eighteenth-Century London', *Social History*, 26/2 (2001): 190–208

Sidenvall, Erik, *The Making of Manhood among Swedish Missionaries in China and Mongolia, c. 1890–c. 1914* (Leiden: Brill, 2009)

Sinclair, Allan, 'The Use of the Neville Lewis Portraits for the Second World War Stamp Series', *Military History Journal*, 11/1 (1998): 1–14

Singletary, Otis A., 'The Negro Militia during Radical Reconstruction', *Military Affairs*, 19/4 (1955): 177–186

Sivasundaram, Sujit, *Nature and the Godly Empire: Science and Evangelical Mission in the Pacific, 1795–1850* (Cambridge: Cambridge University Press, 2005)

Skaggs, David C., 'The Sixty Years' War for the Great Lakes, 1754–1814: An Overview', in D.C. Skaggs and L.L. Nelson (eds), *The Sixty Years' War for the Great Lakes, 1754–1814* (East Lansing: Michigan State University Press, 2001)

Slotkin, Richard, *Regeneration through Violence: The Mythology of the American Frontier, 1600–1800* (Middletown: Wesleyan University Press, 1973)

Slotkin, Richard, *The Fatal Environment: The Myth of the Frontier in the Age of Industrialization, 1800–1890* (New York: Atheneum, 1985)

Slotkin, Richard, *Gunfighter Nation: The Myth of the Frontier in Twentieth Century America* (Norman: University of Oklahoma Press, 1998; 1st edn, 1994)

Small, Stephen, *Political Thought in Ireland 1776–1798: Republicanism, Patriotism and Radicalism* (Oxford: Oxford University Press, 2002)

Smith, Henry N., *Virgin Land: The American West as Symbol and Myth* (Cambridge, MA: Harvard University Press, 1950)

Smits, David D., 'The Frontier Army and the Destruction of the Buffalo: 1865–1883', *Western Historical Quarterly*, 25/3 (1994): 312–338

Smyth, D.H., 'The Volunteer Movement in Ulster: Background and Development, 1745–85', unpublished PhD thesis, Queen's University of Belfast, 1974

Soltow, Lee, 'Kentucky Wealth at the End of the Eighteenth Century', *Journal of Economic History*, 43/3 (1983): 617–633

Somerville, Christopher, *Our War: How The British Commonwealth Fought the Second World War* (London: Weidenfeld and Nicolson, 1998)

Sontag, Susan, *On Photography* (New York: Farrar, Straus and Giroux, 1977; 1st edn, 1973)

Spalding, Henry S., *Suvóroff* (London: Chapman and Hall, 1890)

Spear, Thomas, 'Neo-traditionalism and the Limits of Invention in British Colonial Africa', *Journal of African History*, 44/1 (2003): 3–27

Speece, Mark, 'Aspects of Economic Dualism in Oman, 1830–1930', *International Journal of Middle East Studies*, 21/4 (1989): 495–515

Spierenburg, Pieter (ed.), *Men and Violence: Gender, Honour, and Rituals in Modern Europe and America* (Columbus: Ohio State University Press, 1998)

Spiers, Edward, *The Late Victorian Army, 1868–1902* (Manchester: University of Manchester Press, 1992)

Sramek, Joseph, '"Face Him like a Briton": Tiger Hunting, Imperialism, and British Masculinity in Colonial India, 1800–1875', *Victorian Studies*, 48/4 (2006): 659–680

Stanley, Brian, '"Hunting for Souls": The Missionary Pilgrimage of George Sherwood Eddy', in P.N. Holtrop and H. McLeod (eds), *Missions and Missionaries* (Woodbridge: The Boydell Press, 2000)

Stansky, Keith, '"So These Folks Are Aggressive": An Orientalist Reading of "Afghan Warlord"', *Security Dialogue*, 40/1 (2009): 73–94

Stapleton, Timothy, *African Police and Soldiers in Colonial Zimbabwe, 1923–80* (Rochester, NY: University of Rochester Press, 2011)

Starkey, Armstrong, *European and Native American Warfare, 1675–1815* (London: UCL Press, 1998)

Stedman Jones, Gareth, 'The "Cockney" and the Nation', in D. Feldman and G. Stedman Jones (eds), *Metropolis London: Histories and Representations since 1800* (London: Routledge, 1989)

Steinhart, Edward I., *Black Poachers, White Hunters: A Social History of Hunting in Colonial Kenya* (Oxford: James Currey, 2006)

Steward, Dick, *Duels and the Roots of Violence in Missouri* (Columbia: University of Missouri Press, 2000)

Stone, David, 'Imperialism and Sovereignty: The League of Nations' Drive to Control the Global Arms Trade', *Journal of Contemporary History*, 35/2 (2000): 213–230

Stone, Jay, and Erwin Schmidl, *The Boer War and Military Reforms* (Lanham: University Press of America, 1988)

Storey, William K., *Guns, Race and Power in Colonial South Africa* (Cambridge: Cambridge University Press, 2008)

Straight, Bilinda, 'Killing God: Extraordinary Moments in the Colonial Encounter', *Current Anthropology*, 49/5 (2008): 837–860.

Streets, Heather, *Martial Races: The Military, Race and Masculinity in British Imperial Culture, 1857–1914* (Manchester: Manchester University Press, 2004)

Sundkler, Bengt, and Christopher Steed, *A History of the Church in Africa* (Cambridge: Cambridge University Press, 2000)

Tahmassebbi, Stefan, 'Gun Control and Racism', *George Mason University Civil Rights Law Journal*, 2/1 (1991): 67–100

Taylor, Anthony, '"Pig-Sticking Princes": Royal Hunting, Moral Outrage, and the Republican Opposition to Animal Abuse in Nineteenth- and Early Twentieth-Century Britain', *History*, 89/293 (2004): 30–48

Temple, B.A., and I.D. Skennerton, *A Treatise on the British Military Martini: The Martini-Henry, 1869–1900* (Burbank: B.A. Temple, 1983)

Teute, Fredrika J., 'Land, Liberty, and Labor in the Post-Revolutionary Era: Kentucky as the Promised Land', unpublished PhD thesis, Johns Hopkins University, 1988

Thompson, F.M.L., 'Landowners and the Rural Community', in G. Mingay (ed.), *The Victorian Countryside* (2 vols, London: Routledge and Kegan Paul, 1981)

Thompson, T. Jack, 'The Origins, Migration, and Settlement of the Northern Ngoni', *Society of Malawi Journal*, 34/1 (1981): 6–35

Tosh, John, *Manliness and Masculinities in Nineteenth-Century Britain: Essays on Gender, Family and Empire* (New York: Pearson Longman, 2005)

Tosh, John, *A Man's Place: Masculinity and the Middle-class Home in Victorian England* (Yale: Yale University Press, 2007)

Townsend, John, 'Some Reflections on the Career of Sir Percy Cox', *Asian Affairs*, 24/3 (1993): 259–272

Townshend, Charles, *The British Campaign in Ireland, 1919–1921* (London and New York: Oxford University Press, 1975)

Townshend, Charles, 'The IRA and the Development of Guerrilla Warfare, 1916–21', *English Historical Review*, 94/371 (1979): 318–345

Townshend, Charles, *Political Violence in Ireland: Government and Resistance since 1848* (Oxford: Clarendon, 1983)

Townshend, Charles, *Easter 1916: The Irish Rebellion* (London: Allen Lane, 2005)

Turner, Frederick J., *The Frontier in American History* (New York: Holt, 1921)

Tylden, G., 'A Study in Attack: Majuba 27 February 1881', *Journal of the Society for Army Historical Research*, 39/157 (1961): 27–36

Vail, Leroy, 'The Making of the "Dead North": A Study of the Ngoni Rule in Northern Malawi, *c.* 1855–1907', in J.B. Peires (ed.), *Before and after Shaka: Papers in Nguni History* (Grahamstown: Institute of Social and Economic Research, Rhodes University, 1981)

van der Waag, Ian, 'The Union Defence Force between the Two World Wars, 1919–1940', *Scientia Militaria*, 30/2 (2000): 183–219

van Onselen, Charles, 'The Role of Collaborators in the Rhodesian Mining Industry, 1900–1935 ', *African Affairs*, 72/289 (1973): 401–418

Vellut, Jean-Luc, 'Garenganze/Katanga–Bié–Benguela and beyond: The Cycle of Rubber and Slaves at the Turn of the 20th Century', *Portuguese Studies Review*, 19/1–2 (2011): 133–152

Vernon, Ken, *Penpricks: The Drawing of South Africa's Political Battlelines* (Cape Town: Spearhead Press, 2000)

Walls, Andrew, *The Missionary Movement in Christian History: Studies in the Transmission of the Faith* (Maryknoll: Orbis Books, 1996)

Ward, Matthew C., *Breaking the Backcountry: The Seven Years' War in Virginia and Pennsylvania, 1754–1765* (Pittsburgh: University of Pittsburgh Press, 2003)

Ward, Matthew C., 'The American Militias: "The Garnish of the Table"', in R. Chickering and S. Förster (eds), *War in an Age of Revolution, 1775–1815* (Cambridge: Cambridge University Press, 2010)

Warren, Louis S., *The Hunter's Game: Poachers and Conservationists in Twentieth Century America* (New Haven: Yale University Press, 1997)

Webster, Andrew, 'Making Disarmament Work: The Implementation of the International Disarmament Provisions in the League of Nations Covenant, 1919–1925', *Diplomacy and Statecraft*, 16/3 (2005): 551–569

Webster, Andrew, 'From Versailles to Geneva: The Many Forms of Interwar Disarmament', *Journal of Strategic Studies*, 29/2 (2006): 225–246

Wedon, Carolyn, *Inheritors of the Spirit: Mary White Ovington and the Founding of the NAACP* (New York: Wiley, 1999)

Wessels, Andre, *Lord Roberts and the War in South Africa 1899–1902* (Stroud: Army Records Society, 2000)

Wheatley, Michael, *Nationalism and the Irish Party: Provincial Ireland, 1910–1916* (Oxford: Oxford University Press, 2005)

Whisker, James B., *Arms Makers of Colonial America* (Selinsgrove: Susquehanna University Press, 1992)

White, Gavin, 'Firearms in Africa: An Introduction', *Journal of African History*, 12/2 (1971):173–184

White, Lynn Jr, 'The Historical Roots of our Ecological Crisis', *Science*, 155/3767 (1967): 1203–1207

Whittam, John, *The Politics of the Italian Army 1861–1918* (London: Croom Helm, 1977)

Williams, Robert F., *Negroes with Guns* (Detroit: Wayne State University Press, 1993; 1st edn, 1962)

Williams-Myers, Albert J., 'The Nsenga of Central Africa: Political and Economic Aspects of Clan History, 1700 to the Late Nineteenth Century', unpublished PhD thesis, UCLA, 1978

Williamson, Joel, *After Slavery: The Negro in South Carolina during Reconstruction, 1861–1877* (Chapel Hill: University of North Carolina Press, 1965)

Wilson, Timothy, *Frontiers of Violence: Conflict and Identity in Ulster and Upper Silesia, 1918–1922* (Oxford: Oxford University Press, 2010)

Witchard, Anne, '"A Fatal Freshness": Mid-Victorian Suburbophobia', in L. Phillips and A. Witchard (eds), *London Gothic: Place, Space and the Gothic Imagination* (London: Continuum, 2010)

Wonders, Karen, 'Hunting Narratives of the Age of Empire', *Environment and History*, 11/2 (2005): 269–291

Worman, Charles G., *Gunsmoke and Saddle Leather: Firearms in the Nineteenth-Century American West* (Albuquerque: University of New Mexico Press, 2005)

Wright, John, 'Making the James Stuart Archive', *History in Africa*, 23 (1996): 333–350

Wright, John, 'Turbulent Times: Political Transformations in the North and East, 1760s–1830s', in B.K. Mbenga et al. (eds), *The Cambridge History of South Africa. Volume I: From Early Times to 1885* (Cambridge: Cambridge University Press, 2010)

Wright, Tim, *The History of the Northern Rhodesia Police* (Bristol: British Empire & Commonwealth Museum, 2001)

Yorke, Edmund, 'Isandlwana, 1879: Reflections on the Ammunition Controversy', *Journal of the Society for Army Historical Research*, 72 (1994): 205–218

Yorke, Edmund, *Rorke's Drift, 1879* (Stroud: Tempus, 2001)

Zirinsky, Michael, 'Imperial Power and Dictatorship: Britain and the Rise of Reza Shah, 1921–1926', *International Journal of Middle East Studies*, 24/4 (1992): 639–663

Index